TROUBLESHOOTING HYDRAULIC COMPONENTS

USING LEAKAGE PATH ANALYSIS METHODS

A PRACTICAL GUIDE

Rory S. McLaren

TROUBLESHOOTING HYDRAULIC COMPONENTS
USING LEAKAGE PATH ANALYSIS METHODS

First printing .... December 1993
Second printing .... February 1996
Third printing .... November 2001
Fourth printing .... September 2005
Fifth printing .... January 2012
Sixth printing .... September 2014
Seventh printing .... August 2018
Eighth printing .... October 2022

Library of Congress Catalog Card Number: 93-91821 - ISBN No. 0-9639619-1-8

Printed in the United States of America

Troubleshooting Hydraulic Components

This textbook, along with a complete set of PowerPoint™ Cd-Rom, cutaway models, and trainers, form the basis for a hands-on troubleshooting workshop which is offered on a regular basis in Salt Lake City, Utah. It is also available as an in-house workshop.
Most of the procedures outlined in this textbook are covered during these workshops.

For more detailed information regarding these workshops and/or training aids available for purchase, please contact:

FLUID POWER TRAINING INSTITUTE™
2170 South 3140 West
West Valley City, UT 84119

PH:　(801) 908-5456
FAX:　(801) 908-5734
http://www.fpti.org
e-mail: info@fpti.org

LIABILITY DISCLAIMER

IMPORTANT NOTE: All the test procedures outlined in this textbook have been carefully analyzed. Safety has been vigorously stressed and the importance of training as an **absolute** prerequisite has been outlined.
We assume no liability for persons who willfully proceed with, deviate in any way from, or utilize sub-standard diagnostic equipment, in the performance of any or all of these procedures.
If there is any question whatsoever regarding personal safety or the validity of a test procedure, contact the respective component or system manufacturer.

This manual is dedicated with love and gratitude:
to my wife, Christine Gaenor McLaren.
Her contributions to this work extend far beyond those one
would expect from a colleague, an intellectual companion,
a friend, and wife;
to my children, Gary, Daryl, and Kirsty-Jane;
to my dear friend Patrick Luers;
and to the memory of my twin brother, Craig.
He was the best friend I ever had.

contents

chapter five

Procedures for Testing Industrial Directional Control Valves

chapter six

Directional Control Valve Conversion Procedures

chapter seven

Procedures for Testing Mobile Directional Control Valves

chapter eight

Procedures for Testing Hydraulic Cylinders

chapter nine

Procedures for Testing Hydraulic Motors

chapter ten

Procedures for Testing Cartridge Valves

chapter eleven

Procedures for Testing Flow Control Valves

chapter twelve

Procedures for Testing Check Valves

chapter thirteen

Safety with Accumulators

Preface

This manual is a collection of test procedures covering hydraulic components found in industrial and mobile hydraulic systems. It is intended to demonstrate safe and effective troubleshooting procedures and to be a reference for experienced equipment and plant maintenance technicians and engineers.

To eliminate confusing cross-reference, each test procedure is totally independent. Reference to other test procedures is only given, where necessary, to determine root-cause analysis.

I have written this manual from two points of view: new system start-up problems and existing system problems. In addition, a series of troubleshooting root-cause analysis charts has been included as a step-by-step guide to solving problems.

Troubleshooting Hydraulic Components Using Leakage Path Analysis Methods is a culmination of my many years of field troubleshooting experience and the experience gained while operating a hydraulic component test stand. I would like to thank the many people I have had the opportunity to train over the years for giving me thought-provoking questions to answer regarding test procedures which forced me to conduct many research projects.

If I have overlooked a test procedure or if a procedure does not seem to work on a specific component, I would appreciate hearing from readers so that additional information can be included in future editions.

Safety is the key to working with hydraulic systems. The degree of safety can only be appreciated through education to principles and laws pertaining to hydraulic power transmission. The procedures outlined in this manual are intended only for technicians who have received formal training in hydraulics.

I hope you find this book of invaluable use.

Rory S. McLaren -

chapter one

Before You Proceed!

Safety First!

General

The test procedures outlined in this manual are intended for plant and equipment maintenance and engineering technicians who have received formal training in fluid power systems and components.

The training curriculum should have covered the following:
- a. Fluid power principles and laws.
- b. Anatomy of the component(s) being tested.
- c. Theory of operation of the component(s) being tested.
- d. An introduction to flow, pressure, speed and temperature-measuring devices.
- e. How to use flow, pressure, speed and temperature-measuring devices.
- f. Pressure/velocity relationship.

Safety glasses

-WARNING-   **Do not work on or around hydraulic systems without wearing safety glasses which conform to ANSI Z87.1-1989 standard.**

Do not be fooled by the external appearance of a hydraulic system. There could be hundreds of gallons per minute at thousands of pounds per square inch within the system. Since oil seeks the path of least resistance, it is constantly seeking ways to escape to atmosphere. All that is needed is a loose or incorrect connector, an incorrectly installed or damaged "O" ring, a pinhole in a porous valve body or hose, and the resultant velocity could literally amputate a limb or knock out an eye!

The potential for injury as a result of hot oil should also be considered. In certain instances, it is necessary to wear "full-face" protection when working on hydraulic systems.

What Is Leakage Path Analysis (LPA)?

Introduction

All hydraulic components have component parts which move relative to one another within a stationary body. Clearances are necessary between the moving parts to allow the components to move, to compensate for expansion and contraction as the component heats up and cools down, and to allow oil to form a lubricating film to prevent galling and to dissipate heat.

Smaller clearances provide less leakage. The lower the leakage, the more efficient the component. Consequently, higher efficiency means less heat generation within the system.

Low leakage rates are critical in machine tool, aerospace, and industrial process equipment applications. It is not uncommon to see radial clearances in hydraulic components as low as 2 - 15 microns. In low pressure mobile and agricultural applications where control repeatability precision and filtration are not as critical, higher internal leakage is tolerable.

Manufacturers naturally invest heavily in complex test equipment to test the efficiency of their products. They also have extensive research and development departments which study ways to improve manufacturing processes to keep component leakage to a minimum.

Leakage Path Analysis was developed to give maintenance and plant engineering technicians the ability to study component leakage without cost-prohibitive test stands while maintaining a high degree of accuracy.

The overall objective of Leakage Path Analysis is to:
        a. Enhance workplace safety.
        b. Halt the unnecessary removal of components (process-of-elimination).
        c. Reduce operating costs.
        d. Increase production.

Leakage Path Analysis (LPA)

Since it is not practical or feasible for the end user to duplicate the rigorous tests conducted by manufacturers, I felt it necessary to bring to the shop floor level a simple means of testing leakage in hydraulic components.

After testing hydraulic components on my own test bench for seven years, I determined what the average acceptable leakage is in most general hydraulic components. I was also able to determine what excessive leakage levels are, and at what level leakage affects the operation of the machine. The data presented in this guide is empirical.

Since the most difficult components to test appeared to be directional control valves, pressure control valves, and check valves, which coincidentally, also had the lowest leakage rates of any of the hydraulic components, I developed two simple methods of testing these and other components, while at the same time attaining a high degree of accuracy.

Mobile Directional Valves

It does not take long to become proficient at testing directional control valves using leakage path analysis methods if you practice on new valves, or valves which are operating normally. This method is especially useful if you are trying to isolate a problem in a mobile directional valve with integral cylinder port relief valves, anti-cavitation valves, and/or load check valves.

Industrial Directional Valves

Isolating leakage paths in industrial, sub-plate or manifold mounted, directional control valves and "sandwich" circuit modules is simplified using the techniques outlined in this manual.

Cartridge Valves

Complex cartridge valve manifolds have the reputation of being the most challenging to trouble-shoot. The methods outlined in this manual make the task of troubleshooting cartridge valves simple and effective.

Diagnostic Equipment

Pressure, flow, speed, and temperature need to be controlled and monitored in all hydraulic systems. Periodic adjustment is necessary to keep equipment operating at peak performance.

Diagnostic equipment provides a "window" into the hydraulic system to scientifically monitor pressure, flow, speed, and temperature.

Always follow these basic guidelines when using diagnostic equipment:

a. Diagnostic equipment must be in good operating condition and must be calibrated and serviced according to the manufacturer's recommendations.

b. Diagnostic equipment must **meet** or **exceed** the normal pressure and/or flow requirements of the system being tested.

c. Have the manufacturer's representative provide training for new diagnostic equipment.

d. Never modify diagnostic equipment or use it for anything other than its intended purpose.

e. Read all the respective manufacturer's operating and safety instructions before using the equipment.

Typical applications for diagnostic equipment include:

Safety   Prevent plant and equipment maintenance technicians from conducting "open-to-atmosphere" testing procedures.

System Start-up   To conduct "zero-fault" component start-up procedures.

Proactive Maintenance   To monitor individual component performance.

Research and Development   To test and record performance levels of systems/components in new equipment.

Troubleshooting   To help plant and equipment maintenance technicians identify and correct problems.

Diagnostic equipment can be classified either as "essential" or "non-essential."

Essential
Those used by plant and equipment maintenance technicians who support proactive maintenance functions and diagnostic evaluations on a daily basis for pressure, flow, speed, and temperature monitoring.

Non-essential
Those used by manufacturers in research and development to perform specialized testing such as: oil analysis, peak pressure analysis, and horsepower computations.

Flow Meter

"The flow meter is to a fluid power technician what a stethoscope is to a doctor!"

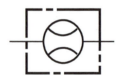
There are many different types of flow meters available to suit most maintenance budgets. They range from inexpensive in-line types to sophisticated electronic units.

(approximate cost: $200.00 - $1500.00)

In-line Mechanical Flow Meter

Advantages:
     a. Relatively inexpensive.
     b. Inexpensive to maintain or repair.
     c. Lightweight.
     d. Suitable for monitoring a wide range of liquids and gases.
     e. Accuracy within 5% of full scale.
     f. Repeatability within 1%.
     g. Do not require flow straighteners or special piping.
     h. Relatively insensitive to shock and vibration.

    i. Good viscosity stability.
    j. Direct reading.
    k. No electrical connections.

Disadvantages:
    a. Laboratory level accuracy sometimes not achievable.
    b. Uni-directional flow monitoring only.
    c. No permanent record of flow rate.
    d. No load generating capability.
    e. Transmission lines have to be interrupted to install.

Portable Digital Electronic Hydraulic Flow Meters (approximate cost: $1800.00-$4000.00)

Advantages:
    a. Laboratory level accuracy.
    b. Turbine flow sensors offer rapid response.
    c. Integral load generating capability.
    d. Compact.
    e. Bi-directional flow monitoring.
    f. Interface with PC's.
    g. Integral flow, pressure and temperature monitoring.
    h. Can record short pressure spikes as fast as 0.6 millisecond.
    i. Can record pressure differential.
    j. Sensor readings can be sent to a memory or printer.
    k. Alphanumeric or graph printer output.

Disadvantages:
    a. Initial investment.
    b. Expensive to maintain.
    c. Requires frequent calibration.
    d. Can usually only be serviced and calibrated by the manufacturer.
    e. Extremely sensitive to mishandling.
    f. Very limited pressure transducer and printer compatibility.
    g. Extremely contaminant sensitive.
    h. Physical size and weight.

Pressure Gauges

A pressure gauge is a device which measures pressure. They are available in two different configurations: Bourdon tube type, and plunger type.

(approximate cost: Test Gauge Kit - $350.00 - $450.00)

Bourdon Tube-Type Pressure Gauge - A Bourdon tube gauge consists of a needle and dial face with any number of calibrations. The needle is attached via a linkage to a flexible metal coiled tube, called a Bourdon tube. The inside diameter of the Bourdon tube is connected to the system pressure.

As system pressure increases, the Bourdon tube tends to straighten out. This is due to the difference in areas between the inside and outside diameters of the tube. This action causes the needle to indicate the appropriate pressure on the face of the dial.

Bourdon tube gauges are precision instruments with accuracies ranging from 0.1% to 3.0% of full scale. They are used for laboratory level testing and for precise pressure adjustments.

Plunger Pressure Gauge - A plunger gauge consists of a small cylinder or plunger opposed by a bias spring. The plunger is connected to a needle which points to a calibrated scale. As the pressure in the system increases, the plunger moves against the opposing force of the bias spring. This movement causes the needle, attached to the plunger, to indicate the appropriate pressure on the dial. Plunger gauges are used primarily as an economical means of measuring pressure.

Pyrometer

Hydraulic systems are not 100% efficient. Inefficiency in the form of heat is prevalent in all hydraulic systems. Even the most well-engineered systems can anticipate some percentage of their input horsepower to turn into heat. The level of heat generated in a hydraulic system must be frequently monitored and strictly controlled.

(approximate cost: $275.00 - $500.00)

There is a comprehensive range of hand-held digital pyrometers used for monitoring temperature. Specially designed probes are available for immersion into liquids, for measuring surface temperatures, and for ambient or air flow temperature monitoring.

Tachometer

Microprocessor-controlled, digital hand-held, non-contact tachometers will directly display RPM (revolutions per minute) measurements sensed from an easy-to-aim visible light beam directed at a rotating shaft or pulley. They will measure speeds as low as 6 RPM, and some offer expanded memory functions. They are highly reliable and require minimum maintenance. Some manufacturers offer an adapter for direct contact operation.

(approximate cost: $125.00 - $500.00)

Accumulator Precharge Assembly

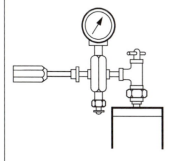

The gas precharge assembly consists of a manifold block with accumulator valve adapter, "T" bar handle to open valve in accumulator valve stem, bleeder valve, charging valve stem, and heavy duty 3000 PSI (207 bar) gauge.
It also includes a 6 ft (2 meter) high pressure, steel-wire braid-reinforced synthetic rubber hose with connections for accumulator and gas cylinder connection.

(approximate cost: $140.00 - $250.00)

MicroLeak

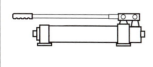

The simple, hand-held MicroLeak analyzer proved to be an indispensable, yet inexpensive, tool for testing hydraulic components with impressive results. A miniature power unit mounted atop a mobile hand cart proved to be more functional for leakage path testing in plants with large numbers of hydraulic components. The mobile cart can be dispatched swiftly to a suspect component. The component can be tested in-situ, or removed for immediate leakage path analysis.

(approximate cost: $130.00 - $300.00)

Safety First!

None of the test procedures outlined in this textbook can or should be conducted without the appropriate diagnostic equipment and proper training.

If you do not have the proper test equipment and/or do not know how to use it, **DO NOT** attempt to troubleshoot hydraulic components.

We highly recommend that you attend a course of training in hydraulics before servicing, repairing, or troubleshooting hydraulic components or systems.

Hydraulic Symbols Used in This Guide

PRESSURE CONTROL

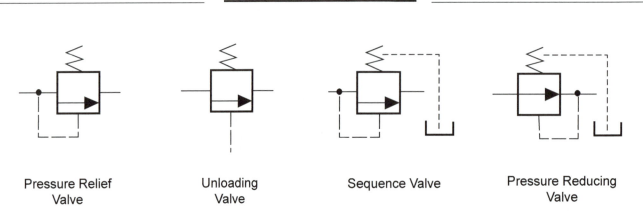

Pressure Relief
Valve

Unloading
Valve

Sequence Valve

Pressure Reducing
Valve

DIRECTIONAL CONTROL

2-WAY and 3-WAY VALVES

   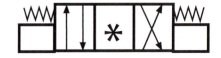

2-Way Normally Open
2-Position

2-Way Normally Closed
2-Position

3-Way
Directional Control

3-Way
Selector

4-WAY VALVES

  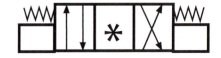

2-Position
Single Actuator

2-Position
Double Actuator

3-Position
Spring Centered

SPOOL CENTERS FOR 3-POSITION VALVES

    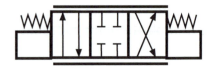

Closed
Center

Tandem
Center

Float
Center

Open
Center

Proportional
Solenoid Valve

PUMPS

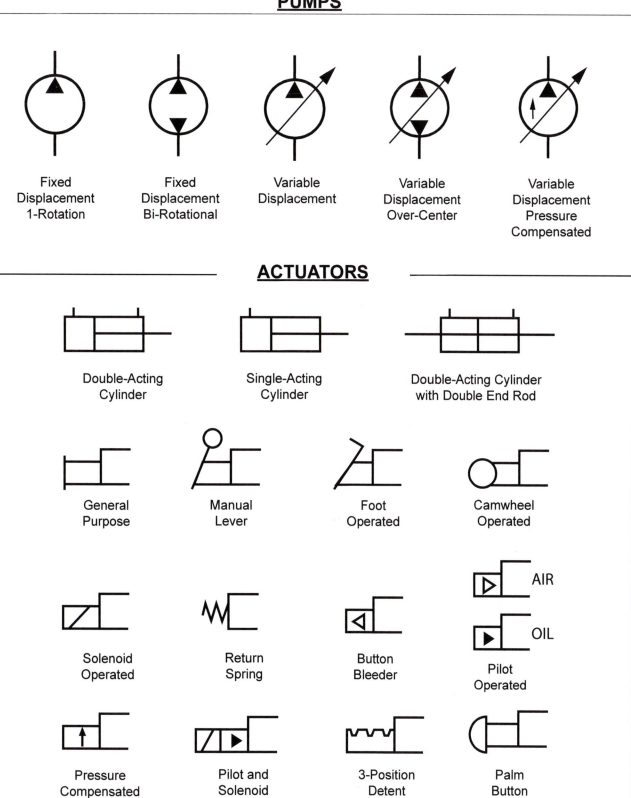

Fixed Displacement 1-Rotation

Fixed Displacement Bi-Rotational

Variable Displacement

Variable Displacement Over-Center

Variable Displacement Pressure Compensated

ACTUATORS

Double-Acting Cylinder

Single-Acting Cylinder

Double-Acting Cylinder with Double End Rod

General Purpose

Manual Lever

Foot Operated

Camwheel Operated

Solenoid Operated

Return Spring

Button Bleeder

AIR

OIL

Pilot Operated

Pressure Compensated

Pilot and Solenoid

3-Position Detent

Palm Button

HYDRAULIC and ELECTRIC MOTORS

Fixed
Displacement
1-Rotation

Fixed
Displacement
Reversible

Variable
Displacement
1-Rotation

Electric
Motor

FLOW CONTROL

Fixed Orifice

Needle Valve

Pressure-Compensated
Flow Control

Check Valve

Pilot to Close
Check Valve

Pilot to Open
Check Valve

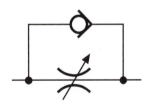

Flow Control Valve

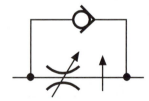

Pressure-Compensated
Flow Control Valve - Restrictor Type

OTHER

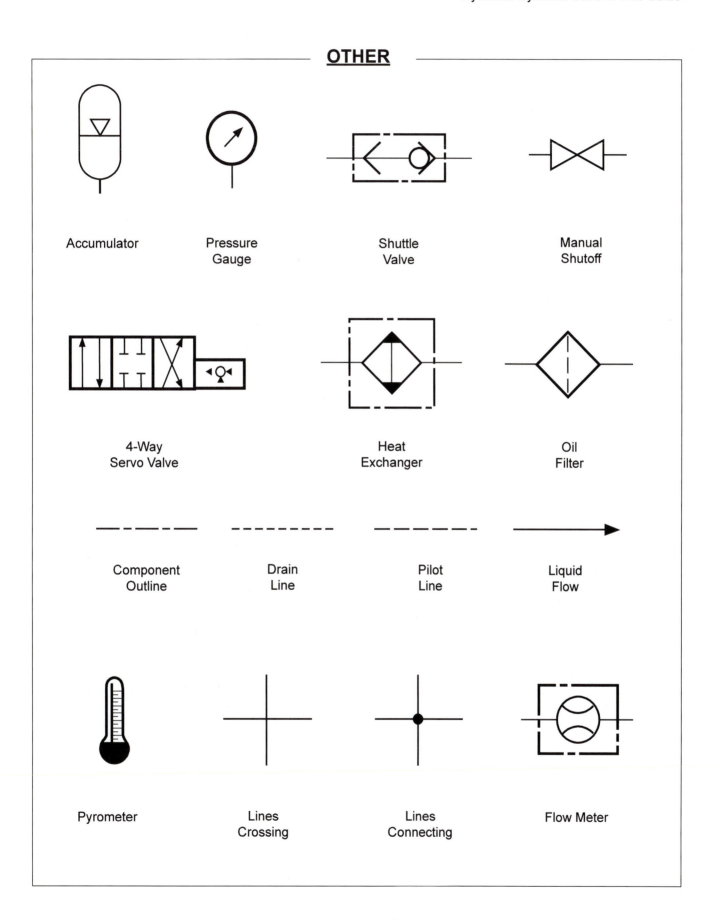

Accumulator	Pressure Gauge	Shuttle Valve	Manual Shutoff
4-Way Servo Valve		Heat Exchanger	Oil Filter
Component Outline	Drain Line	Pilot Line	Liquid Flow
Pyrometer	Lines Crossing	Lines Connecting	Flow Meter

Installing Diagnostic Equipment - Series versus Parallel

Before proceeding, it is important to understand the reference made throughout the textbook regarding the installation of diagnostic equipment.

Series Installation:

EXAMPLE: "Install a flow meter in series with the transmission line at the outlet port of the pump."

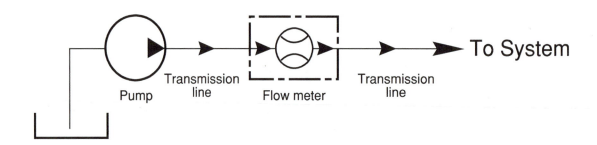

Since the outlet port of the pump is connected to the inlet port of the flow meter, the fluid has no alternative but to pass through each component "in series."

Parallel Installation:

EXAMPLE: "Install a pressure gauge in parallel with the connector at the outlet port of the pump."

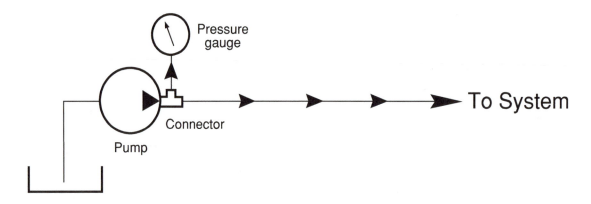

In a parallel circuit the fluid has one or more "alternative" flow paths. It can flow to the gauge or "alternatively" to the system. In a parallel circuit, the fluid will always flow to the path which offers the least resistance.

chapter two

System Start-up Procedures and Troubleshooting

Starting A New Hydraulic Pump - the "Zero-Fault" Method

The following outline provides a step-by-step method for starting a hydraulic pump correctly. Refer to Figure 2-1 for correct placement of diagnostic equipment.

---

1. Connect vacuum gauge (A) in parallel with the pump inlet port transmission line.

Why? It will offer visual indication immediately upon start-up of any problems related to the suction side of the pump. Pump manufacturers provide maximum inlet restriction specifications which should be strictly adhered to. Refer to the pump manufacturer's specification for maximum inlet restriction recommendations.

Some common pump inlet start-up problems:

1.1. Low inlet restriction
   a. No oil.
   b. Low oil level.
   c. Loose connectors, filter, and/or clamps.
   d. Case drain line above the oil - bent axis pumps.
   e. Pump dry - not creating partial vacuum- mounted above oil level.

1.2. High inlet restriction
   a. Pump rotating too fast.
   b. Incorrect oil - viscosity.
   c. Plugged inlet transmission line.
   d. Suction screen incorrectly sized.
   e. Suction filter incorrectly sized.
   f. Suction line incorrectly sized.
   g. Suction hose - not suitable for suction line applications.
   h. Suction hose - kinked or crimped.
   i. Reservoir/pump isolator valve closed or partially closed.
   j. Reservoir/pump isolator valve incorrectly sized.
   k. Ambient temperature extreme - oil viscosity.

NOTE: If the pump inlet port is incorrectly sized, it will not be detected by a vacuum gauge installed in parallel with the pump inlet port connector. If the vacuum or pump inlet restriction is within specification and cavitation persists, check the pump manufacturer's specifications for pump inlet port size relative to pump flow.

---

2. Connect pressure gauge (B) in parallel with the connector at the outlet port of the pump. Where possible, disconnect the pump outlet transmission line, and connect it directly to the inlet port of the return line filter (5) or into the top of the reservoir (if the transmission line is placed directly into the top of the reservoir, it must be securely fastened).

Why?   Pressure control is not always the function of a pressure relief valve in a hydraulic system. Certain types of variable displacement hydraulic pumps have pressure compensators which control pressure in the system. Refer to the manufacturer's specifications for recommended pressure relief valve or compensator settings.

2.1.   Compensators on new or rebuilt pumps are very rarely adjusted to suit a specific system. Compensator adjustment can be made before introducing the pump to the system.

2.2.   Pump performance can be monitored under varying pressure conditions.

2.3.   Pressure compensator adjustments can be made safely and accurately.

2.4.   Pump manufacturer's step-level pressure "run-in" procedures can be systematically followed.

Certain types of unbalanced pumps must be "run-in." Manufacturers provide a step-level pressure start-up recommendation. Step-level simply means that the pressure at the pump outlet must be applied gradually and in specific increments.

It is not uncommon for gear pumps, especially aluminum bodied designs, to experience gear-to-housing contact upon start-up. Some larger pumps will remove as much as a half-a-cup of aluminum on start-up. By running the pump outlet directly into the return line or "slave" filter, none of the excess material will enter the system. This procedure is especially important if high pressure filters are not installed in series with the pump outlet transmission line directly ahead of the pump.

About 90% of rebuilt pumps are either not tested, or are not tested to maximum pressure and flow levels. This is due to either an absence of test capability, or insufficient test-stand horsepower.

---

3. Connect flow meter (D) in series with the pump outlet transmission line.

Why?   Satisfactory actuator speed and system performance is dependent on the amount of oil delivered by the pump under various load conditions. Pump flow must be consistent with pump volumetric efficiency and design requirements. Refer to manufacturer's specifications for recommended pump flow specifications.

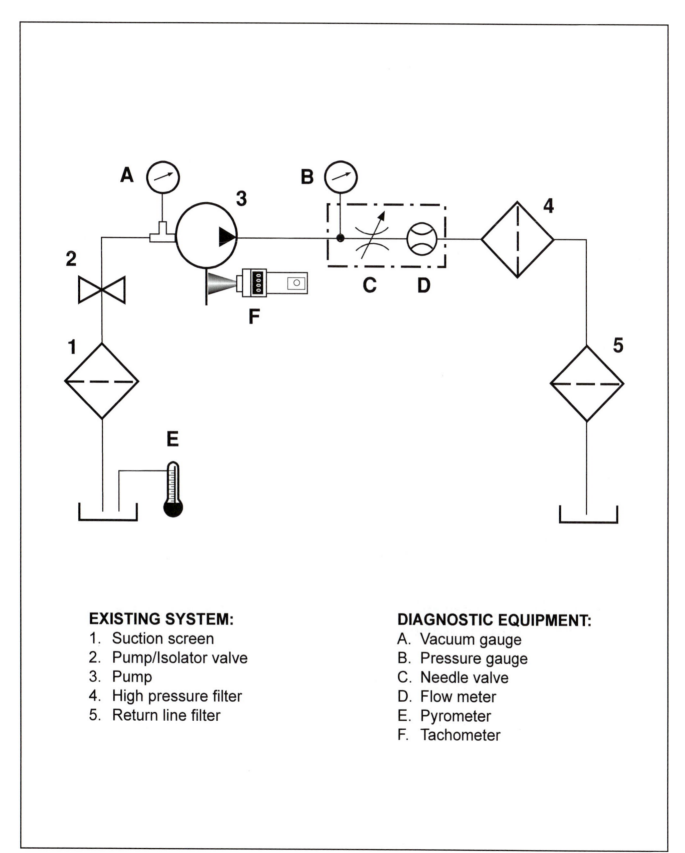

Figure 2-1   Placement of diagnostic equipment

3.1.   The flow meter will:
      a. Indicate if the pump is receiving oil at its inlet port.
      b. Indicate if the correct amount of flow is being delivered to the system.
      c. Indicate an abnormal variation in pump input speed.
      d. Indicate pump efficiency under various load conditions.
      e. Allow you to make flow adjustments on pumps equipped with mechanical flow adjustment options (variable displacement).

---

4. Connect needle valve (C) in series with the pump outlet transmission line.

Why?   Pump efficiency can only be accurately measured when the pump is pumping against a pressure (load). A needle valve can be used to generate an "artificial" load at the pump outlet to check pump flow at various pressure levels.

4.1.   Gradual loads can be introduced at the pump oulet easily without having to load the actuator(s).

4.2.   Manufacturer's start-up procedures can be followed more easily. Certain manufacturers require that pumps be loaded sequentially at step-level pressures for optimum in-service performance.

4.3.   Many new and rebuilt hydraulic pumps are not tested under full pressure and flow conditions. This is primarily due to a lack of test-bench horsepower. Therefore, the prime mover can be used as a "dynamometer" or test bench, as it has the horsepower capability to operate the pump at full flow and pressure levels.

4.4.   Contamination from catastrophic pump start-up failure will not be transported into the system.

4.5.   Warranty claims can be dramatically reduced as a well-documented systematic start-up procedure can rarely be argued against.

---

5. Check the pump input shaft speed using a tachometer.

Why?   Pump flow is directionally proportional to shaft speed. The combined volume of the pumping chambers through one shaft revolution (360°) is called "pump displacement", and is usually expressed in cubic inches per revolution (CIR). An abnormal increase or decrease in shaft speed will have an adverse affect on pump life. Refer to the manufacturer's specification for recommended prime mover speed.

5.1.   High Speed Related Problems
   a. Bearing life is inversely proportional to pump speed. A pump operating at 1800 RPM will have a life expectancy about 50% lower than one running at 1200 RPM.
   b. Excessive speed will result in cavitation because inlet transmission lines may be undersized for the increased velocity, and suction screens may be undersized for the increased flow.
   c. Pump balancing is also critical to shaft speed. Excessive speed may cause vibration due to the increased centrifugal force on unbalanced rotating groups.
   d. Lubrication may be affected by increased centrifugal force "squeezing" the lubrication film out, causing adhesive wear.
   e. Cavitation and vibration will also produce excessive noise.

5.2.   Low Speed Related Problems
   a. Mobile equipment is particularly vulnerable to prime mover speed reductions due to wear in foot throttle to fuel injection pump linkages and cables.
   b. A pump becomes less efficient at lower speeds because of internal leakage which is proportional to system pressure and pump design clearances. Leakage is a higher percentage of reduced flow rates.

Motor Testing

Hydraulic motors typically slow down under load when worn. Historical analysis will determine that the problem is viscosity sensitive, i.e. during cold start-up operation they appear to operate normally, and as the oil temperature increases the motor shaft speed decreases.

---

6. Monitor the oil temperature using a pyrometer.

Why?   Monitoring and maintaining oil at a moderate temperature is essential for reliable machine operation. When hydraulic systems are started up, the oil temperature will increase until it reaches a "levelling-off" temperature. This condition occurs when the heat being generated is equal to the heat being removed by radiation through the walls of the reservoir and/or by a heat exchanger. Refer to the manufacturer's specification for maximum & minimum operating temperatures.

6.1.   Machine tools and plant power units operate at temperatures between 140°F.(60° C.) and 160°F (71°C). Mobile equipment systems are forced to operate at temperatures up to 180°F (82°C).

6.2.   Systems should never be allowed to operate at temperatures above 200°F (93°C). At this temperature and above, oil life is greatly reduced.
   Viscosity may decrease to a point where:
   a. Power loss becomes excessive.
   b. Lubrication loss may result in adhesive wear.
   c. Rubber seals and filter elements may deteriorate rapidly.

6.3.   Studies have shown that oil life decreases by approximately 50% for every 20ºF (-6.6ºC) rise in temperature.

6.4.   A pyrometer should be used on a daily basis to allow the user to become thoroughly familiar with system temperatures under all load conditions. Cold ambient conditions will also contribute to poor pump life. High viscosity will cause excessive pump inlet restriction resulting in cavitation.

Troubleshooting - Root-Cause Analysis

Introduction
Many premature hydraulic component failures are attributed to poor initial start-up procedures. It is therefore essential to implement strict guidelines when starting new systems or introducing new components to existing systems. This will optimize component reliability and maximize component life.

Common system/component start-up pitfalls:

There are some obvious and not so obvious reasons why start-up failures occur:

   a. Accumulated contamination in reservoirs, transmission lines, and components during manufacture or repair.

   b. Tape and liquid-type connector and pipe sealants used to seal connections.

   c. Pieces of rubber from hose assemblies, "O"-rings, and cylinder packings.

   d. Burrs from connectors, tube assemblies, metal fabrication, substandard replacement parts and component machining practices.

   e. Substandard component storage methods, ports not plugged adequately, masking tape and duct tape covering ports, or rags in ports.

   f. Sub-assemblies being "borrowed" from inventoried components leaving the housings open to the elements.

   g. Sub-assemblies being "borrowed" from existing operating systems to conduct "process-of-elimination" diagnostic procedures.

   h. Due to design clearances, certain "fixed clearance" pumps will not "pick-up" oil upon initial start-up. Oil is needed to create a "seal" to initiate a partial vacuum.

   i. Pressure manifold systems which are not equipped with air-bleed valves, can cause the pump to become "air-bound." This condition will prevent the pumping chambers from filling with oil.

j. Pump and motor housings are not filled with oil prior to start-up which can cause excessive bearing temperatures and loss of lubrication.

k. Plastic plugs which are inadvertently left in component ports upon start-up.

l. Incorrect prime mover input speed.

m. Incorrectly installed components.

n. Incorrectly adjusted components.

o. Incorrect components.

p. Incorrect rotation of prime mover.

q. Rags or cloths left in reservoir or reservoir transmission lines during fabrication.

r. Reservoir/pump isolation valve left closed upon start-up.

s. Incorrect oil.

t. Return line, directional control valve, heat exchanger, and associated components sized for pump flow rather than cylinder flow.

u. External pressure control valve drain lines inadvertently plugged.

v. Pump and motor case drain lines incorrectly plumbed, sized, or inadvertently plugged.

w. Incorrect hose type used on suction side of pump.

The start-up guidelines offered in this manual should be followed to avoid these and other start-up pitfalls.

notes

Troubleshooting
Quick-Reference Guide

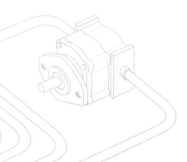

Look for the problem which most closely resembles the condition your system may be experiencing. Refer to the adjacent page number for possible causes and recommended course of action. Where necessary, specific component test procedures are recommended.

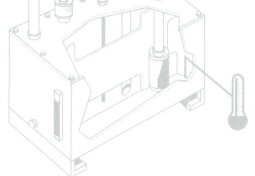

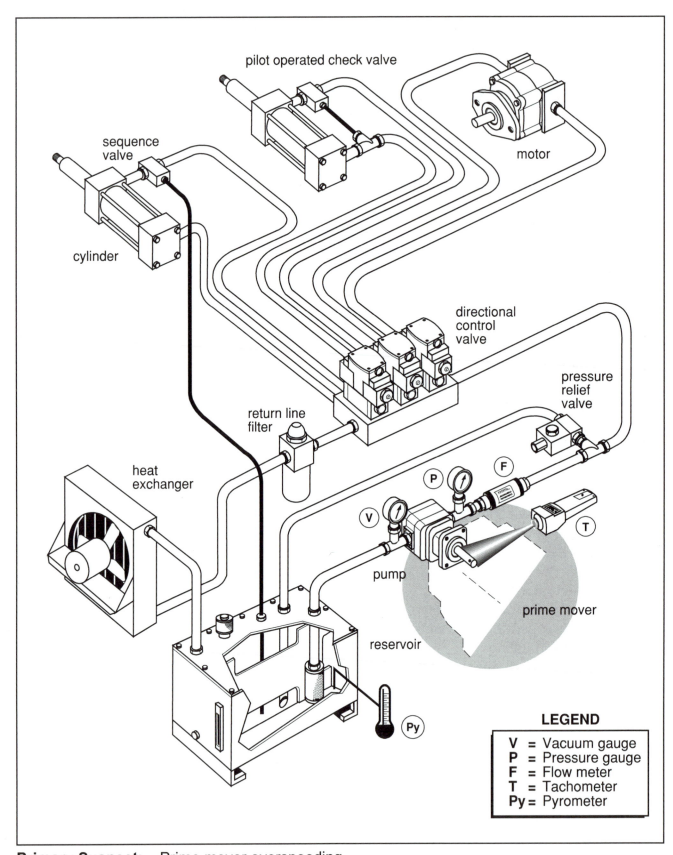

Primary Suspect: Prime mover overspeeding

PROBLEM

High pump input shaft speed

DIAGNOSTIC OBSERVATION

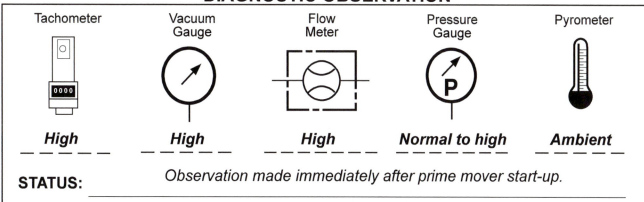

Tachometer	Vacuum Gauge	Flow Meter	Pressure Gauge	Pyrometer
High	*High*	*High*	*Normal to high*	*Ambient*

STATUS: *Observation made immediately after prime mover start-up.*

PRIMARY SUSPECT

Prime mover overspeeding

POSSIBLE CAUSE

Electric motor overspeeding.

Engine overspeeding.

Pump drive gear ratio incorrect.

RECOMMENDED COURSE OF ACTION

Refer to manufacturer's specification for recommended pump input speed. If unable to rectify, submit for application engineering review.

Is pump speed correct?

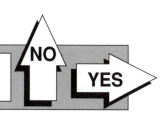

NO

YES

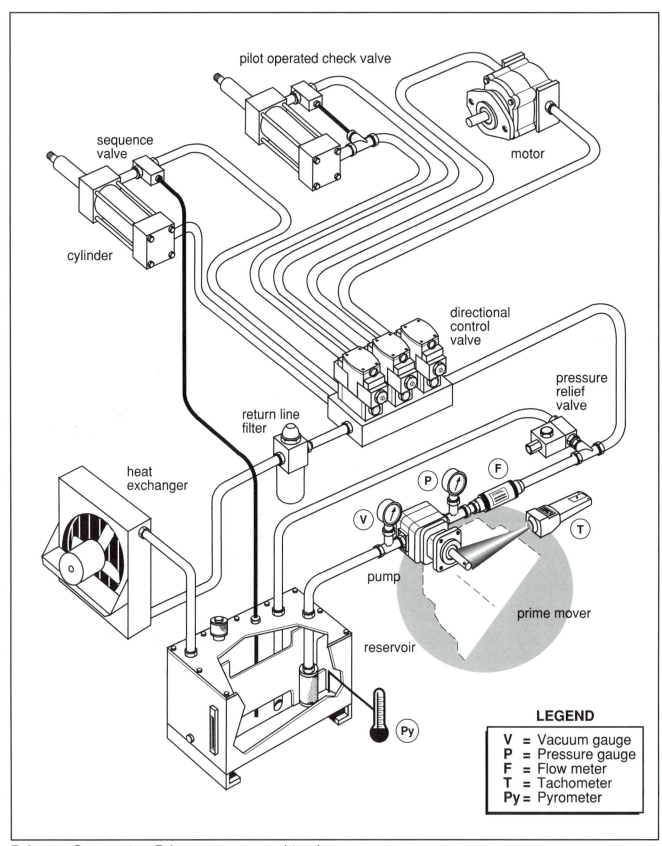

pilot operated check valve

sequence valve

motor

cylinder

directional control valve

pressure relief valve

return line filter

heat exchanger

P

F

V

T

pump

prime mover

reservoir

Py

LEGEND

V	=	Vacuum gauge
P	=	Pressure gauge
F	=	Flow meter
T	=	Tachometer
Py	=	Pyrometer

Primary Suspect: Prime mover speed too low

PROBLEM

Low pump input shaft speed

DIAGNOSTIC OBSERVATION

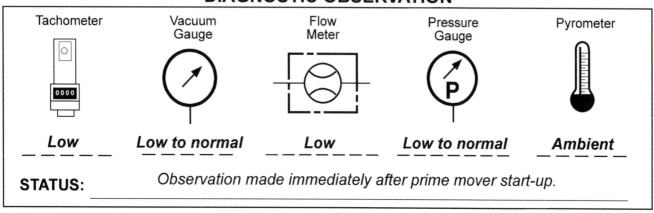

Tachometer	Vacuum Gauge	Flow Meter	Pressure Gauge	Pyrometer
Low	*Low to normal*	*Low*	*Low to normal*	*Ambient*

STATUS: *Observation made immediately after prime mover start-up.*

PRIMARY SUSPECT

Prime mover speed too low

POSSIBLE CAUSE

Electric motor speed too low.

Engine speed too low.

Pump drive gear ratio incorrect.

RECOMMENDED COURSE OF ACTION

Refer to manufacturer's specification for recommended pump input speed. If unable to rectify, submit for application engineering review.

Is pump speed correct?

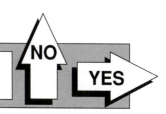

NO

YES

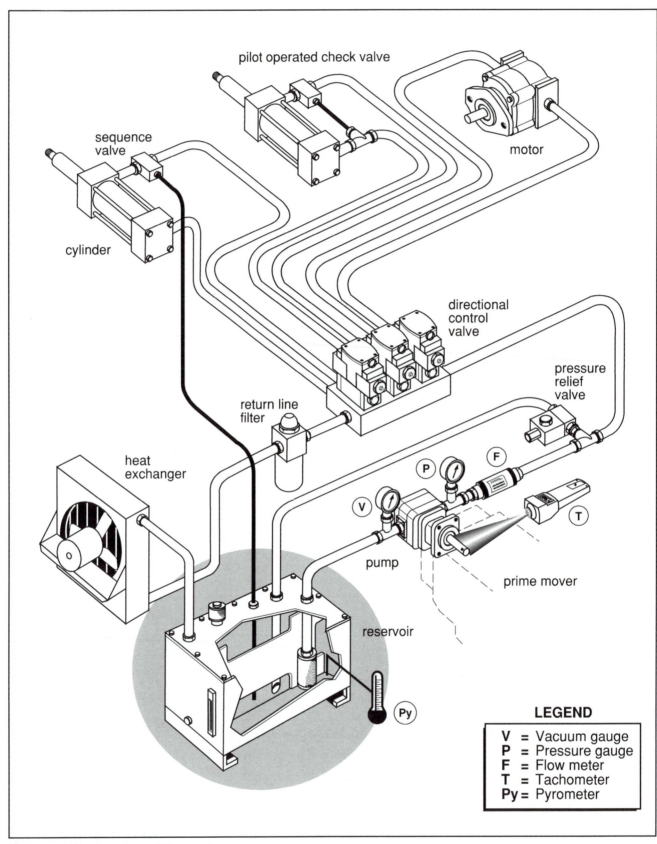

pilot operated check valve

sequence
valve

motor

cylinder

directional
control
valve

pressure
relief
valve

return line
filter

heat
exchanger

P

F

V

pump

prime mover

T

reservoir

Py

LEGEND

V	=	Vacuum gauge
P	=	Pressure gauge
F	=	Flow meter
T	=	Tachometer
Py	=	Pyrometer

Primary Suspect: Oil supply

PROBLEM

No oil at pump inlet

DIAGNOSTIC OBSERVATION

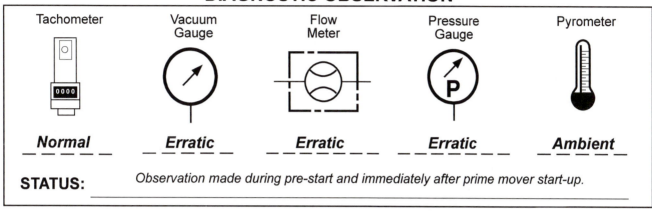

Tachometer	Vacuum Gauge	Flow Meter	Pressure Gauge	Pyrometer
Normal	*Erratic*	*Erratic*	*Erratic*	*Ambient*

STATUS: *Observation made during pre-start and immediately after prime mover start-up.*

PRIMARY SUSPECT

Oil supply

POSSIBLE CAUSE

No pre-start fill.

Transmission lines, actuators, coolers, filters, etc., consume total reservoir capacity upon start-up.

RECOMMENDED COURSE OF ACTION

Fill to correct level with recommended fluid.

Fill to correct level with recommended fluid.

Is oil at correct level?

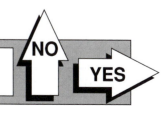

NO

YES

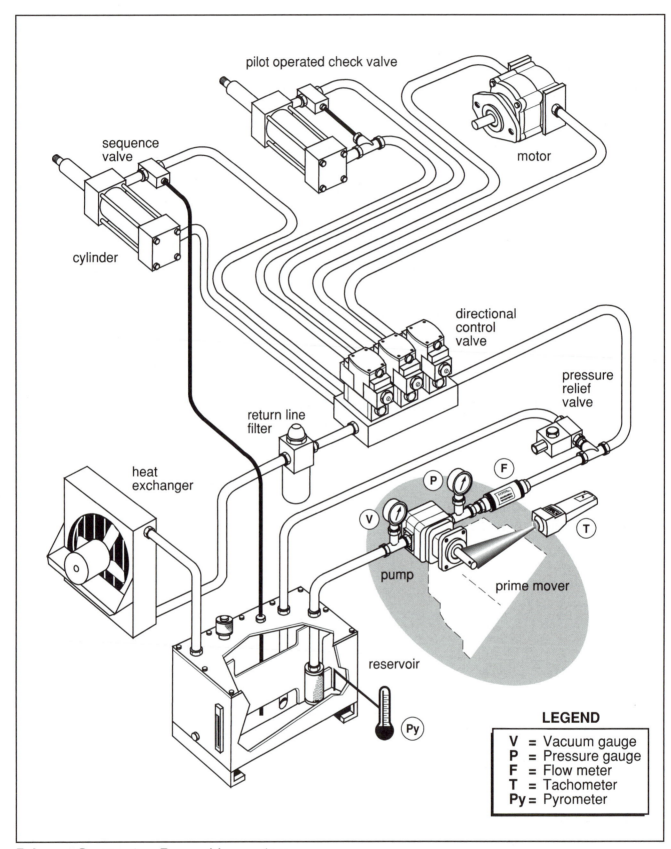

pilot operated check valve

sequence valve

motor

cylinder

directional control valve

pressure relief valve

return line filter

heat exchanger

P

F

V

T

pump

prime mover

reservoir

Py

LEGEND

V	=	Vacuum gauge
P	=	Pressure gauge
F	=	Flow meter
T	=	Tachometer
Py	=	Pyrometer

Primary Suspect: Pump drive system

PROBLEM

Pump not rotating

DIAGNOSTIC OBSERVATION

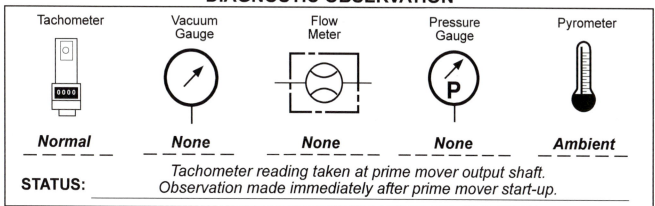

Tachometer	Vacuum Gauge	Flow Meter	Pressure Gauge	Pyrometer
Normal	*None*	*None*	*None*	*Ambient*

STATUS: *Tachometer reading taken at prime mover output shaft. Observation made immediately after prime mover start-up.*

PRIMARY SUSPECT

Pump drive system

POSSIBLE CAUSE

Keyway not installed on pump drive shaft.

Pump shaft not properly coupled.

Spline too small for female connector.

Pump drive shaft too short to engage.

Pump shaft sheared on start-up.

Power-Take-Off (PTO) not engaged.

Drive clutch slipping.

Drive "V"-belt slipping.

RECOMMENDED COURSE OF ACTION

Remove pump and inspect.

Inspect coupling.

Remove and measure.

Remove and measure.

Pump "dead-head" condition on start-up. Replace pump and follow "zero-fault" pump start-up procedure page 2-1.

Engage.

Check clutch adjustment & horsepower capability.

Check "V"-belt adjustment & horsepower capability.

Is pump shaft connected and rotating under all load conditions?

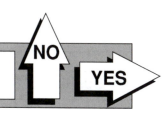

NO

YES

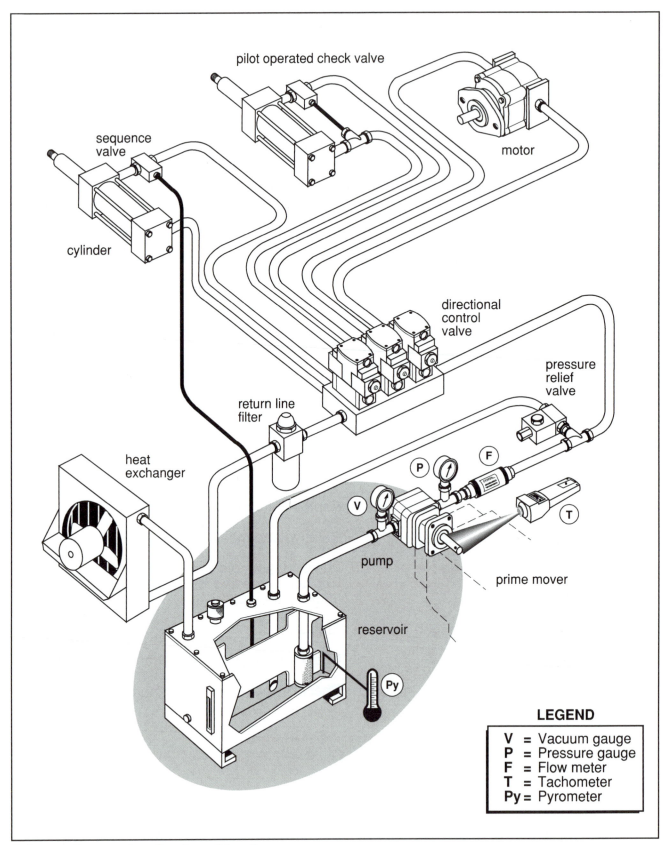

pilot operated check valve

sequence valve

motor

cylinder

directional control valve

pressure relief valve

return line filter

heat exchanger

V

P

F

T

pump

prime mover

reservoir

Py

LEGEND

V	=	Vacuum gauge
P	=	Pressure gauge
F	=	Flow meter
T	=	Tachometer
Py	=	Pyrometer

Primary Suspect: Suction side of pump

PROBLEM

Low pump inlet condition (vacuum)

DIAGNOSTIC OBSERVATION

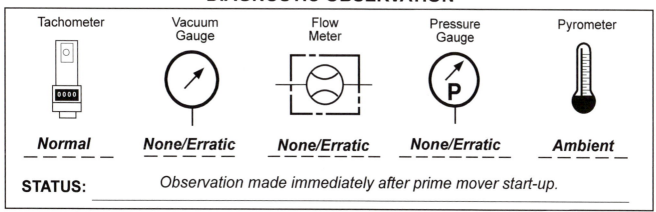

Tachometer	Vacuum Gauge	Flow Meter	Pressure Gauge	Pyrometer
Normal	*None/Erratic*	*None/Erratic*	*None/Erratic*	*Ambient*

STATUS: *Observation made immediately after prime mover start-up.*

PRIMARY SUSPECT

Suction side of pump

POSSIBLE CAUSE	RECOMMENDED COURSE OF ACTION
Air leak at suction side of pump.	Refer to chapter three for root-cause analysis.
Improper rebuild.	Refer to chapter three for root-cause analysis.
Pump not primed - Certain fixed clearance pumps require fluid to seal clearances before partial vacuum can be created.	Fill pumping chambers with fluid.
Pump airbound - If pump inlet is aerated and pump output is flowing to a closed center system, air cannot be evacuated, and the pump will not receive fluid.	Bleed pump outlet port. If necessary install air-bleed valve.
Mechanical flow adjustment on pump in neutral position.	Adjust to recommended flow.

(Continued on next page)

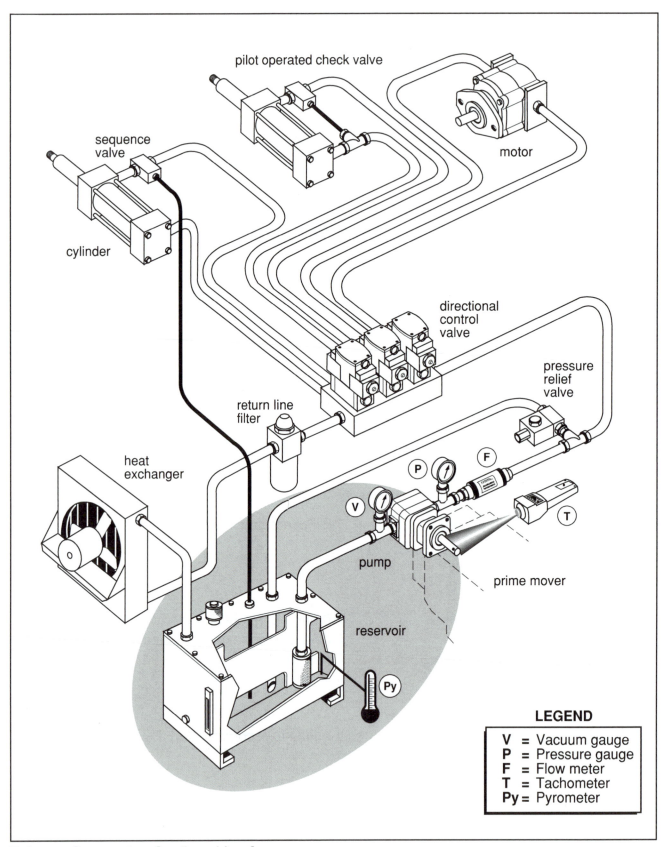

Primary Suspect: Suction side of pump

POSSIBLE CAUSE	RECOMMENDED COURSE OF ACTION
Incorrect pump rotation.	Check rotation on pump.
Multiple pump application (tandem mount) intermediate coupling not installed or broken.	Check and install.
Multiple pump application (tandem mount) rear pump incorrectly assembled.	Remove and change configuration for proper rotation.

Has low inlet condition been rectified?

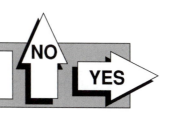

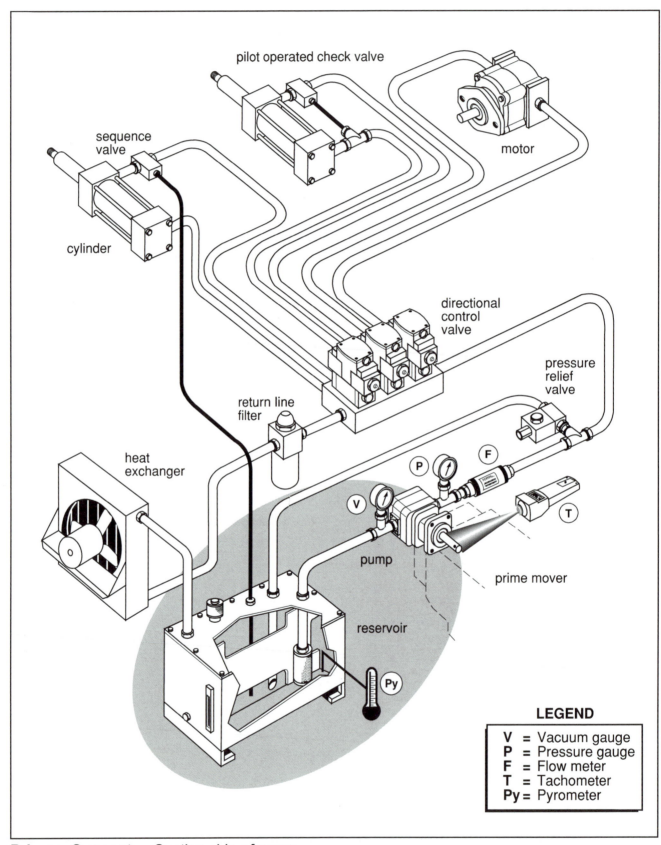

pilot operated check valve

sequence valve

motor

cylinder

directional control valve

pressure relief valve

return line filter

heat exchanger

pump

prime mover

reservoir

LEGEND

V	=	Vacuum gauge
P	=	Pressure gauge
F	=	Flow meter
T	=	Tachometer
Py	=	Pyrometer

Primary Suspect: Suction side of pump

PROBLEM

High pump inlet condition (vacuum)

DIAGNOSTIC OBSERVATION

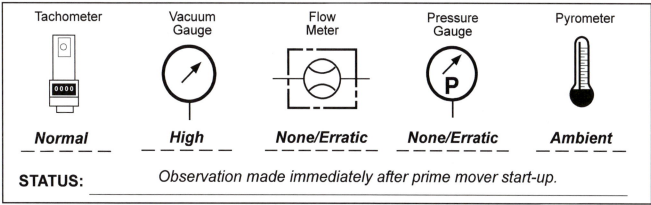

Tachometer	Vacuum Gauge	Flow Meter	Pressure Gauge	Pyrometer
Normal	*High*	*None/Erratic*	*None/Erratic*	*Ambient*

STATUS: *Observation made immediately after prime mover start-up.*

PRIMARY SUSPECT

Suction side of pump

POSSIBLE CAUSE	RECOMMENDED COURSE OF ACTION
Manufacturing debris or cleaning rag in suction line.	Inspect and remove.
Plastic plug in pump suction port (possible with split-flange adapter).	Remove split-flange and inspect.
Suction screen too small.	Submit for application engineering review.
Suction filter application design.	Most open-loop pump manufacturers recommend no suction line filtration other than screens. Refer to specific pump manufacturer's specifications for suction line filtration recommendations. Submit for application engineering review.
Incorrect oil viscosity.	Refer to oil specifications relative to ambient temperature and operating conditions. Submit for application engineering review.

(Continued on next page)

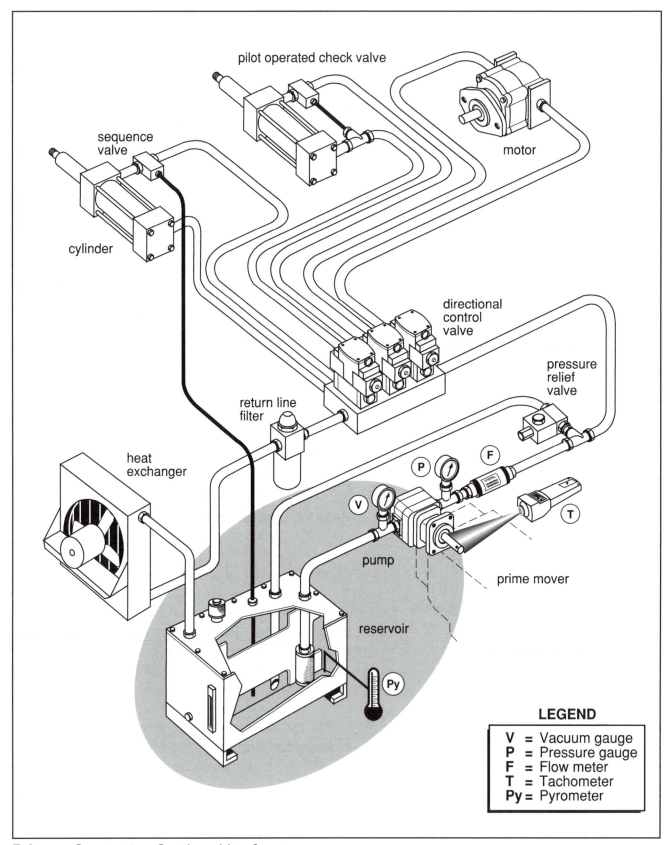

Primary Suspect: Suction side of pump

POSSIBLE CAUSE	RECOMMENDED COURSE OF ACTION
Hose collapses under vacuum (external).	Check hose specification relative to suction line applications. Submit for application engineering review.
Hose collapses under vacuum (internal).	Check hose specification relative to suction line applications. Submit for application engineering review.
Internal hose liner damage due to poor connector installation procedures causing "flap" to behave like a check valve.	Use vacuum gauge to isolate problem. Remove hose and inspect inner wall for damage.
No reservoir breather.	Install breather on reservoir.
Port in pump housing too small.	In most cases this problem cannot be detected at the pump inlet. Check the pump inlet port size relative to system operating temperature and maximum pump flow.
Anti-syphon check valve in suction line incorrectly installed.	Check for "free-flow" arrow on check valve or refer to port markings. Re-install correctly.

**Refer to chapter three
for additional test procedures.**

Has high inlet condition been rectified?

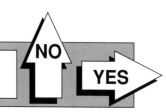

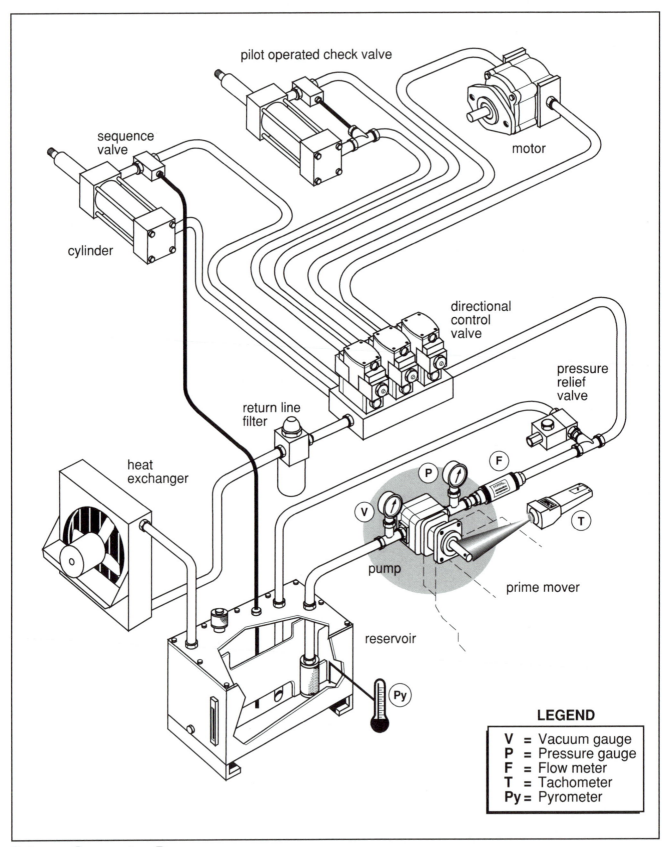

pilot operated check valve

sequence
valve

cylinder

motor

directional
control
valve

pressure
relief
valve

return line
filter

heat
exchanger

P

F

V

T

pump

prime mover

reservoir

Py

LEGEND

V =	Vacuum gauge
P =	Pressure gauge
F =	Flow meter
T =	Tachometer
Py =	Pyrometer

Primary Suspect: Pump

PROBLEM

Low pump flow condition

DIAGNOSTIC OBSERVATION

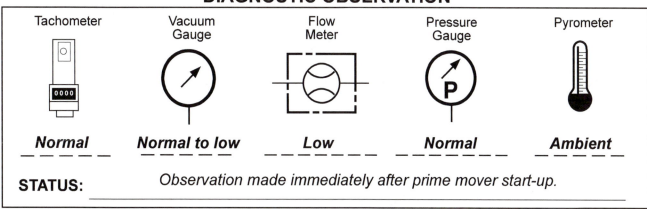

Tachometer	Vacuum Gauge	Flow Meter	Pressure Gauge	Pyrometer
Normal	*Normal to low*	*Low*	*Normal*	*Ambient*

STATUS: *Observation made immediately after prime mover start-up.*

PRIMARY SUSPECT

Pump

POSSIBLE CAUSE	RECOMMENDED COURSE OF ACTION
Incorrect pump. Pump flow not consistent with design criteria.	Replace pump.
Mechanical flow adjustment on pump not adjusted for system flow requirements.	Refer to pump manufacturer's specifications for flow adjustment procedures. Adjust to recommended flow rate.
Variable pump not stroking to "full-flow" position- manual or electronic control.	Check manual control adjustment. Check electronic control adjustment.
Rebuilt pump- Incorrect replacement parts.	Replace pump.

Is pump flow consistent with manufacturers specifications?

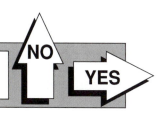

NO

YES

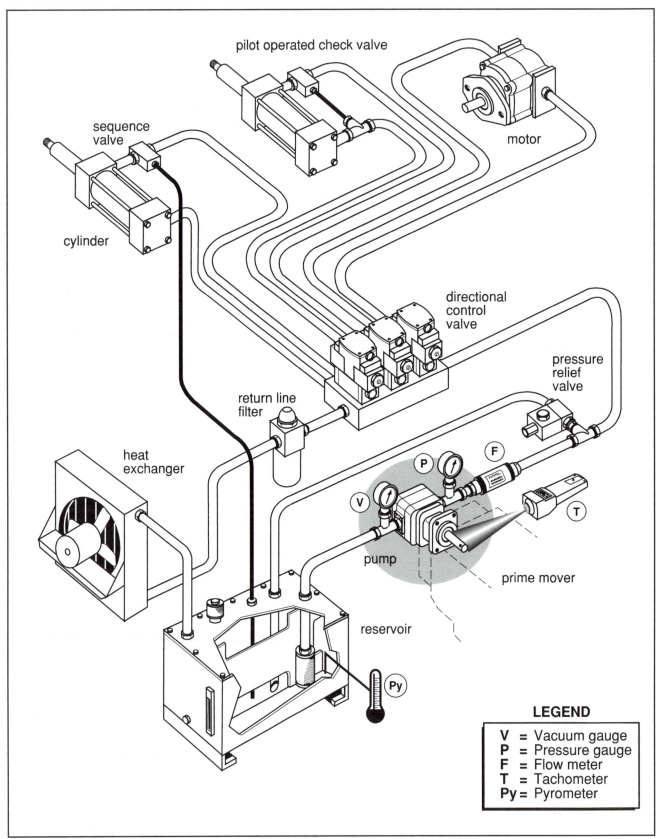

pilot operated check valve

motor

sequence valve

cylinder

directional control valve

pressure relief valve

return line filter

heat exchanger

V

P

F

T

pump

prime mover

reservoir

Py

LEGEND

V	=	Vacuum gauge
P	=	Pressure gauge
F	=	Flow meter
T	=	Tachometer
Py	=	Pyrometer

Primary Suspect: Pump

PROBLEM

High pump flow condition

DIAGNOSTIC OBSERVATION

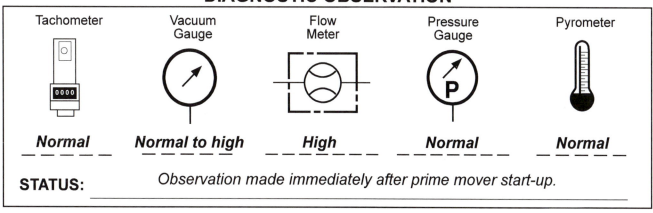

Tachometer	Vacuum Gauge	Flow Meter	Pressure Gauge	Pyrometer
Normal	*Normal to high*	*High*	*Normal*	*Normal*

STATUS: *Observation made immediately after prime mover start-up.*

PRIMARY SUSPECT

Pump

POSSIBLE CAUSE	RECOMMENDED COURSE OF ACTION
Incorrect pump. Pump flow not consistent with design criteria.	Replace pump.
Mechanical flow adjustment on pump not adjusted for system flow requirements.	Refer to pump manufacturer's specifications for flow adjustment procedures. Adjust to recommended flow rate.
Variable pump control stroking too far -- manual or electronic control.	Refer to pump manufacturer's specifications for maximum and minimum flow adjustment procedures. Adjust to recommended maximum flow rate.
Rebuilt pump -- incorrect replacement parts.	Replace pump.

Is pump flow consistent with manufacturers specifications?

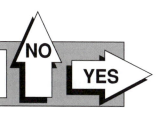

NO

YES

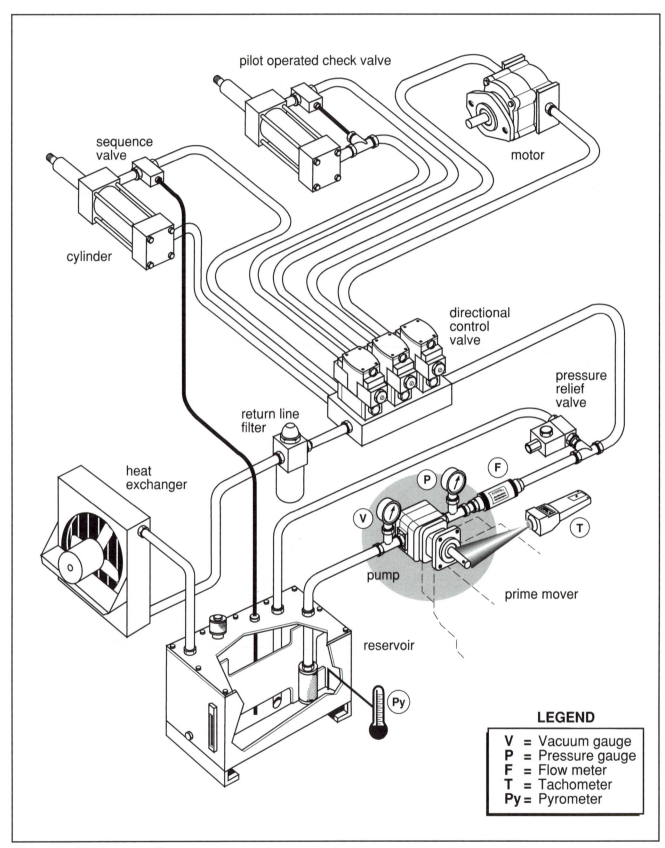

pilot operated check valve

sequence
valve

cylinder

motor

directional
control
valve

return line
filter

pressure
relief
valve

heat
exchanger

pump

prime mover

reservoir

LEGEND

V	=	Vacuum gauge
P	=	Pressure gauge
F	=	Flow meter
T	=	Tachometer
Py	=	Pyrometer

Primary Suspect: Pump

PROBLEM

Pump flow decreases as pressure in the system increases

DIAGNOSTIC OBSERVATION

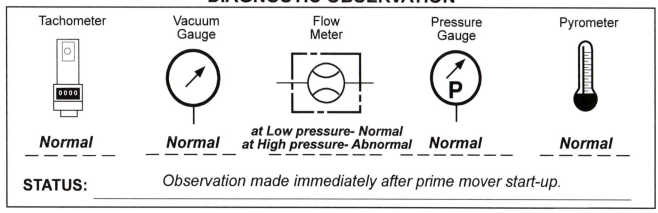

Tachometer	Vacuum Gauge	Flow Meter	Pressure Gauge	Pyrometer
Normal	*Normal*	at Low pressure- Normal at High pressure- Abnormal	*Normal*	*Normal*

STATUS: *Observation made immediately after prime mover start-up.*

PRIMARY SUSPECT

Pump

POSSIBLE CAUSE

Pressure-compensated vane pump-- direct-operated. Pressure set below value of compensator spring setting causing pump to destroke prematurely.

Pump operating below maximum rated flow. Internal leakage is a larger percentage of reduced flow rates.

RECOMMENDED COURSE OF ACTION

Adjust pressure compensator back to original setting or exchange compensator spring to comply with compensator pressure setting.

Test pump at maximum flow output (at maximum speed).

Could be caused by:
1. Reducing flow (variable speed input) to "fit" flow meter which is undersized for pump being tested.
2. Insufficient test horsepower. Low flow, high pressure test procedure on pump with low volumetric efficiency.

Is pump flow consistent with volumetric efficiency relative to pump type?

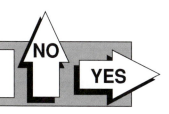

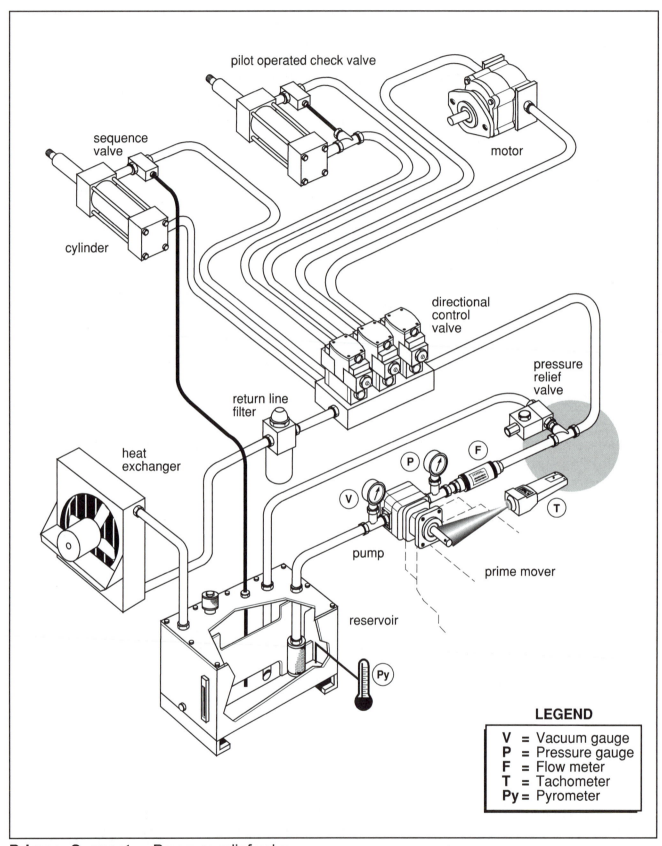

sequence
valve

pilot operated check valve

motor

cylinder

directional
control
valve

pressure
relief
valve

return line
filter

heat
exchanger

V

P

F

T

pump

prime mover

reservoir

Py

LEGEND

V	=	Vacuum gauge
P	=	Pressure gauge
F	=	Flow meter
T	=	Tachometer
Py	=	Pyrometer

Primary Suspect: Pressure relief valve

PROBLEM

Actuator fails to respond in either direction when the directional control valve is activated

DIAGNOSTIC OBSERVATION

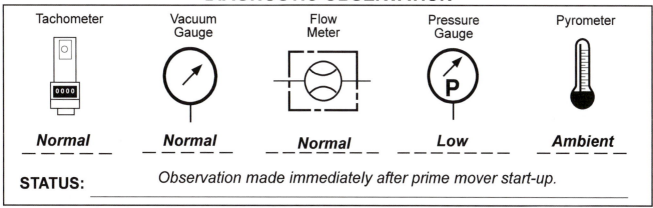

Tachometer	Vacuum Gauge	Flow Meter	Pressure Gauge	Pyrometer
Normal	*Normal*	*Normal*	*Low*	*Ambient*

STATUS: *Observation made immediately after prime mover start-up.*

PRIMARY SUSPECT

Pressure relief valve

NOTE: Flow may reduce if flow meter is installed downstream of the pressure relief valve.

POSSIBLE CAUSE	RECOMMENDED COURSE OF ACTION
Pressure adjustment too low.	Adjust to recommended pressure.
Orifice in pilot-operated relief valve is plugged with debris from initial system start-up.	Service or replace valve. Refer to chapter four for root-cause analysis.
Transmission lines connected to incorrect ports.	Check port reference on valve and reconnect.
Remote pilot port (pilot-operated) connected to tank.	Plug remote pilot port and adjust relief valve to recommended pressure.

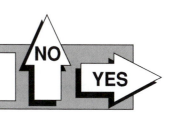

Is pressure relief valve applied, operating, and adjusted according to design specifications?

NO

YES

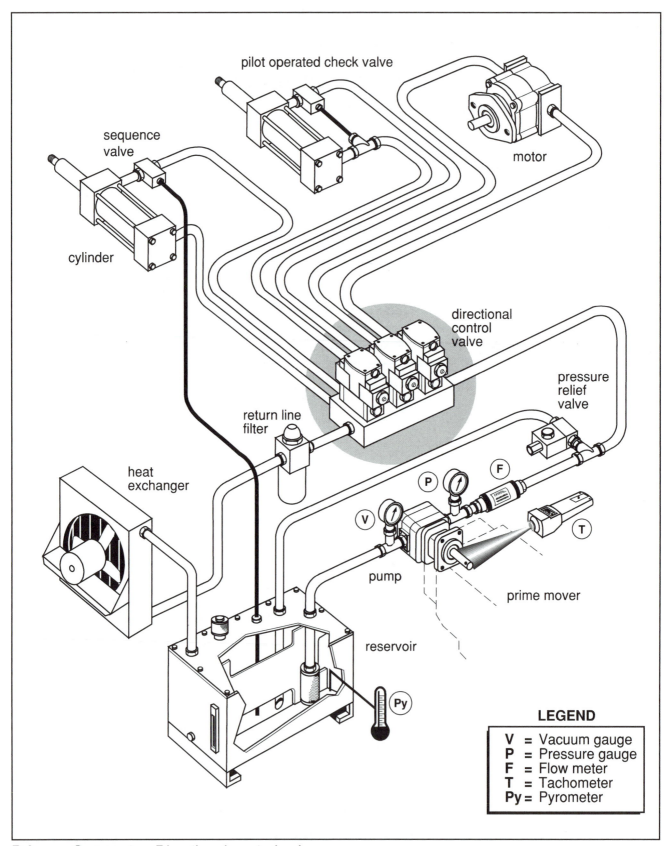

pilot operated check valve

sequence valve

cylinder

motor

directional control valve

pressure relief valve

return line filter

heat exchanger

P

F

V

T

pump

prime mover

reservoir

Py

LEGEND

V =	Vacuum gauge
P =	Pressure gauge
F =	Flow meter
T =	Tachometer
Py =	Pyrometer

Primary Suspect: Directional control valve

PROBLEM

Actuator fails to respond in either direction when the directional control valve is activated

DIAGNOSTIC OBSERVATION

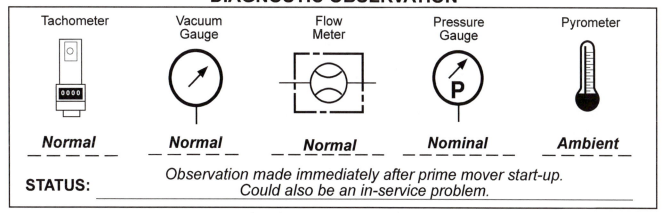

Tachometer	Vacuum Gauge	Flow Meter	Pressure Gauge	Pyrometer
Normal	*Normal*	*Normal*	*Nominal*	*Ambient*

STATUS: *Observation made immediately after prime mover start-up. Could also be an in-service problem.*

PRIMARY SUSPECT

Directional control valve

POSSIBLE CAUSE	RECOMMENDED COURSE OF ACTION
Pilot pressure - Oil.	Refer to chapter five for industrial valve root-cause analysis. Refer to chapter seven for mobile valve root-cause analysis.
Pilot pressure - Air.	Refer to chapter five for industrial valve root-cause analysis. Refer to chapter seven for mobile valve root-cause analysis.
Solenoid wiring.	Check wiring and voltage.
Directional control valve spool jammed -- solenoid-controlled or air-piloted industrial valve. Will occur in valves with low solenoid forces or low pilot pressures.	Loosen and re-torque mounting bolts. Check sub-plate mounting surface for flatness. Check for valve sub-plate distortion from tapered pipe connectors.
Directional control valve spool jammed -- solenoid-controlled or air-piloted mobile valve. Will occur in valves with low solenoid forces or low pilot pressures.	Loosen and re-torque tie-rod bolts. Check for valve body distortion from tapered pipe fittings.

(Continued on next page)

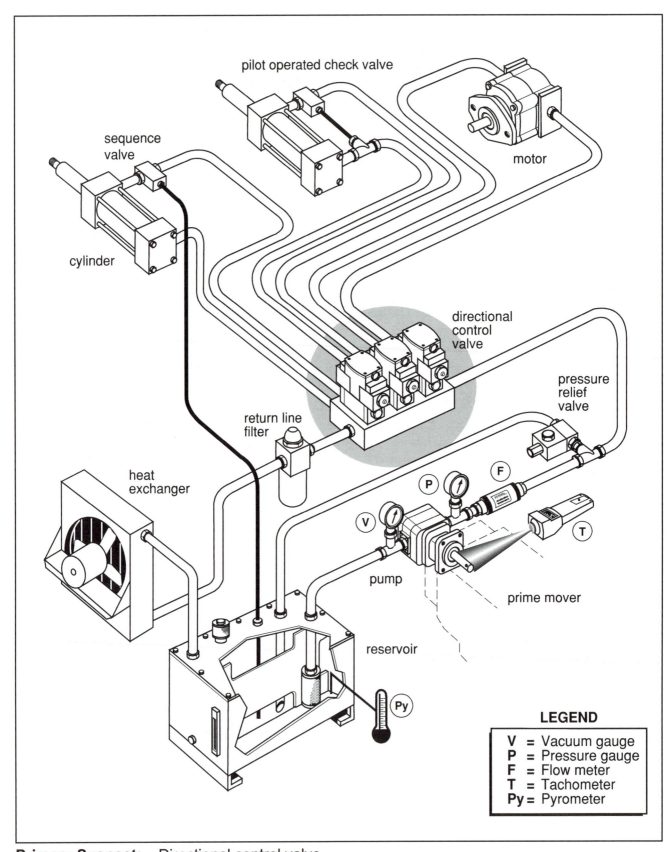

pilot operated check valve

motor

sequence
valve

cylinder

directional
control
valve

pressure
relief
valve

return line
filter

heat
exchanger

pump

prime mover

reservoir

LEGEND

V	=	Vacuum gauge
P	=	Pressure gauge
F	=	Flow meter
T	=	Tachometer
Py	=	Pyrometer

Primary Suspect: Directional control valve

POSSIBLE CAUSE	RECOMMENDED COURSE OF ACTION
Parallel manifold in place of series manifold.	Check part number on manifold. Submit for application engineering review.
No back-pressure check valve.	Check and install if missing. Submit for application engineering review.

Is Directional Control Valve operating according to design specifications?

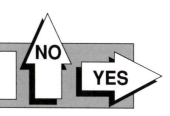

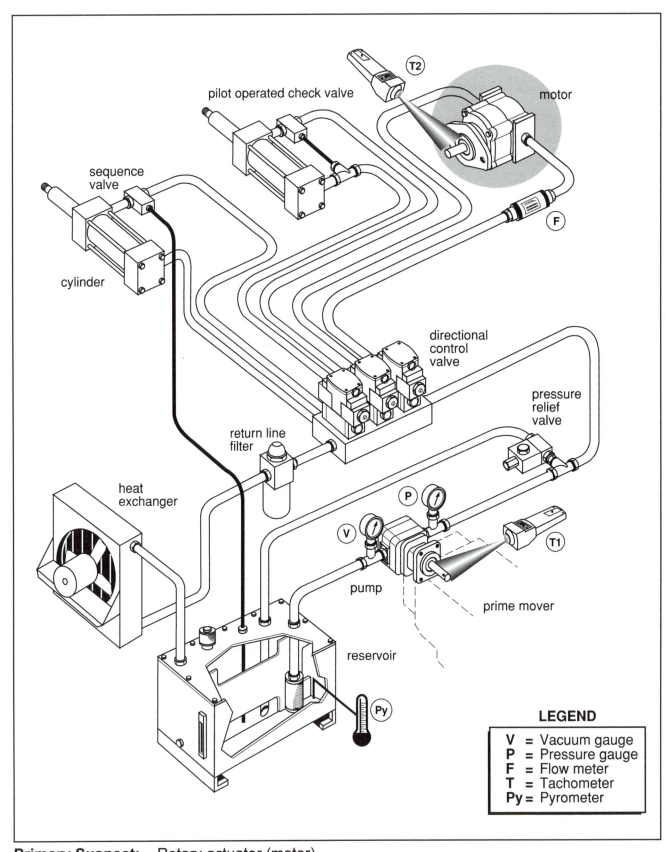

pilot operated check valve

T2

motor

sequence valve

F

cylinder

directional control valve

pressure relief valve

return line filter

heat exchanger

P

V

T1

pump

prime mover

reservoir

Py

LEGEND

V	=	Vacuum gauge
P	=	Pressure gauge
F	=	Flow meter
T	=	Tachometer
Py	=	Pyrometer

Primary Suspect: Rotary actuator (motor)

PROBLEM

Actuator fails to respond in either direction when the directional control valve is activated

DIAGNOSTIC OBSERVATION

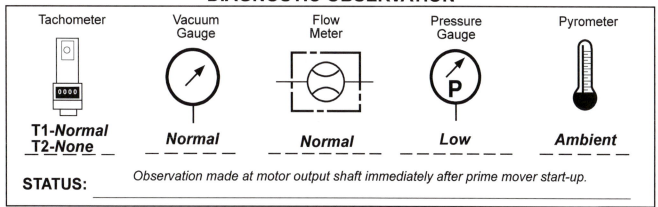

Tachometer	Vacuum Gauge	Flow Meter	Pressure Gauge	Pyrometer
T1-*Normal* T2-*None*	*Normal*	*Normal*	*Low*	*Ambient*

STATUS: *Observation made at motor output shaft immediately after prime mover start-up.*

PRIMARY SUSPECT

Rotary actuator (motor)

POSSIBLE CAUSE

Motor incorrectly assembled.

Motor drive shaft not coupled.

Motor drive shaft keyway not installed.

Motor drive shaft too short.

RECOMMENDED COURSE OF ACTION

Refer to chapter nine for root-cause analysis.

Remove and inspect.

Remove and inspect.

Remove and inspect.

Has proper operation of the motor been determined?

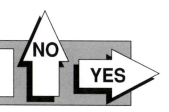

NO

YES

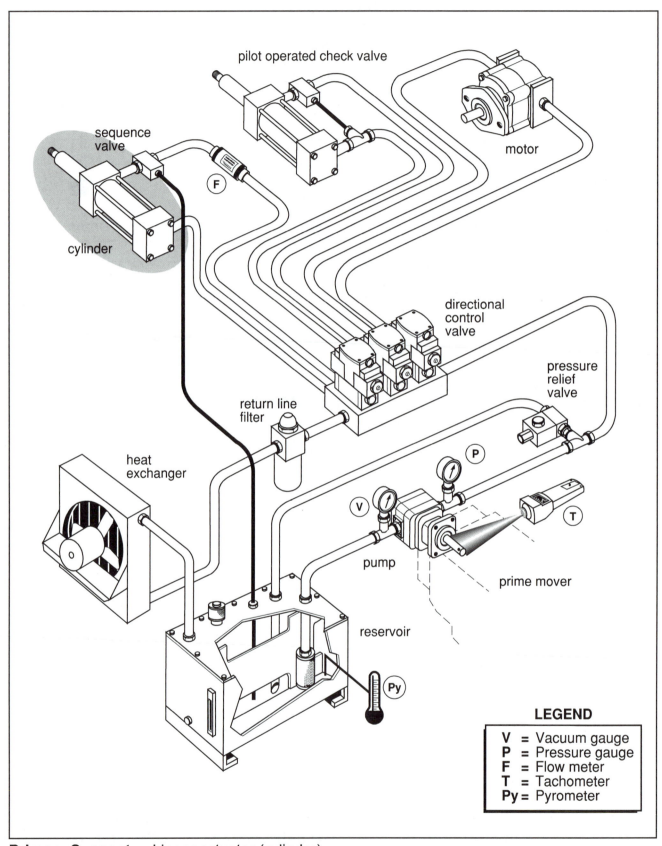

Primary Suspect: Linear actuator (cylinder)

PROBLEM

Actuator fails to respond in either direction when the directional control valve is activated

DIAGNOSTIC OBSERVATION

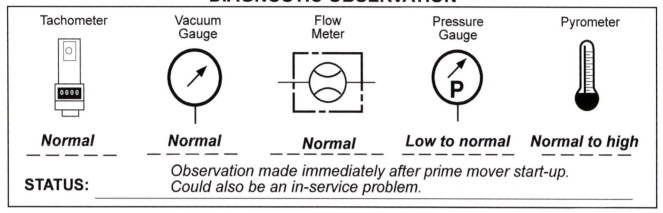

Tachometer	Vacuum Gauge	Flow Meter	Pressure Gauge	Pyrometer
Normal	*Normal*	*Normal*	*Low to normal*	*Normal to high*

STATUS: *Observation made immediately after prime mover start-up. Could also be an in-service problem.*

PRIMARY SUSPECT

Linear actuator (cylinder)

POSSIBLE CAUSE	RECOMMENDED COURSE OF ACTION
Seal damage during assembly.	Refer to chapter eight for root-cause analysis.
Bore/piston clearance.	Refer to chapter eight for root-cause analysis.
Seal worn.	Refer to chapter eight for root-cause analysis.

Has proper operation of cylinder been determined?

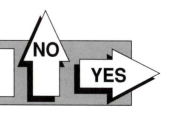

NO

YES

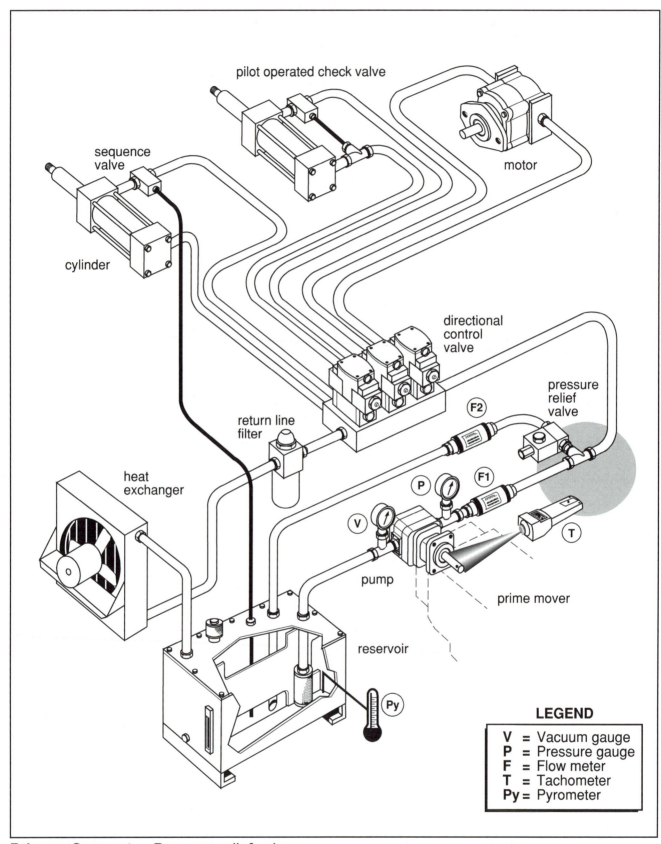

pilot operated check valve

sequence valve

motor

cylinder

directional control valve

pressure relief valve

return line filter

F2

heat exchanger

P **F1**

V

pump

prime mover

T

reservoir

Py

LEGEND

V	=	Vacuum gauge
P	=	Pressure gauge
F	=	Flow meter
T	=	Tachometer
Py	=	Pyrometer

Primary Suspect: Pressure relief valve

PROBLEM

Erratic actuator operation

DIAGNOSTIC OBSERVATION

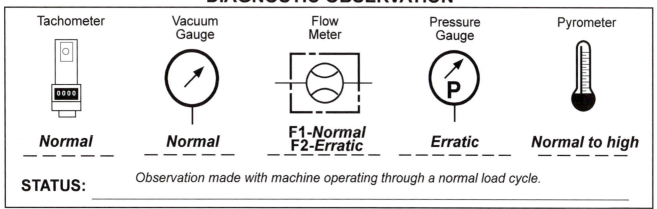

Tachometer	Vacuum Gauge	Flow Meter	Pressure Gauge	Pyrometer
Normal	*Normal*	**F1-Normal** **F2-Erratic**	*Erratic*	*Normal to high*

STATUS: *Observation made with machine operating through a normal load cycle.*

PRIMARY SUSPECT

Pressure relief valve

POSSIBLE CAUSE	**RECOMMENDED COURSE OF ACTION**
Direct-operated relief valve -- early "cracking".	Replace with pilot-operated relief valve.
Direct-operated relief valve -- spring ratio.	Exchange for spring with correct pressure ratio.
Relief valve set too low.	Adjust to recommended pressure.
Relief valve application -- insufficient pressure to overcome initial load acceleration.	Submit for application engineering review.
High vent spring not installed in pilot-operated relief valve. Unable to generate pilot pressure.	Install high vent spring.
Pressure relief valve worn or damaged.	Refer to chapter four for root-cause analysis.

Has proper application, operation and adjustment of relief valve been determined?

NO

YES

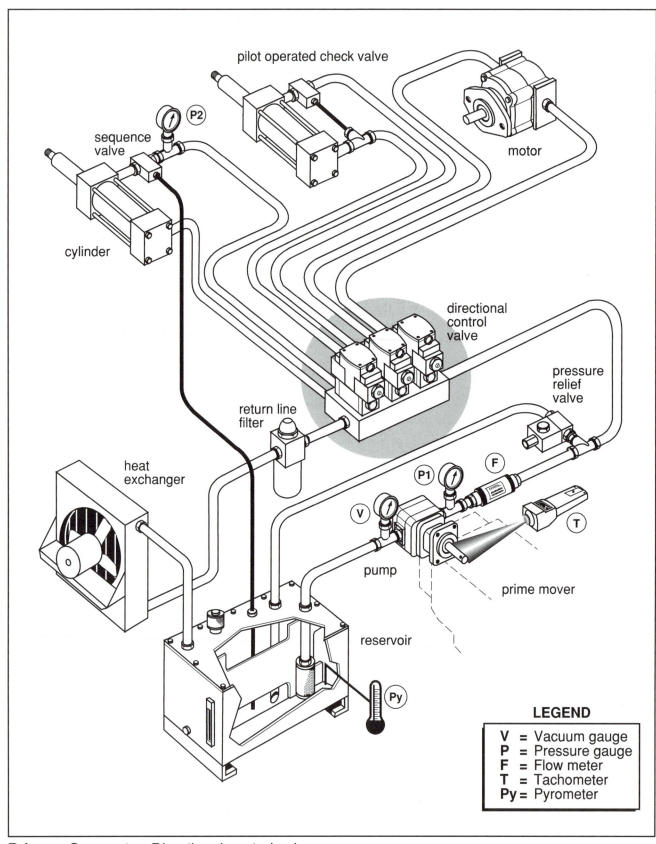

pilot operated check valve

(P2)

sequence
valve

cylinder

motor

directional
control
valve

pressure
relief
valve

return line
filter

heat
exchanger

(P1) (F)

(V)

(T)

pump

prime mover

reservoir

(Py)

LEGEND

V	=	Vacuum gauge
P	=	Pressure gauge
F	=	Flow meter
T	=	Tachometer
Py	=	Pyrometer

Primary Suspect: Directional control valve

PROBLEM

Erratic actuator operation

DIAGNOSTIC OBSERVATION

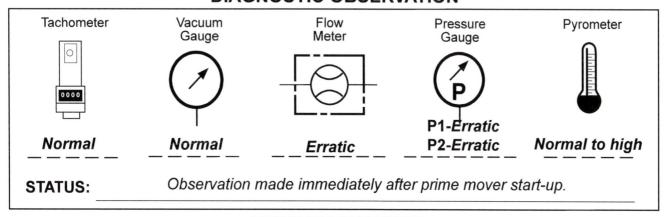

Tachometer	Vacuum Gauge	Flow Meter	Pressure Gauge	Pyrometer
Normal	*Normal*	*Erratic*	P1-*Erratic* P2-*Erratic*	*Normal to high*

STATUS: *Observation made immediately after prime mover start-up.*

PRIMARY SUSPECT

Directional control valve

POSSIBLE CAUSE

Valve sizing -- not sized correctly for pump flow or flow intensification from linear actuator applications.

Pilot pressure (air or oil) insufficient to control main spool.

Back-pressure check valve not installed.

Valve not configured correctly for application.

Oil contamination -- system not flushed correctly on start-up.

Solenoid voltage too low.

Excessive return-line back pressure.

Directional control valve spool "jamming."

RECOMMENDED COURSE OF ACTION

Refer to "loop-drop" pressure data in manufacturer's specifications. Take into account rod-to-bore ratio in linear actuator applications.

Refer to chapters five and seven for root-cause analysis.

Refer to chapter five for root-cause analysis.

Refer to chapter six for correct valve configuration procedures.

Replace or service valve and flush system.

Refer to machine specifications for correct solenoid voltage. Replace solenoid or complete valve.

Refer to chapter five for root-cause analysis.

Loosen and re-torque mounting bolts and check alignment of subplate.

Has proper application and operation of directional control valve been determined?

NO

YES

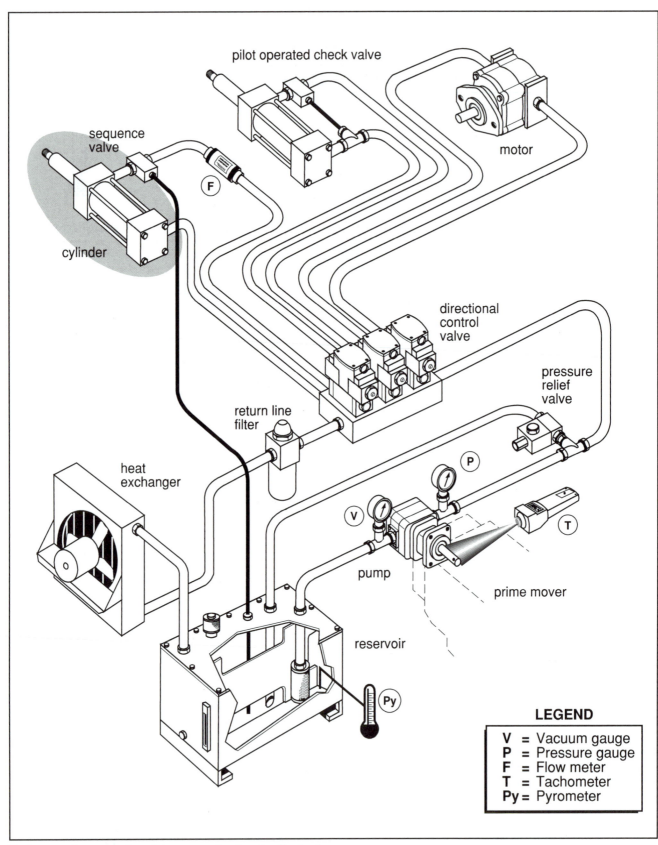

pilot operated check valve

motor

sequence valve

F

cylinder

directional control valve

pressure relief valve

return line filter

heat exchanger

P

V

T

pump

prime mover

reservoir

Py

LEGEND

V = Vacuum gauge
P = Pressure gauge
F = Flow meter
T = Tachometer
Py = Pyrometer

Primary Suspect: Linear actuator (cylinder)

PROBLEM

Erratic actuator operation

DIAGNOSTIC OBSERVATION

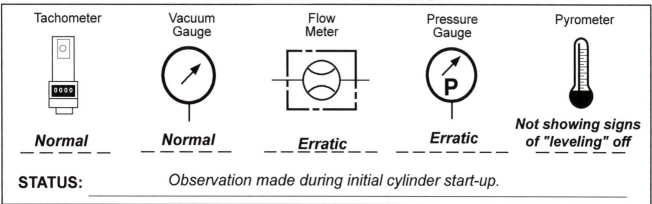

Tachometer	Vacuum Gauge	Flow Meter	Pressure Gauge	Pyrometer
Normal	*Normal*	*Erratic*	*Erratic*	*Not showing signs of "leveling" off*

STATUS: *Observation made during initial cylinder start-up.*

PRIMARY SUSPECT

Linear actuator (cylinder)

POSSIBLE CAUSE	RECOMMENDED COURSE OF ACTION
Cylinder tube damaged, retarding piston/rod movement.	Refer to chapter eight for root-cause analysis.
Cylinder seals too tight, retarding piston/rod movement.	Refer to chapter eight for root-cause analysis.
Excessive load.	Check pressure relief valve setting and compare to pressure when load is being moved. Submit for application engineering review.
Mechanical binding in rotary or linear load.	Check mechanical aspect of machine for binding, and repair.
Cylinder seal damage during assembly or due to storage methods.	Refer to chapter eight for root-cause analysis.
Telescoping cylinder - taking load in late stages - area reduction causing erratic operation.	Check pressure as cylinder "stages." If pressure approaches relief valve setting in late stages, submit for application engineering review.

(Continued on next page)

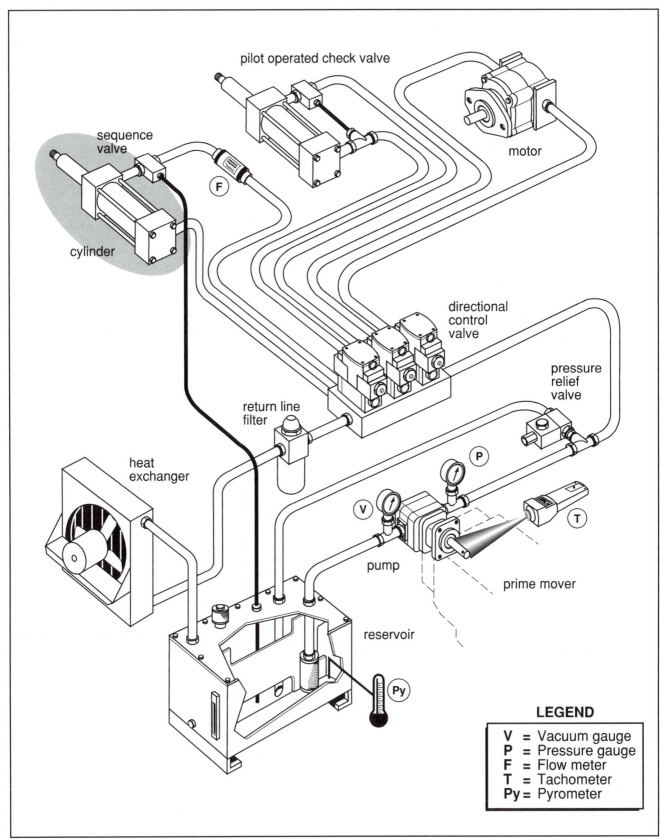

Primary Suspect: Linear actuator (cylinder)

POSSIBLE CAUSE	RECOMMENDED COURSE OF ACTION
Cylinder operating geometry -- increasing or decreasing angles causes erratic operation.	Submit for application engineering review.
Air in cylinder.	Follow "air-bleed" procedure.
Meter-in flow control circuit with no positive load on output.	Normal variation. If troublesome, change to meter-out flow control. Refer to application engineering data regarding pressure intensification.
Excessive transmission line length. In certain cases, actuator distance, from directional control valve, exceeds design considerations.	Submit for application engineering review.
Pilot-operated check valve.	Pilot pressure fluctuating due to fluctuating load. Submit for application engineering review. Check pilot ratios.

Has proper application and operation of the cylinder been determined?

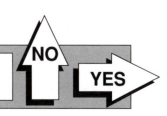

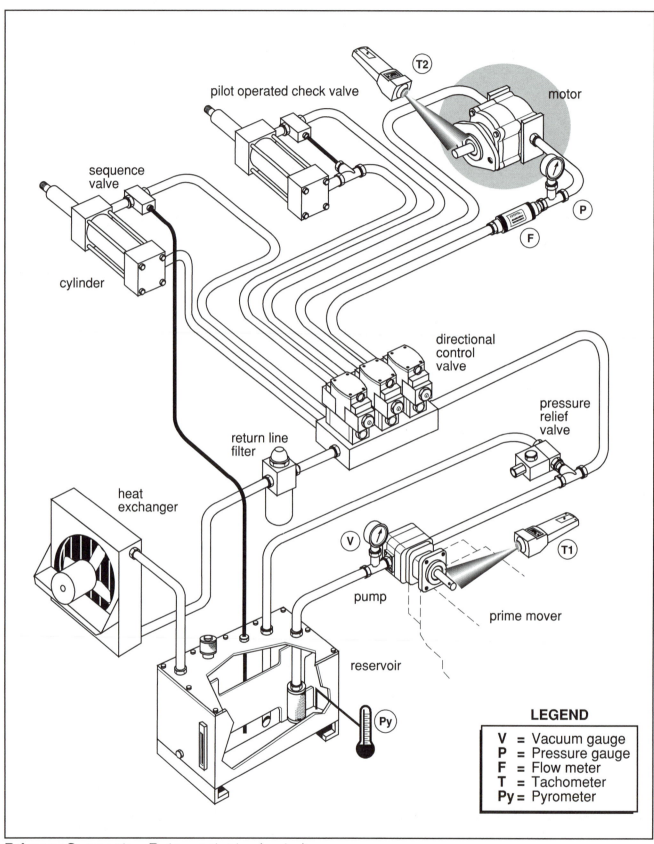

pilot operated check valve

T2

motor

sequence valve

P

F

cylinder

directional control valve

pressure relief valve

return line filter

heat exchanger

V

T1

pump

prime mover

reservoir

Py

LEGEND

V	=	Vacuum gauge
P	=	Pressure gauge
F	=	Flow meter
T	=	Tachometer
Py	=	Pyrometer

Primary Suspect: Rotary actuator (motor)

PROBLEM

Erratic actuator operation

DIAGNOSTIC OBSERVATION

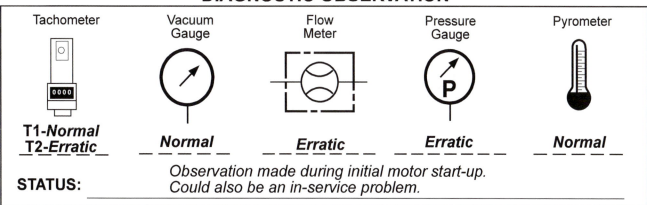

Tachometer	Vacuum Gauge	Flow Meter	Pressure Gauge	Pyrometer
T1-*Normal* T2-*Erratic*	*Normal*	*Erratic*	*Erratic*	*Normal*

STATUS: *Observation made during initial motor start-up.*
Could also be an in-service problem.

PRIMARY SUSPECT

Rotary actuator (motor)

POSSIBLE CAUSE	RECOMMENDED COURSE OF ACTION
Operating speed too slow.	Check type of motor. Refer to manufacturer's specification for lowest operating speed.
Meter-in flow control will cause motor shaft speed to fluctuate if there is no shaft load.	If troublesome, change to meter-out flow control. Check motor manufacturer's specification for shaft seal pressure recommendations.
Excessive load.	Check system pressure during "load" cycle. If pressure approaches relief valve setting, submit problem for application engineering review.
Inconsistent load - causing pressure fluctuations.	Check system pressure during "load" cycle. If pressure is erratic due to load changes, submit problem for application engineering review.
Mechanical binding caused by misaligned or oval sprockets (chain drive).	Check mechanical aspect of machine for binding, and repair.
Motor brakes (spring apply/hydraulic release).	Check pilot brake release pressure.
Motor worn or damaged.	Refer to chapter nine for root-cause analysis.

Has proper application and operation of the motor been determined?

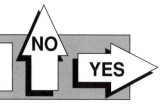

NO

YES

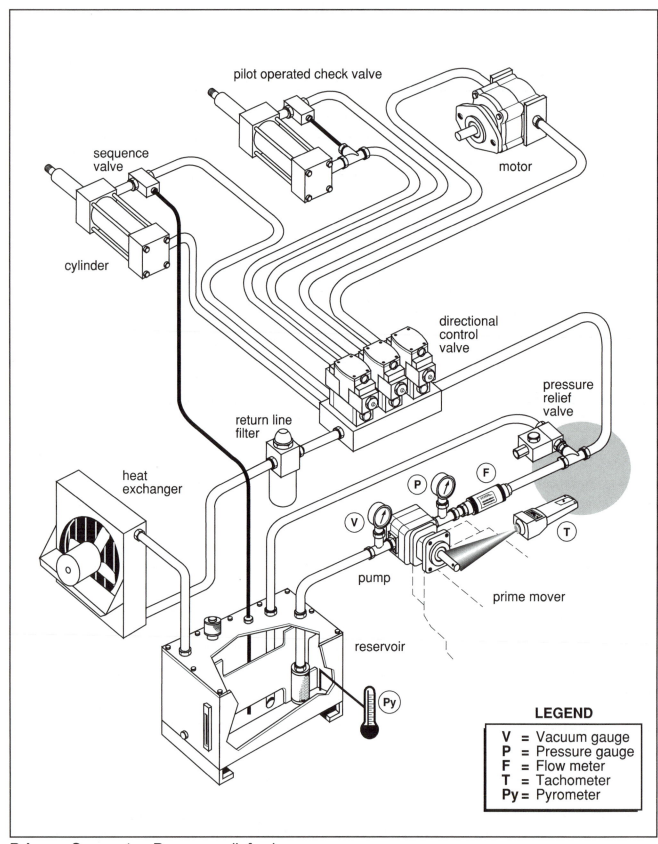

Primary Suspect: Pressure relief valve

PROBLEM

Actuator(s) operate too slowly

DIAGNOSTIC OBSERVATION

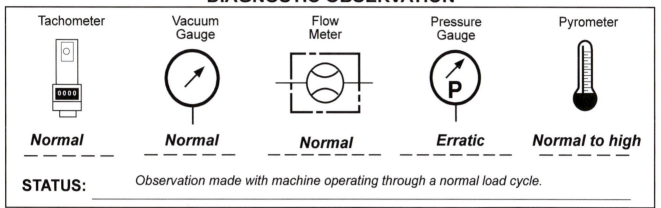

Tachometer	Vacuum Gauge	Flow Meter	Pressure Gauge	Pyrometer
Normal	*Normal*	*Normal*	*Erratic*	*Normal to high*

STATUS: *Observation made with machine operating through a normal load cycle.*

PRIMARY SUSPECT

Pressure relief valve

POSSIBLE CAUSE

Pressure relief valve -- set too low.

Direct-operated relief valve -- early "cracking."

Cross-port relief valve -- set too low.

Cylinder port relief valve -- set too low.

Pressure relief valve worn or damaged.

RECOMMENDED COURSE OF ACTION

Adjust to recommended pressure.

Replace with pilot-operated relief valve.
Submit for application engineering review.

Adjust to recommended pressure.

Adjust to recommended pressure.

Refer to chapter four for root-cause analysis.

Has proper application, operation and adjustment of the relief valve been determined?

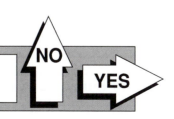

NO

YES

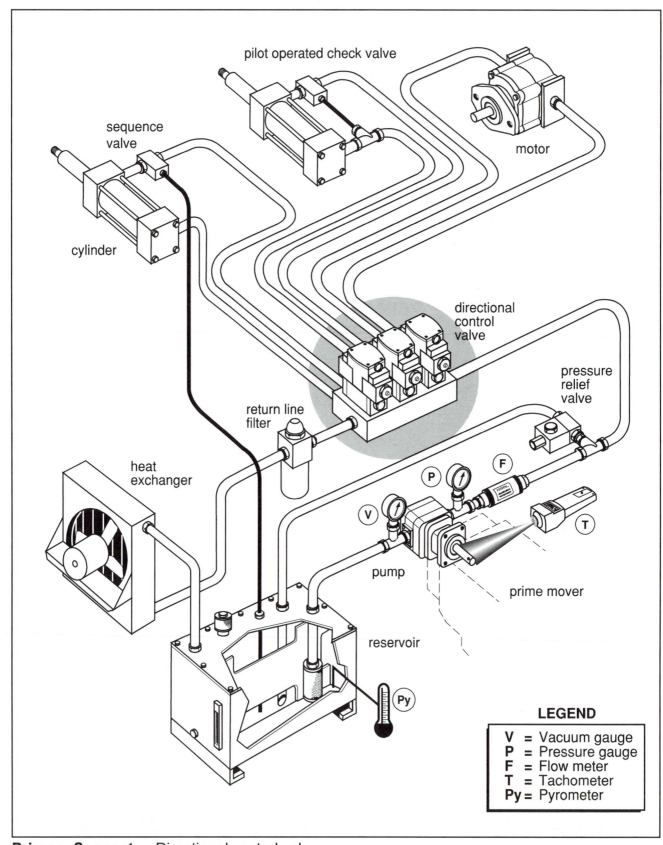

pilot operated check valve

sequence
valve

motor

cylinder

directional
control
valve

pressure
relief
valve

return line
filter

heat
exchanger

P

F

V

T

pump

prime mover

reservoir

Py

LEGEND

V	=	Vacuum gauge
P	=	Pressure gauge
F	=	Flow meter
T	=	Tachometer
Py	=	Pyrometer

Primary Suspect: Directional control valve

PROBLEM

Actuator(s) operate too slowly

DIAGNOSTIC OBSERVATION

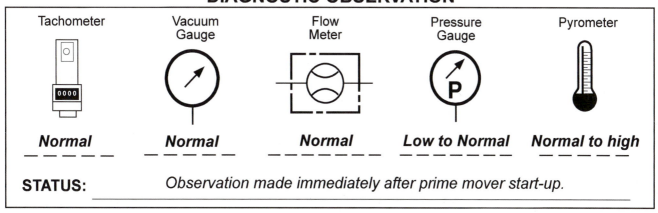

Tachometer	Vacuum Gauge	Flow Meter	Pressure Gauge	Pyrometer
Normal	*Normal*	*Normal*	*Low to Normal*	*Normal to high*

STATUS: *Observation made immediately after prime mover start-up.*

PRIMARY SUSPECT

Directional control valve

POSSIBLE CAUSE	RECOMMENDED COURSE OF ACTION
Spool binding -- retaining bolts or tie-rods too tight.	Loosen and re-tighten to correct torque.
Pilot pressure too low (air or oil).	Adjust pilot pressure to recommended level.
Internal pilot plug not installed with external pilot configuration.	Conduct external pilot conversion. Refer to chapter six for valve conversion procedures.
Spool travel limiter adjustment not set correctly.	Adjust according to design specifications.
Pilot chokes not adjusted correctly.	Adjust pilot chokes according to design specifications.

Has proper application, operation and adjustment of directional control valve been determined?

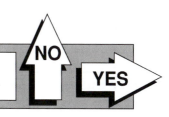

NO

YES

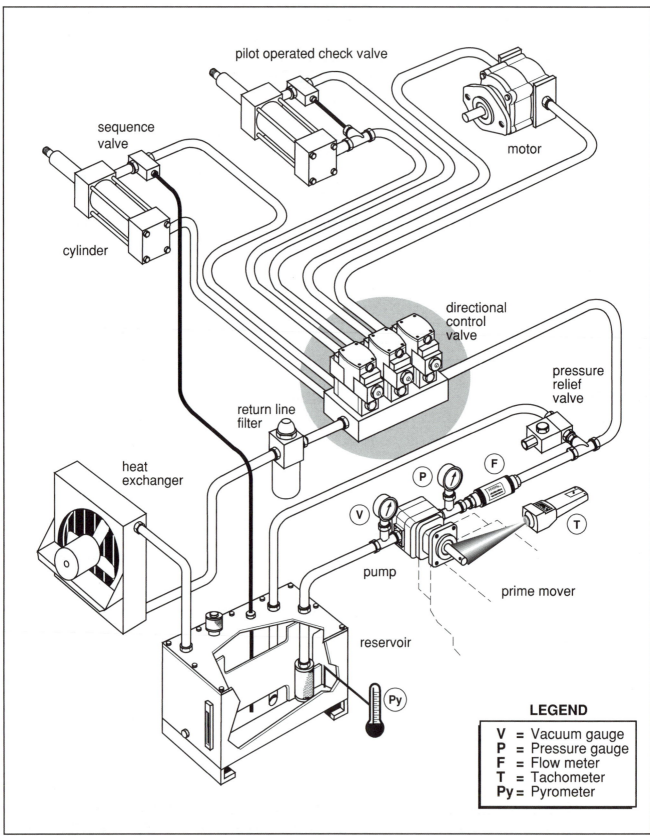

pilot operated check valve

sequence valve

motor

cylinder

directional control valve

pressure relief valve

return line filter

heat exchanger

P

F

V

pump

T

prime mover

reservoir

Py

LEGEND

V	= Vacuum gauge
P	= Pressure gauge
F	= Flow meter
T	= Tachometer
Py	= Pyrometer

Primary Suspect: Directional control valve

PROBLEM

Actuator(s) operate in one direction only

DIAGNOSTIC OBSERVATION

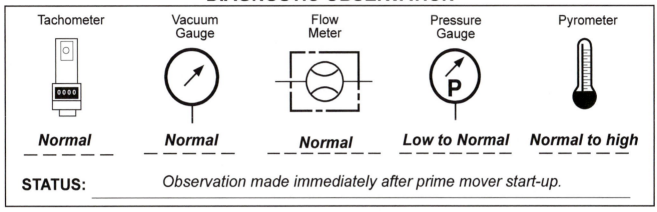

Tachometer	Vacuum Gauge	Flow Meter	Pressure Gauge	Pyrometer
Normal	*Normal*	*Normal*	*Low to Normal*	*Normal to high*

STATUS: *Observation made immediately after prime mover start-up.*

PRIMARY SUSPECT

Directional control valve

POSSIBLE CAUSE

Solenoid malfunction -- not wired correctly.

Cylinder port relief valve (integral) -- set too low.

Pilot pressure on one side of valve only.

Pilot choke closed on one side of valve.

RECOMMENDED COURSE OF ACTION

Check wiring. Correct if necessary.

Adjust to recommended pressure.

Check pilot pressure valve. Refer to chapters five and seven for root-cause analysis.

Adjust pilot choke according to manufacturer's specification.

Has proper application, operation and adjustment of directional control valve been determined?

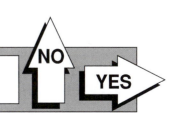

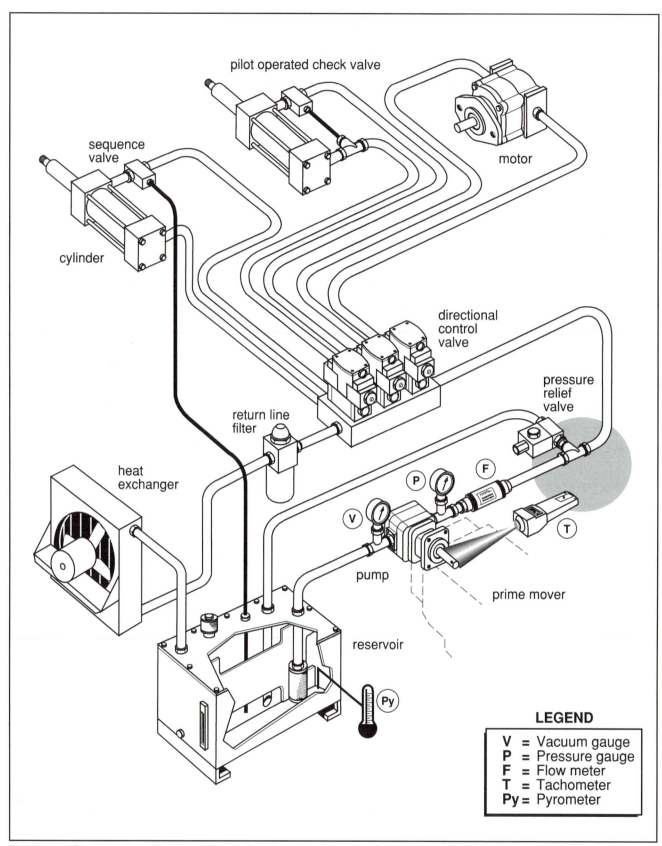

pilot operated check valve

sequence
valve

cylinder

motor

directional
control
valve

pressure
relief
valve

return line
filter

heat
exchanger

P **F**

V

pump

T

prime mover

reservoir

Py

LEGEND

V	**=**	Vacuum gauge
P	**=**	Pressure gauge
F	**=**	Flow meter
T	**=**	Tachometer
Py	**=**	Pyrometer

Primary Suspect: Pressure relief valve

PROBLEM

System overheating

DIAGNOSTIC OBSERVATION

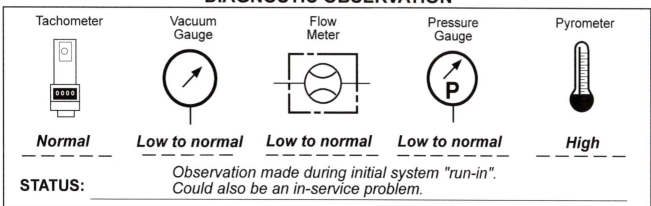

Tachometer	Vacuum Gauge	Flow Meter	Pressure Gauge	Pyrometer
Normal	*Low to normal*	*Low to normal*	*Low to normal*	*High*

STATUS: *Observation made during initial system "run-in". Could also be an in-service problem.*

PRIMARY SUSPECT

Pressure relief valve

POSSIBLE CAUSE	**RECOMMENDED COURSE OF ACTION**
Set too low.	Adjust to recommended pressure.
Set too high.	Adjust to recommended pressure.
Direct-operated relief valve -- early cracking.	Replace with pilot-operated relief valve. Submit for application engineering review.
Worn or damaged.	Refer to chapter four for root-cause analysis.
Oil passing through relief valve constantly. System modification or applications engineering oversight. Inherent design problem.	Submit for application engineering review.

Has proper application and operation of the pressure relief valve been determined?

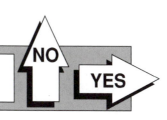

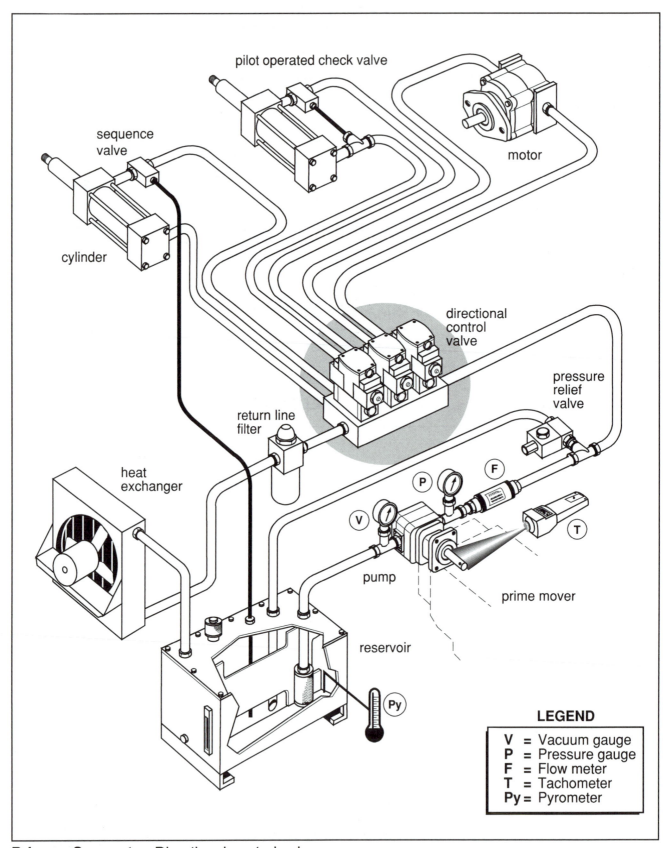

Primary Suspect: Directional control valve

PROBLEM

System overheating

DIAGNOSTIC OBSERVATION

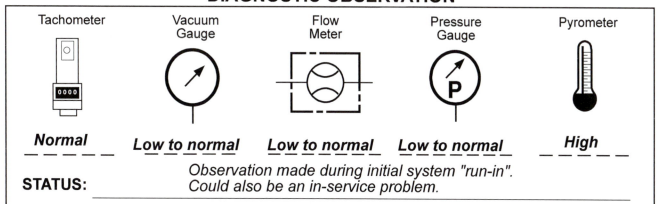

Tachometer	Vacuum Gauge	Flow Meter	Pressure Gauge	Pyrometer
Normal	*Low to normal*	*Low to normal*	*Low to normal*	*High*

STATUS: *Observation made during initial system "run-in". Could also be an in-service problem.*

PRIMARY SUSPECT

Directional control valve

POSSIBLE CAUSE	RECOMMENDED COURSE OF ACTION
Worn or damaged.	Refer to chapters five and seven for root-cause analysis.
Integral cylinder port relief valve set too low.	Adjust to recommended pressure.
Integral cylinder port relief valve worn or damaged.	Refer to chapter seven for root-cause analysis.
Anti-cavitation valve worn or damaged.	Refer to chapter seven for root cause analysis.
Incorrectly sized.	Submit for application engineering review.
Internal pilot configuration with external pilot application.	Refer to chapter six for correct conversion procedure.
Circuit module inter-port leakage.	Refer to chapter five for root-cause analysis.
Spool not centering.	Refer to chapter five or seven for root-cause analysis.

Has proper application and operation of the directional control valve been determined?

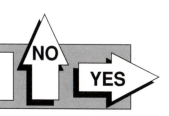

NO

YES

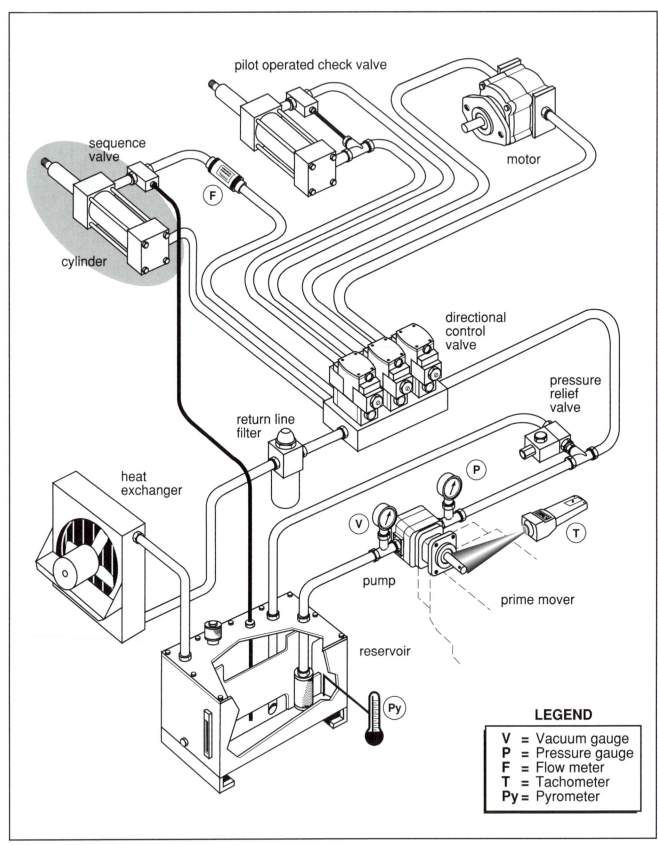

pilot operated check valve

sequence valve

motor

cylinder

F

directional control valve

pressure relief valve

return line filter

heat exchanger

P

V

T

pump

prime mover

reservoir

Py

LEGEND

V	=	Vacuum gauge
P	=	Pressure gauge
F	=	Flow meter
T	=	Tachometer
Py	=	Pyrometer

Primary Suspect: Linear actuator (cylinder)

PROBLEM

System overheating

DIAGNOSTIC OBSERVATION

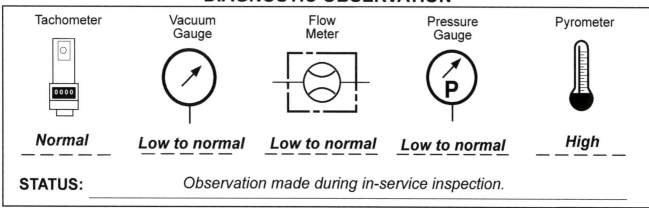

Tachometer	Vacuum Gauge	Flow Meter	Pressure Gauge	Pyrometer
Normal	*Low to normal*	*Low to normal*	*Low to normal*	*High*

STATUS: *Observation made during in-service inspection.*

PRIMARY SUSPECT

Linear actuator (cylinder)

POSSIBLE CAUSE	**RECOMMENDED COURSE OF ACTION**
Seal worn or damaged.	Refer to chapter eight for root-cause analysis.
Cylinder bore damage.	Refer to chapter eight for root-cause analysis.
Seal too tight causing partial flow across the relief valve.	Remove load and conduct pressure test at the cylinder port.
Integral pilot-operated check valve damaged or worn causing cross-port leakage.	Refer to chapter four for root-cause analysis.
Piston loose on rod. Retaining nut or screw loose.	Refer to chapter seven for root-cause analysis.

Has proper application and operation of the cylinder been determined?

NO

YES

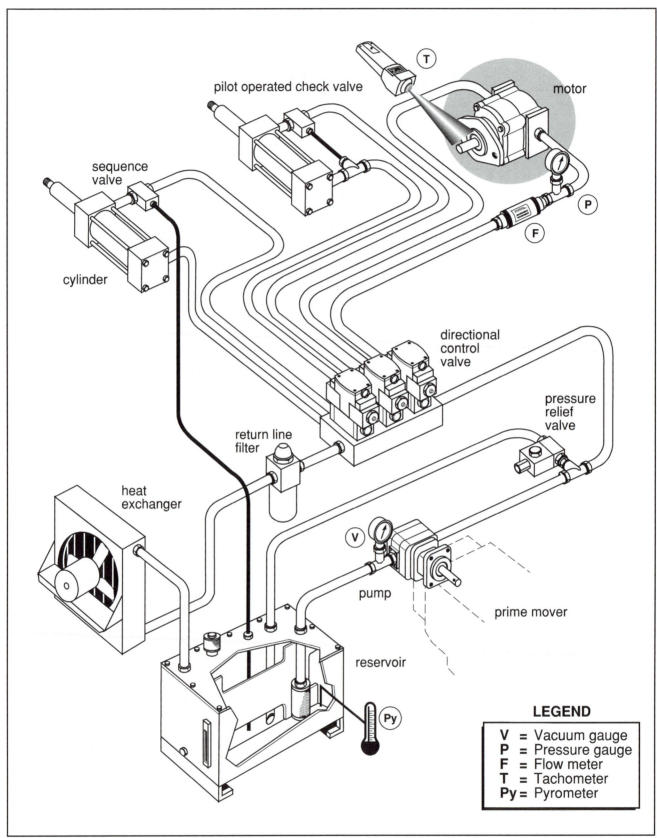

Primary Suspect: Rotary actuator (motor)

PROBLEM

System overheating

DIAGNOSTIC OBSERVATION

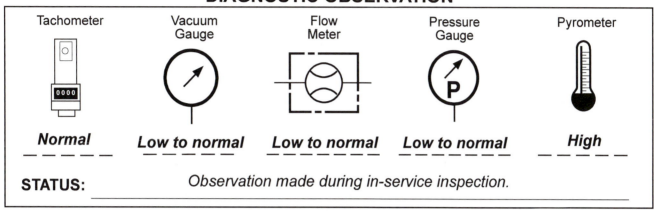

Tachometer	Vacuum Gauge	Flow Meter	Pressure Gauge	Pyrometer
Normal	*Low to normal*	*Low to normal*	*Low to normal*	*High*

STATUS: *Observation made during in-service inspection.*

PRIMARY SUSPECT

Rotary actuator (motor)

POSSIBLE CAUSE	RECOMMENDED COURSE OF ACTION
Motor worn or damaged.	Refer to chapter nine for root-cause analysis.
Motor operating below minimum rated speed.	Submit for application engineering review.
Sub-standard case drain-line installation prevents motor case from filling up (externally drained motor).	Connect case drain-line to the top port in the motor housing. Submit for application engineering review.
Motor improperly rebuilt. Problem coincides with installation of new or rebuilt motor.	Refer to chapter nine for root-cause analysis.
Integral anti-cavitation valve worn or damaged causing cross-port leakage.	Refer to chapter nine for root-cause analysis.

Has proper application and operation of the motor been determined?

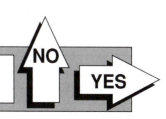

NO

YES

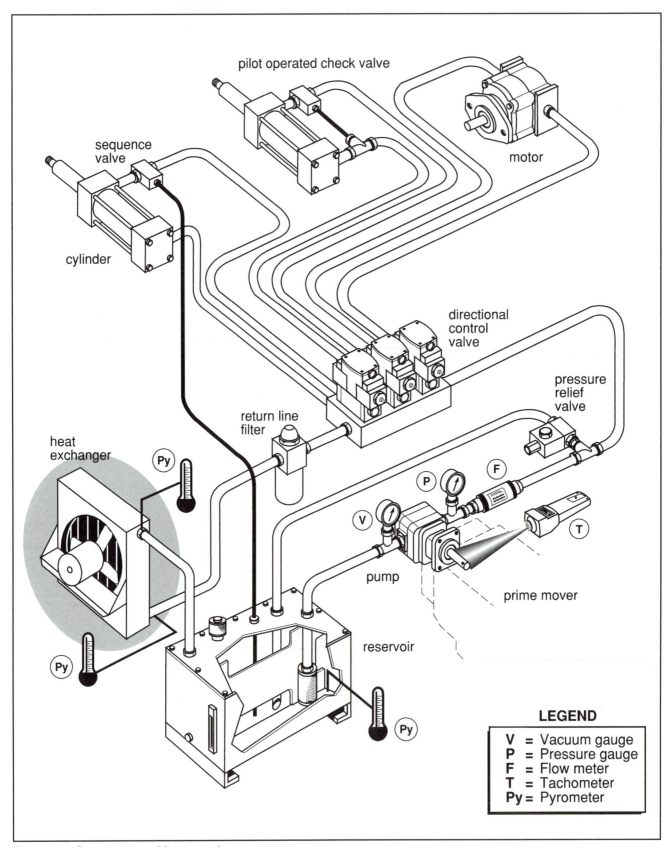

pilot operated check valve

sequence
valve

motor

cylinder

directional
control
valve

pressure
relief
valve

return line
filter

heat
exchanger

Py

P

F

V

T

pump

prime mover

reservoir

Py

Py

LEGEND

V	=	Vacuum gauge
P	=	Pressure gauge
F	=	Flow meter
T	=	Tachometer
Py	=	Pyrometer

Primary Suspect: Heat exchanger

PROBLEM

System overheating

DIAGNOSTIC OBSERVATION

Tachometer	Vacuum Gauge	Flow Meter	Pressure Gauge	Pyrometer
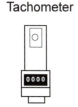		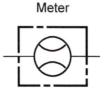		
Normal	*Low to normal*	*Low to normal*	*Low to normal*	*High*

STATUS: *Observation made during initial system "run-in". Could also be an in-service problem.*

PRIMARY SUSPECT

Heat exchanger

POSSIBLE CAUSE	RECOMMENDED COURSE OF ACTION
Water/Oil Type: Water supply closed off.	Open or connect water supply.
Insufficient water supply.	Measure water supply. Must be at least 30% of hydraulic pump flow.
Incorrectly sized.	Submit for application engineering review.
Scale formation in cooler tubes.	Test temperature drop across heat exchanger. If inadequate, service or replace.
Thermostat malfunction.	Test thermostat. Service or replace.
Air/Oil Type: Cooling fins plugged with dirt.	Clean.
Too many layers of paint.	Paint causes a heat shield. Remove paint or replace.
Incorrectly mounted. Inlet and outlet ports are at the bottom of the heat exchanger. Heat exchanger mounted above the oil reservoir.	Rotate heat exchanger to a position which locates the inlet port at the bottom and the outlet port at the top.

Is temperature drop across heat exchanger within design specifications?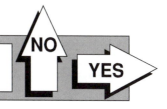

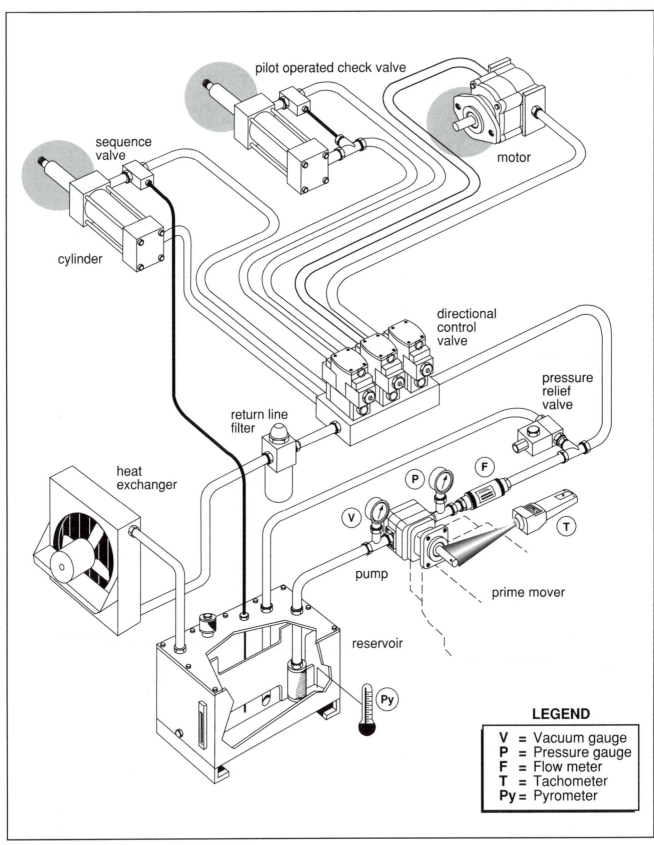

Primary Suspect: Actuator output

PROBLEM

System overload- Mechanical

DIAGNOSTIC OBSERVATION

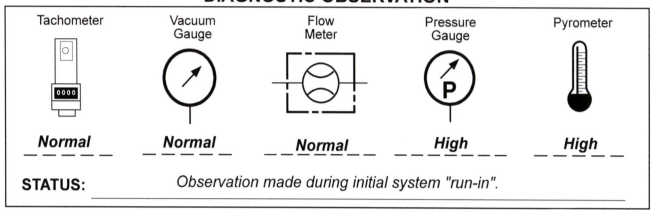

Tachometer	Vacuum Gauge	Flow Meter	Pressure Gauge	Pyrometer
Normal	*Normal*	*Normal*	*High*	*High*

STATUS: *Observation made during initial system "run-in".*

PRIMARY SUSPECT

Actuator output

POSSIBLE CAUSE	**RECOMMENDED COURSE OF ACTION**
Excessive load. Full system pressure (pressure relief valve setting) recorded. Load does not move.	Reduce load. Submit for application engineering review.
Mechanical bind in operation of machine.	Remove the load and operate the actuator through a normal "no-load" cycle. Observe pressure at actuator port. If there is an abnormal variation in pressure, conduct an evaluation of the mechanical aspect of the machine.
Motor application with integral, spring-apply hydraulic-release brake. Worn or damaged.	Check brake release pressure. Service or replace brake assembly.

Has proper application and operation of the mechanical aspect of the machine been determined?

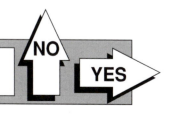

NO

YES

notes

chapter three

Procedures for Testing Hydraulic Pumps

Internally Drained Pump
Direct-Access Test Procedure

What will this test procedure accomplish?

All hydraulic pumps have component parts which move in relation to one another separated by a small oil-filled clearance. These components are generally loaded toward one another by forces related to system pressure, surface area, and springs or seals.

Theoretically, a pump delivers an amount of fluid equal to its displacement during each revolution. The actual output is reduced because of internal leakage. As pressure increases, the leakage also increases causing a decrease in volumetric efficiency. Volumetric efficiency is equal to the actual output divided by the theoretical output, and is expressed as a percentage.

The amount of leakage is influenced by four factors:
1. Pressure difference across the clearances;
2. Oil viscosity;
3. Temperature; and
4. Size of clearance.

Internal leakage can either flow from the outlet port back to the inlet port (internal drain), or from the outlet port into the pump case (external drain). If a pump is externally drained, a separate port, which is generally smaller in diameter than the inlet and outlet ports, is provided in the pump case. A case drain-line is connected to the drain port to transport the leakage directly back to the fluid reservoir.

Generally, positive displacement hydraulic pumps achieve volumetric efficiencies of between 80% - 97%. The flow curves on the graph (Figure 3-2) represent generally acceptable pressure/ flow levels for piston, vane, and gear pumps.

The volumetric efficiency or leakage of an internally drained pump can be monitored with a flow meter installed in series with the pump outlet port transmission line. However, it is important to note that a reliable flow test can only be made if the pump is operating under load (pressure). Some internally drained pumps include: internal gear, external gear, and balanced vane.

The most common symptoms of problems associated with excessive pump leakage are:
a. A loss of actuator speed as the fluid temperature increases.
b. A moderate to high increase in the operating temperature of the fluid.
c. Actuator stalling at low pump speed -- variable speed prime mover applications.

This test procedure will determine:

    a. If pump input speed is consistent with design specification.
    b. If pump inlet restriction is consistent with design specification.
    c. If pressure relief valve setting is consistent with design specification.
    d. The amount of leakage across the ports of an internally drained pump.

 **-WARNING-** | **Do not work on or around hydraulic systems without wearing safety glasses which conform to ANSI Z87.1-1989 standard.**

YOU WILL NEED THE FOLLOWING DIAGNOSTIC EQUIPMENT
TO CONDUCT THIS TEST PROCEDURE

CAUTION! *The pressure and flow ratings of the diagnostic equipment which will be used to conduct this test procedure must be equal to, or greater than, the pressure and flow ratings of the system being tested. Refer to the system schematic for recommended pressure and flow data.*

1.	Pyrometer	4.	Pressure gauge
2.	Vacuum gauge	5.	Needle Valve
3.	Tachometer	6.	Flow meter

PREPARATION

To conduct this test safely and accurately, refer to Figure 3-1 for correct placement of diagnostic equipment.
To record the test data, make a copy of the test worksheet on page 3-8 (Figure 3-3).

TEST PROCEDURE

Step 1.   Shut the prime mover off.

Step 2.   Lock the electrical system out or tag the keylock switch.

Step 3.   Observe the system pressure gauge. Release any residual pressure trapped in the system by an accumulator, counterbalance or pilot-operated check valve, suspended load on an actuator, intensifier, or a pressurized reservoir.

Step 4.   Install flow meter (6) in series with the transmission line at the outlet port of the pump. The flow meter must be equipped with needle valve (5) to generate an "artificial" load.

NOTE: *If necessary, connect a length of hose to the outlet port of the flow meter and route it directly to the reservoir. Fasten it securely. If the existing transmission line is steel pipe or tubing, replace it with a suitable hose to conduct the test procedure.*

1. Pyrometer
2. Vacuum gauge
3. Tachometer
4. Pressure gauge
5. Needle valve
6. Flow meter

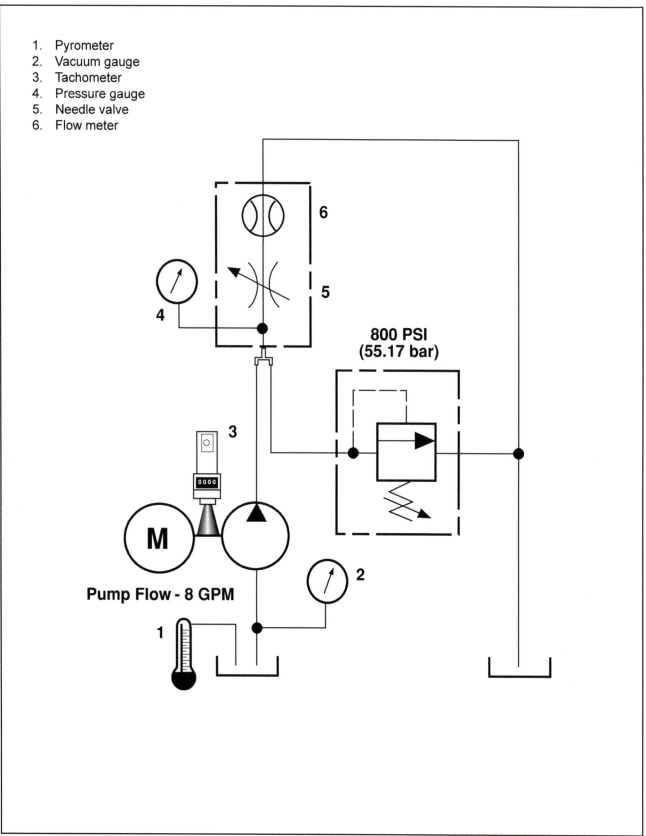

Figure 3-1   Placement of diagnostic equipment

Step 5.   Open needle valve (5) fully (turn counter-clockwise).

Step 6.   Install pressure gauge (4) in parallel with the connector at the inlet port of flow meter (6).

Step 7.   Install vacuum gauge (2) in parallel with the connector at the inlet port of the pump.

Step 8.   Start the prime mover. Inspect the diagnostic equipment connectors for leaks.

NOTE: *If catastrophic pump failure is suspected, do not run the pump for longer than is absolutely necessary to determine its condition. If metal fragments are found in transmission lines, pump outlet port, return filter, and/or reservoir, DO NOT restart the pump. Replace the pump and conduct a post-catastrophic failure start-up procedure.*

Step 9.   Allow the system to warm up to approximately 130º F (54º C). Observe pyrometer (1).

NOTE: *For accurate test results, check and adjust (if necessary) the system's main pressure relief valve. Set it according to the manufacturer's specification.*

Step 10.   Operate the prime mover at maximum governed speed (Conduct Step 11 through Step 16 with the prime mover operating at maximum governed speed).

Step 11.   Record on the test worksheet, the pump "no-load" input shaft speed, indicated on tachometer (3). Adjust to the manufacturer's specification if necessary.

Step 12.   Record on the test worksheet, the pump inlet restriction indicated on vacuum gauge (2).

NOTE: *If the pump inlet restriction is within specification, continue with the test procedure. If the pump inlet restriction is inconsistent with the design specification, refer to the "Pump Cavitation Test Procedure" outlined in this chapter. Continue with this test procedure when the pump inlet restriction is within design specification.*

Step 13.   Operate the pump at "no-load." Record on the test worksheet:
            a. Flow indicated on flow meter (6).
            b. Pressure indicated on pressure gauge (4).
            c. Fluid temperature indicated on pyrometer (1).

Step 14.   Gradually load the pump by restricting the flow with needle valve (5) (turn clockwise). Stop when the pressure reaches 100 PSI (6.9 bar) more than the "no-load" pressure. Enter this data on the test worksheet.

Step 15.   Continue increasing the pressure in increments of 100 PSI (6.9 bar), plotting the flow and pressure at each 100 PSI (6.9 bar) increment on the test worksheet. Stop the test procedure when the value of the system's main pressure relief valve setting is reached.

Step 16. With the prime mover operating at "full-load" (full pump flow passing over the pressure relief valve), record on the test worksheet:
  a. Prime mover speed indicated on tachometer (3).
  b. Fluid temperature indicated on pyrometer (1).

Step 17. Open needle valve (5) fully (turn counter-clockwise) and reduce the "artificial" load on the pump.

Step 18. Shut the prime mover off and analyze the test results.

Step 19. At the conclusion of this test procedure remove the diagnostic equipment. Reconnect the transmission lines and tighten the connectors securely.

Step 20. Start the prime mover and inspect the connectors for leaks.

NOTE: *If the main pressure relief valve is in parallel with the circuit during this test procedure, place the flow meter in series with the transmission line at the outlet port of the relief valve and repeat the test procedure. A relief valve malfunction can cause problems similar to pump-related problems with regard to leakage.*

ANALYZING THE TEST RESULTS

1. **Diagnostic Observation: "full-load" cycle:** Flow meter (6) indicates a nominal flow decrease as the pressure drop across the pump ports increases. The flow decrease does not exceed the anticipated loss relative to the volumetric efficiency rating of the pump, and a nominal decrease in prime mover speed.

    The operating temperature of the fluid indicated on pyrometer (1) remains within design specification.

    Prime mover speed indicated on tachometer (3) remains within design specification.

    Pump inlet restriction indicated on vacuum gauge (2) remains within design specification.

    Diagnosis: The pump appears to be in satisfactory operating condition.

2. **Diagnostic Observation: "full-load" cycle:** Flow meter (6) indicates a moderate to high flow decrease as the pressure drop across the pump ports increases. The flow decrease does not exceed 30% of the theoretical or "no-load" flow (See **NOTE:** on page 3-7).

    Pyrometer (1) indicates a moderate increase in the operating temperature of the fluid.

    Prime mover speed indicated on tachometer (3) remains within design specification.

    Pump inlet restriction indicated on vacuum gauge (2) remains within design specification.

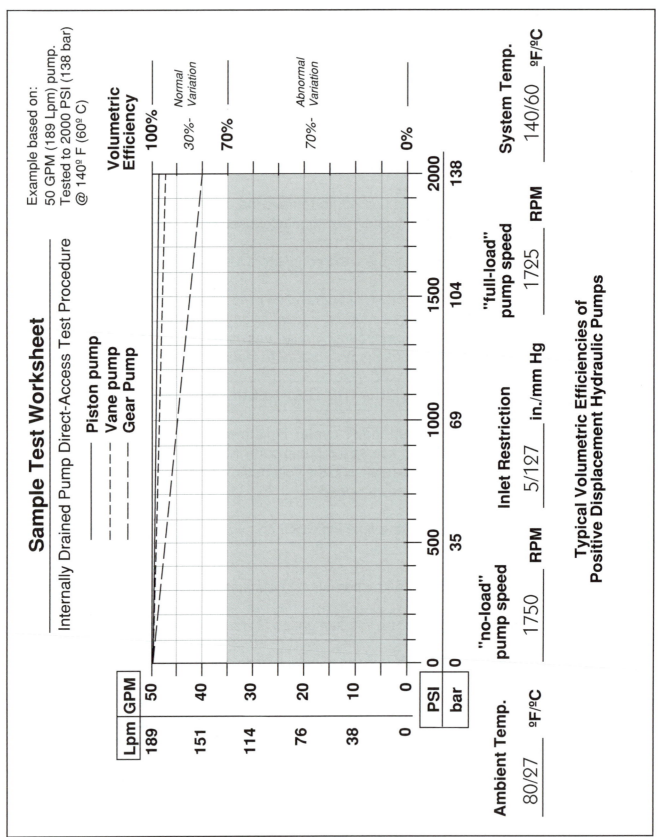

Figure 3-2   Sample Test Worksheet - Internally drained pump direct-access test procedure

Diagnosis: The condition of the pump may be marginal. If there is an abnormal variation (reduction) in actuator speed which causes an unacceptable production loss and/or a marked increase in the operating temperature of the fluid, the pump should be replaced.

3. **Diagnostic Observation: "full-load" cycle:** Flow meter (6) indicates a progressive flow loss as the pressure drop across the pump ports increases. The flow loss exceeds 30% of the theoretical or "no-load" flow rating of the pump.

NOTE: *Component manufacturers rarely publish data on when the useful life of a component expires. The decision on when to remove a component is generally left to the discretion of the end-user.*

I conducted an extensive study on pump leakage versus useful life and determined that if the leakage approaches or exceeds approximately 30% of the theoretical (no-load) flow, at full-load (maximum rated pressure) it should be replaced.

Some pumps will appear to operate normally above this figure, others will show problems well below this figure. However, the side effects of excessive leakage are detrimental to the well-being of the entire hydraulic system. It could cause a significant increase in the operating temperature of the fluid, and/or a loss in production.

A well-managed, pro-active maintenance program will help establish the maximum leakage levels of your hydraulic components.

Pyrometer (1) indicates a progressive increase in the operating temperature of the fluid which does not appear to "level-off."

Prime mover speed indicated on tachometer (3) remains within design specification.

Pump inlet restriction indicated on vacuum gauge (2) remains within design specification.

Diagnosis: There appears to be excessive leakage across the ports in the pump.

A flow decrease of this magnitude will invariably cause an unacceptable loss in actuator speed. It could also cause a marked increase in the operating temperature of the fluid. If neglected, elevated fluid temperatures could lead to a rapid degradation of the entire hydraulic system.

Replace the pump. Refer to chapter two for the "zero-fault" pump start-up procedures.

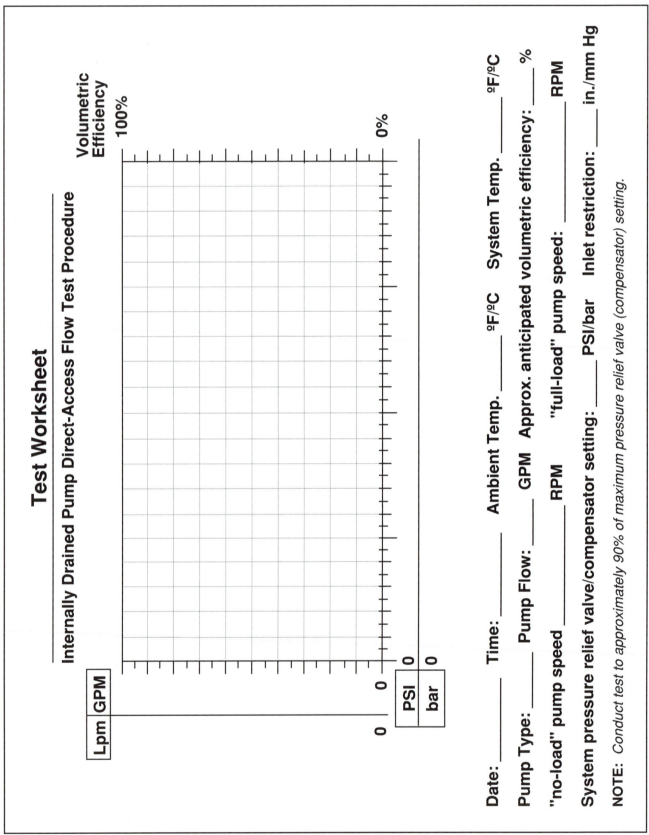

Figure 3-3   Test Worksheet - Internally drained pump direct-access test procedure

Internally Drained Pump
In-Circuit Test Procedure

What will this test procedure accomplish?

All hydraulic pumps have component parts which move in relation to one another separated by a small oil-filled clearance. These components are generally loaded toward one another by forces related to system pressure, surface area, and springs or seals.

Theoretically, a pump delivers an amount of fluid equal to its displacement during each revolution. The actual output is reduced because of internal leakage. As pressure increases, the leakage also increases causing a decrease in volumetric efficiency. Volumetric efficiency is equal to the actual output divided by the theoretical output, and is expressed as a percentage.

The amount of leakage is influenced by four factors:
1. Pressure difference across the clearances;
2. Oil viscosity;
3. Temperature; and
4. Size of clearance.

Internal leakage can either flow from the outlet port back to the inlet port (internal drain), or from the outlet port into the pump case (external drain). If a pump is externally drained, a separate port, which is generally smaller in diameter than the inlet and outlet ports, is provided in the pump case. A case drain-line is connected to the drain port to transport the leakage directly back to the fluid reservoir.

Generally, positive displacement hydraulic pumps achieve volumetric efficiencies of between 80% - 97%. The flow curves on the graph (Figure 3-9) represent generally acceptable pressure/ flow levels for piston, vane, and gear pumps.

The volumetric efficiency or leakage of an internally drained pump can be monitored with a flow meter installed in series with the pump outlet port transmission line. However, it is important to note that a reliable flow test can only be made if the pump is operating under load (pressure). Some internally drained pumps include: internal gear, external gear, and balanced vane.

The most common symptoms of problems associated with excessive pump leakage are:
   a. A loss of actuator speed as the fluid temperature increases.
   b. A moderate to high increase in the operating temperature of the fluid.
   c. Actuator stalling at low pump speed-- variable speed prime mover applications.

This test procedure will determine:
   a. If pump input speed is consistent with design specification.
   b. If pump inlet restriction is consistent with design specification.
   c. If pressure relief valve setting is consistent with design specification.
   d. The amount of leakage across the ports of an internally drained pump.

-WARNING- Do not work on or around hydraulic systems without wearing safety glasses which conform to ANSI Z87.1-1989 standard.

YOU WILL NEED THE FOLLOWING DIAGNOSTIC EQUIPMENT TO CONDUCT THIS TEST PROCEDURE

CAUTION! *The pressure and flow ratings of the diagnostic equipment which will be used to conduct this test procedure must be equal to, or greater than, the pressure and flow ratings of the system being tested. Refer to the system schematic for recommended pressure and flow data.*

1. Flow meter
2. Pressure gauge
3. Tachometer
4. Vacuum gauge
5. Pyrometer

PREPARATION

To conduct this test safely and accurately, refer to Figure 3-4 for correct placement of diagnostic equipment.

To record the test data, make a copy of the test worksheet on page 3-53 (Figure 3-19).

TEST PROCEDURE

Step 1. Shut the prime mover off.

Step 2. Lock the electrical system out or tag the keylock switch.

Step 3. Observe the system pressure gauge. Release any residual pressure trapped in the system by an accumulator, counterbalance or pilot-operated check valve, suspended load on an actuator, intensifier, or a pressurized reservoir.

Step 4. Install flow meter (1) in series with the transmission line at the outlet port of the pump.

Step 5. Install pressure gauge (2) in parallel with the connector at the inlet port of flow meter (1).

Step 6. Install vacuum gauge (4) in parallel with the connector at the inlet port of the pump.

Step 7. Start the prime mover. Inspect the diagnostic equipment connectors for leaks.

NOTE: *If catastrophic pump failure is suspected, do not run the pump for longer than is absolutely necessary to determine its condition. If metal fragments are found in transmission lines, pump outlet port, in-line or return filters, and/or reservoir, DO NOT restart the pump. Replace the pump and conduct a post-catastrophic failure start-up procedure.*

1. Flow meter
2. Pressure gauge
3. Tachometer
4. Vacuum gauge
5. Pyrometer

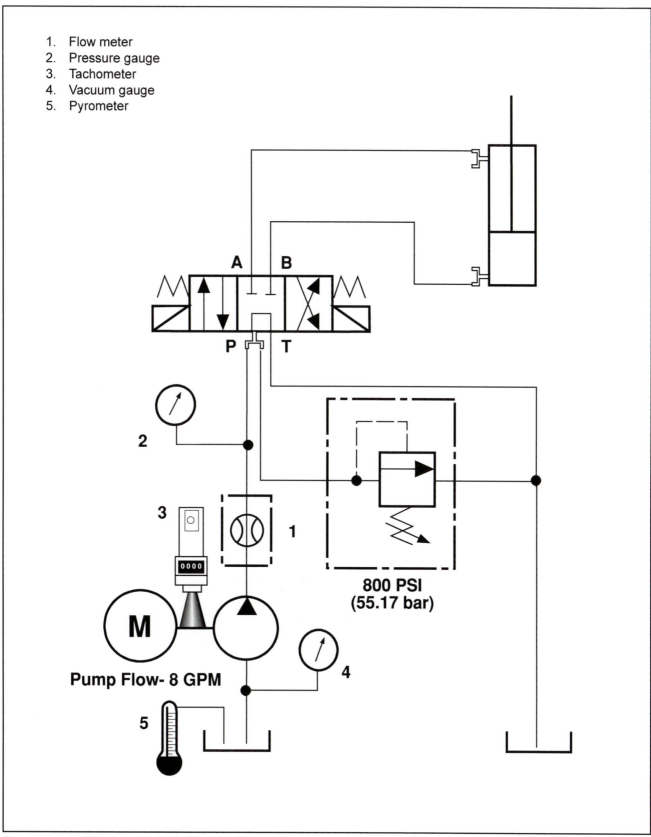

Figure 3-4   Placement of diagnostic equipment

Step 8. Allow the system to warm up to approximately 130º F.(54º C.) Observe pyrometer (5).

NOTE: *For accurate test results, check and adjust (if necessary) the system's main pressure relief valve. Set it according to the manufacturer's specification.*

Step 9. Operate the prime mover at maximum governed speed. (Conduct Step 10 through Step 13 with the prime mover operating at maximum governed speed).

Step 10. Record on the test worksheet, the pump "no-load" input shaft speed indicated on tachometer (3). Adjust it to the manufacturer's specification if necessary.

Step 11. Record on the test worksheet, the pump inlet restriction indicated on vacuum gauge (4).

NOTE: *If the pump inlet restriction is within specification, continue with the test procedure. If the pump inlet restriction is inconsistent with the design specification, refer to the "Pump Cavitation Test Procedure" outlined in this chapter. Continue with this test procedure only if the pump inlet restriction is within specification.*

Step 12. Operate the pump at "no-load". Record on the test worksheet:
  a. Flow indicated on flow meter (1).
  b. Pressure indicated on pressure gauge (2).
  c. Fluid temperature indicated on pyrometer (5).

Step 13. Operate the pump at "full-load" (highest pressure recorded during a normal load cycle). Record on the test worksheet:
  a. Flow indicated on flow meter (1).
  b. Pressure indicated on pressure gauge (2).
  c. Prime mover speed indicated on tachometer (3).
  d. Fluid temperature indicated on pyrometer (5).

Step 14. Shut the prime mover off and analyze the test results.

Step 15. At the conclusion of this test procedure, remove the diagnostic equipment. Reconnect the transmission lines and tighten the connectors securely.

Step 16. Start the prime mover and inspect the connectors for leaks.

NOTE: *If the main pressure relief valve is in parallel with the circuit during this test procedure, place the flow meter in series with the transmission line at the outlet port of the relief valve and repeat the test procedure. A relief valve malfunction can cause problems similar to pump-related problems with regard to leakage.*

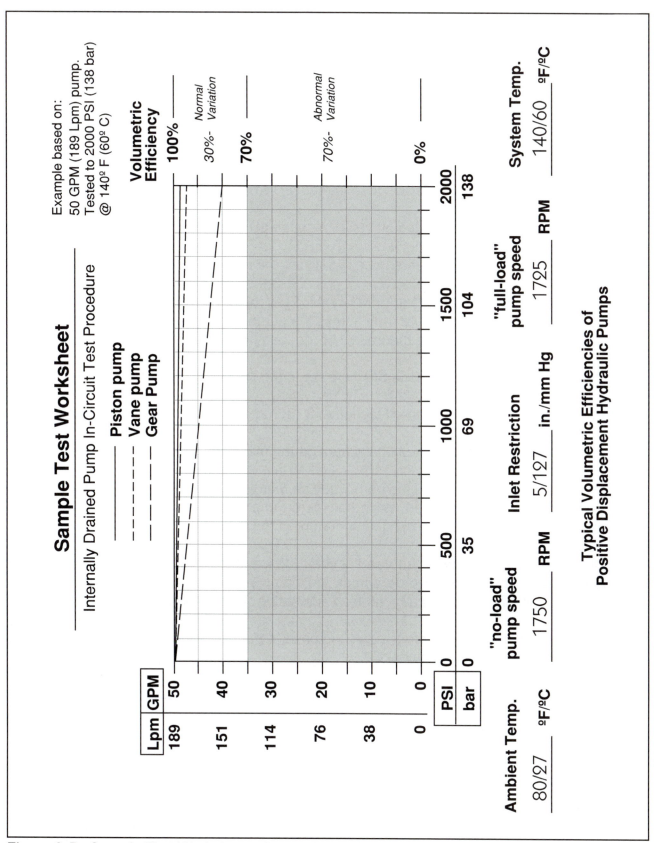

Figure 3-5 Sample Test Worksheet - Internally drained pump in-circuit test procedure

ANALYZING THE TEST RESULTS

1. **Diagnostic Observation: "full-load" cycle:** Flow meter (1) indicates a nominal flow decrease as the pressure drop across the pump ports increases. The flow decrease does not exceed the anticipated loss relative to the volumetric efficiency rating of the pump and a nominal decrease in prime mover speed.

    The operating temperature of the fluid indicated on pyrometer (5) remains within design specification.

    Prime mover speed indicated on tachometer (3) remains within design specification.

    Pump inlet restriction indicated on vacuum gauge (4) remains within design specification.

    Diagnosis: The pump appears to be in satisfactory operating condition.

2. **Diagnostic Observation: "full-load" cycle:** Flow meter (1) indicates a moderate to high flow decrease as the pressure drop across the pump ports increases. The flow decrease does not exceed 30% of the theoretical or "no-load" flow (See **NOTE:** on page 3-7).

    Pyrometer (5) indicates a moderate increase in the operating temperature of the fluid.

    Prime mover speed indicated on tachometer (3) remains within design specification.

    Pump inlet restriction indicated on vacuum gauge (4) remains within design specification.

    Diagnosis: The condition of the pump may be marginal. If there is an abnormal variation (reduction) in actuator speed which causes an unacceptable production loss, and/or a marked increase in the operating temperature of the fluid, the pump should be replaced.

3. **Diagnostic Observation: "full-load" cycle:** Flow meter (1) indicates a progressive flow decrease as the pressure drop across the pump ports increases. The flow decrease exceeds 30% of the theoretical or "no-load" flow rating of the pump (See **NOTE:** on page 3-7).

    Pyrometer (5) indicates a progressive increase in the operating temperature of the fluid which does not appear to "level-off."

    Prime mover speed indicated on tachometer (3) remains within design specification.

    Pump inlet restriction indicated on vacuum gauge (4) remains within design specification.

    Diagnosis: There appears to be excessive leakage across the ports in the pump.

    A flow decrease of this magnitude will invariably cause an unacceptable loss in actuator speed. It could also cause a marked increase in the operating temperature of the fluid. If neglected, elevated fluid temperatures could lead to a rapid degradation of the entire hydraulic system.

Externally Drained Pump
Direct-Access Test Procedure

What will this test procedure accomplish?

All hydraulic pumps have component parts which move in relation to one another separated by a small oil-filled clearance. These components are generally loaded toward one another by forces related to system pressure, surface area, and springs or seals.

Theoretically, a pump delivers an amount of fluid equal to its displacement during each revolution. The actual output is reduced because of internal leakage. As pressure increases, the leakage also increases causing a decrease in volumetric efficiency. Volumetric efficiency is equal to the actual output divided by the theoretical output, and is expressed as a percentage.

The amount of leakage is influenced by four factors:
1. Pressure difference across the clearances;
2. Oil viscosity;
3. Temperature; and
4. Size of clearance.

Internal leakage can either flow from the outlet port back to the inlet port (internal drain), or from the outlet port into the pump case (external drain). If a pump is externally drained, a separate port, which is generally smaller in diameter than the inlet and outlet ports, is provided in the pump case. A case drain-line is connected to the drain port to transport the leakage directly back to the fluid reservoir.

Generally, positive displacement hydraulic pumps achieve volumetric efficiencies of between 80% - 97%. The flow curves on the graph (Figure 3-7) represent generally acceptable pressure/flow levels for piston, vane, and gear pumps.

The volumetric efficiency or leakage of an externally drained pump can be monitored with a flow meter installed in series with the case drain line. However, it is important to note that a reliable flow test can only be made if the pump is operating under load (pressure). Some externally drained pumps include: axial piston, bent axis, radial piston and unbalanced vane.

The most common symptoms of problems associated with excessive pump leakage are:
   a. A loss of actuator speed as the fluid temperature increases.
   b. A moderate to high increase in the operating temperature of the fluid.
   c. Actuator stalling at low pump speed - variable speed prime mover applications.
   d. Increase in case pressure which could result in shaft seal leakage.

This test procedure will determine:
   a. If pump input speed is consistent with design specification.
   b. If pump inlet restriction is consistent with design specification.

c. If pressure relief valve or compensator setting is consistent with design specification.

d. The amount of leakage across the ports of an externally drained pump.

 **-WARNING-** Do not work on or around hydraulic systems without wearing safety glasses which conform to ANSI Z87.1-1989 standard.

YOU WILL NEED THE FOLLOWING DIAGNOSTIC EQUIPMENT TO CONDUCT THIS TEST PROCEDURE

CAUTION! *The pressure and flow ratings of the diagnostic equipment which will be used to conduct this test procedure must be equal to, or greater than, the pressure and flow ratings of the system being tested. Refer to the system schematic for recommended pressure and flow data.*

1.	Pressure gauge	4.	Flow meter
2.	Needle valve	5.	Vacuum gauge
3.	Tachometer	6.	Pyrometer

PREPARATION

To conduct this test safely and accurately, refer to Figure 3-6 for correct placement of diagnostic equipment.

To record the test data, make a copy of the test worksheet on page 3-54 (Figure 3-20).

TEST PROCEDURE

Step 1. Shut the prime mover off.

Step 2. Lock the electrical system out or tag the keylock switch.

Step 3. Observe the system pressure gauge. Release any residual pressure trapped in the system by an accumulator, counterbalance or pilot-operated check valve, suspended load on an actuator, intensifier, or a pressurized reservoir.

Step 4. Install flow meter (4) in series with the case drain-line at the case drain port of the pump.

WARNING! <u>*DO NOT*</u> *plug the case drain port in the pump housing. Plugging the case drain port could cause shaft seal or case seal failure. It could also cause catastrophic pump case failure.*

Step 5. Install needle valve (2) in series with the transmission line at the outlet port of the pump. Open the needle valve fully (turn counter-clockwise).

NOTE: *The outlet port of the needle valve must be connected directly to the reservoir. It may be necessary to construct a length of hose to connect the outlet port of the needle valve to the reservoir.*

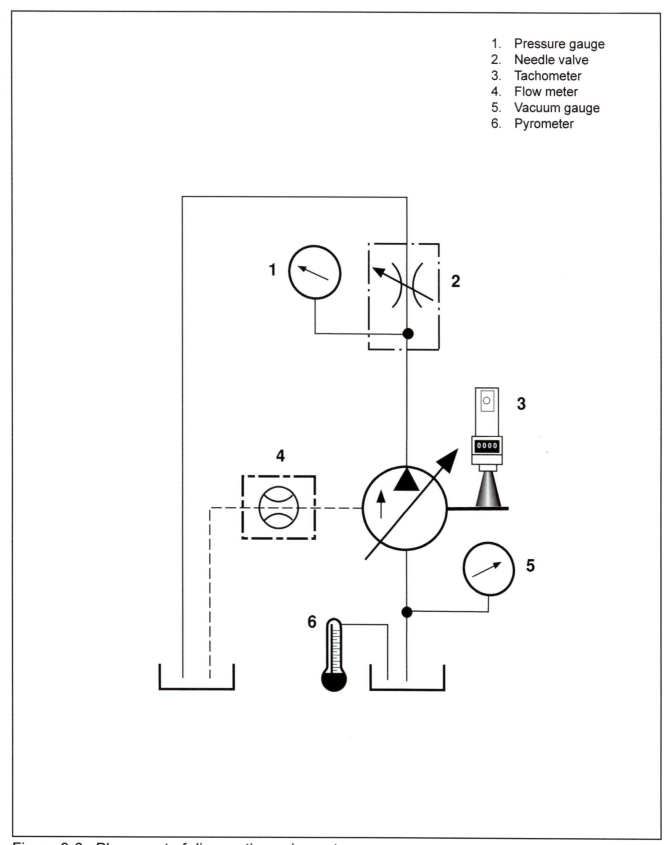

1. Pressure gauge
2. Needle valve
3. Tachometer
4. Flow meter
5. Vacuum gauge
6. Pyrometer

Figure 3-6   Placement of diagnostic equipment

Step 6.   Install pressure gauge (1) in parallel with the connector at the outlet port of the pump.

Step 7.   Install vacuum gauge (5) in parallel with the connector at the inlet port of the pump.

Step 8.   Start the prime mover. Inspect the diagnostic equipment connectors for leaks.

NOTE: *If catastrophic pump failure is suspected, do not run the pump for longer than is absolutely necessary to determine its condition. If metal fragments are found in transmission lines, pump outlet port, in-line or return filters, and/or reservoir, DO NOT restart the pump. Replace the pump and conduct a post-catastrophic failure start-up procedure.*

Step 9.   Allow the system to warm up to approximately 130º F. (54º C.). Observe pyrometer (6).

NOTE: For accurate test results, check and adjust (if necessary) the pressure relief valve or pump compensator setting. Set it according to the manufacturer's specification.

Step 10.   Operate the prime mover at maximum governed speed. (Conduct Step 11 through Step 16 with the prime mover operating at maximum governed speed).

Step 11.   Record on the test worksheet the pump "no-load" input shaft speed indicated on tachometer (3). Adjust it to the manufacturer's specification if necessary.

Step 12.   Record on the test worksheet the pump inlet restriction indicated on vacuum gauge (5).

NOTE: *If the pump inlet restriction is within specification, proceed with the test procedure. If the pump inlet restriction is inconsistent with the design specification, refer to the "Pump Cavitation Test Procedure" outlined in this chapter. Continue with this test procedure when the pump inlet restriction is within specification.*

Step 13.   Record on the test worksheet the "no-load" case drain flow indicated on flow meter (4).

Step 14.   Create an "artificial" load at the pump outlet port by gradually closing needle valve (2) (turn clockwise). Record the case drain flow at each 100 PSI (6.9 bar) increment on the test worksheet.

NOTE: *Conduct the test up to the value of the system's main pressure relief valve or pump compensator pressure setting.*

Step 15.   Record on the test worksheet the prime mover "full-load" speed indicated on tachometer (3).

Step 16.   Open needle valve (2) fully (turn counter-clockwise) and reduce the "artificial" load on the pump.

Step 17.   Shut the prime mover off and analyze the test results.

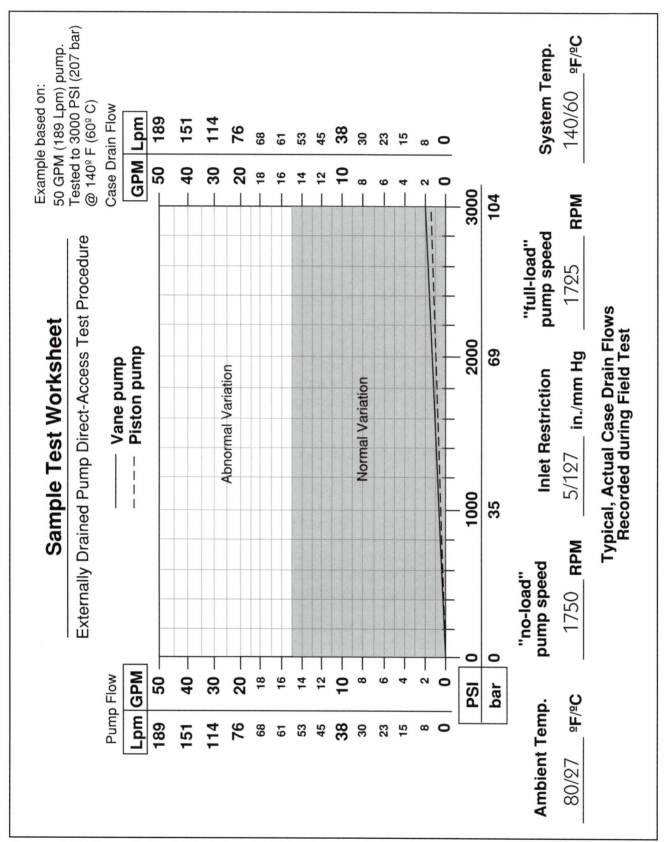

Figure 3-7   Sample Test Worksheet - Externally drained pump direct-access test procedure

Step 18. At the conclusion of this test procedure, remove the diagnostic equipment. Reconnect the transmission lines and tighten the connectors securely.

Step 19. Start the prime mover and inspect the connectors for leaks.

ANALYZING THE TEST RESULTS

1. **Diagnostic Observation: "full-load" cycle:** Flow meter (4) indicates a nominal increase in case drain flow as the pressure drop across the pump ports increases. The flow increase does not exceed the anticipated increase relative to the volumetric efficiency of the pump.

   The operating temperature of the fluid indicated on pyrometer (6) remains within design specification.

   Prime mover speed indicated on tachometer (3) remains within design specification.

   Inlet restriction indicated on vacuum gauge (5) remains within design specification.

   Diagnosis: The pump appears to be in satisfactory operating condition.

2. **Diagnostic Observation: "full-load" cycle:** Flow meter (4) indicates a moderate to high increase in case drain flow as the pressure drop across the pump ports increases. The flow increase does not exceed 30% of the theoretical or "no-load" flow.
   (See **NOTE:** on page 3-7).

   Pyrometer (6) indicates a moderate increase in the operating temperature of the fluid.

   Prime mover speed , indicated on tachometer (3), remains within design specification.

   Pump inlet restriction, indicated on vacuum gauge (5), remains within design specification.

   Diagnosis: The condition of the pump may be marginal. If there is an abnormal variation (reduction) in actuator speed which causes an unacceptable production loss, and/or a marked increase in the operating temperature of the fluid, the pump should be replaced.

3. **Diagnostic Observation: "full-load" cycle:** Flow meter (4) indicates a progressive flow increase as the pressure drop across the pump ports increases. The flow increase exceeds 30% of the theoretical or "no-load" flow (See **NOTE:** on page 3-7).

   Pyrometer (6) indicates a progressive increase in the operating temperature of the fluid which does not appear to "level-off."

   Prime mover speed, indicated on tachometer (3), remains within design specification.

   Pump inlet restriction, indicated on vacuum gauge (5), remains within design specification.

   Diagnosis: A case-drain flow increase of this magnitude will invariably cause an unacceptable loss in actuator speed. It could also cause a marked increase in the temperature of the fluid. If neglected, elevated fluid temperatures could lead to the degradation of the entire hydraulic system.

Externally Drained Pump
In-Circuit Test Procedure

What will this test procedure accomplish?

All hydraulic pumps have component parts which move in relation to one another separated by a small oil-filled clearance. These components are generally loaded toward one another by forces related to system pressure, surface area, and springs or seals.

Theoretically, a pump delivers an amount of fluid equal to its displacement during each revolution. The actual output is reduced because of internal leakage. As pressure increases, the leakage also increases, causing a decrease in volumetric efficiency. Volumetric efficiency is equal to the actual output divided by the theoretical output, and is expressed as a percentage.

The amount of leakage is influenced by four factors:
1. Pressure difference across the clearances;
2. Oil viscosity;
3. Temperature; and
4. Size of clearance.

Internal leakage can either flow from the outlet port back to the inlet port (internal drain), or from the outlet port into the pump case (external drain). If a pump is externally drained, a separate port which is generally smaller in diameter than the inlet and outlet ports, is provided in the pump case. A case drain-line is connected to the drain port to transport the leakage directly back to the fluid reservoir.

Generally, positive displacement hydraulic pumps achieve volumetric efficiencies of between 80% - 97%. The flow curves on the graph (Figure 3-9) represent generally acceptable pressure/flow levels for piston, vane, and gear pumps.

The volumetric efficiency or leakage of an externally drained pump can be monitored with a flow meter installed in series with the case drain line. However, it is important to note that a reliable flow test can only be made if the pump is operating under load (pressure). Some externally drained pumps include: axial piston, bent axis, radial piston and unbalanced vane.

The most common symptoms of problems associated with excessive pump leakage are:
a. A loss of actuator speed as the fluid temperature increases.
b. A moderate to high increase in the operating temperature of the fluid.
c. Actuator stalling at low pump speed-- variable speed prime mover applications.
d. Increase in case pressure which could result in shaft seal leakage.

This test procedure will determine:
 a. If pump input speed is consistent with design specification.
 b. If pump inlet restriction is consistent with design specification.
 c. If pressure relief valve setting is consistent with design specification.
 d. The amount of leakage across the ports of an externally drained pump.

 **-WARNING-** | **Do not work on or around hydraulic systems without wearing safety glasses which conform to ANSI Z87.1-1989 standard.**

YOU WILL NEED THE FOLLOWING DIAGNOSTIC EQUIPMENT
TO CONDUCT THIS TEST PROCEDURE

CAUTION! The pressure and flow ratings of the diagnostic equipment which will be used to conduct this test procedure must be equal to, or greater than, the pressure and flow ratings of the system being tested. Refer to the system schematic for recommended pressure and flow data.

 1. Pressure gauge   4. Vacuum gauge
 2. Tachometer   5. Pyrometer
 3. Flow meter

PREPARATION

To conduct this test safely and accurately, refer to Figure 3-8 for correct placement of diagnostic equipment.
To record the test data, make a copy of the test worksheet on page 3-55 (Figure 3-21).

TEST PROCEDURE

Step 1.   Shut the prime mover off.

Step 2.   Lock the electrical system out or tag the keylock switch.

Step 3.   Observe the system pressure gauge. Release any residual pressure trapped in the system by an accumulator, counterbalance or pilot-operated check valve, suspended load on an actuator, intensifier, or a pressurized reservoir.

Step 4.   Install flow meter (3) in series with the case drain-line at the case drain port of the pump.

Step 5.   Install pressure gauge (1) in parallel with the connector at the outlet port of the pump.

Step 6.   Install vacuum gauge (4) in parallel with the connector at the inlet port of the pump.

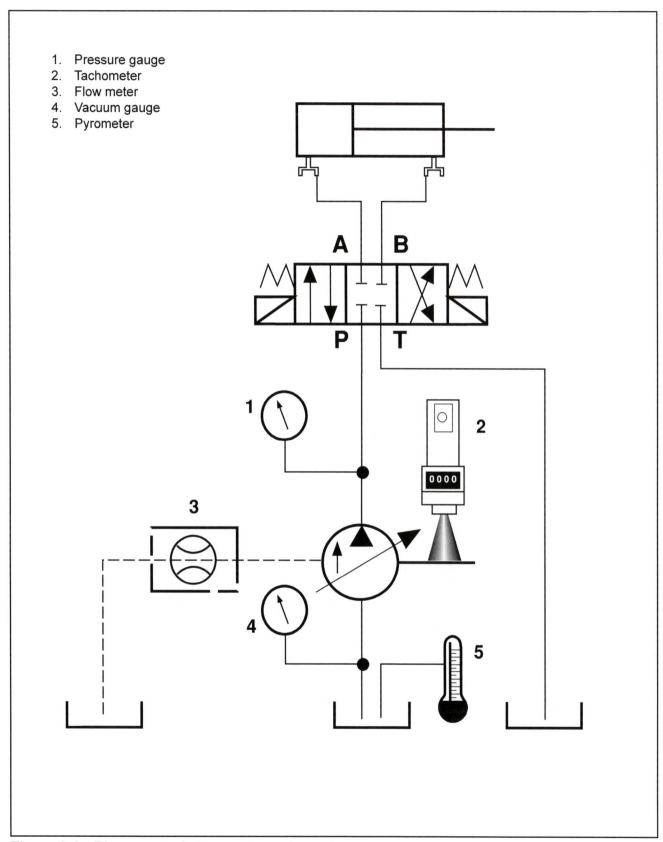

1. Pressure gauge
2. Tachometer
3. Flow meter
4. Vacuum gauge
5. Pyrometer

Figure 3-8   Placement of diagnostic equipment

Step 7.   Start the prime mover. Inspect the diagnostic equipment connectors for leaks.

NOTE: If catastrophic pump failure is suspected, do not run the pump for longer than is absolutely necessary to determine its condition. If metal fragments are found in transmission lines, pump outlet port, in-line or return filters, and/or reservoir, **DO NOT** restart the pump. Replace the pump and conduct a post-catastrophic failure start-up procedure.

Step 8.   Allow the system to warm up to approximately 130º F (54º C). Observe pyrometer (5).

NOTE: *For accurate test results, check and adjust (if necessary) the pressure relief valve or pump compensator setting. Set it according to the manufacturer's specification.*

Step 9.   Operate the prime mover at maximum governed speed. (Conduct Step 10 through Step 13 with the prime mover operating at maximum governed speed).

Step 10.   Record on the test worksheet the pump "no-load" input shaft speed indicated on tachometer (2). Adjust it to the manufacturer's specification if necessary.

Step 11.   Record on the test worksheet the pump inlet restriction indicated on vacuum gauge (4).

NOTE: *If the pump inlet restriction is within specification, continue with the test procedure. If the pump inlet restriction is inconsistent with the design specification, refer to the "Pump Cavitation Test Procedure" outlined in this chapter. Continue with this test procedure when the pump inlet restriction is within specification.*

Step 12.   Operate the pump at "no-load". Record on the test worksheet:
      a. Flow indicated on flow meter (3).
      b. Pressure indicated on vacuum gauge (4).
      c. Fluid temperature indicated on pyrometer (5).

Step 13.   Operate the pump at "full-load" (highest pressure recorded during a normal load cycle). Record on the test worksheet:
      a. Pressure indicated on pressure gauge (1).
      b. Prime mover speed indicated on tachometer (2).
      c. Flow indicated on flow meter (3).
      d. Fluid temperature indicated on pyrometer (5).

Step 14.   Shut the prime mover off and analyze the test results.

Step 15.   At the conclusion of this test procedure, remove the diagnostic equipment. Reconnect the transmission lines and tighten the connectors securely.

Step 16.   Start the prime mover and inspect the connectors for leaks.

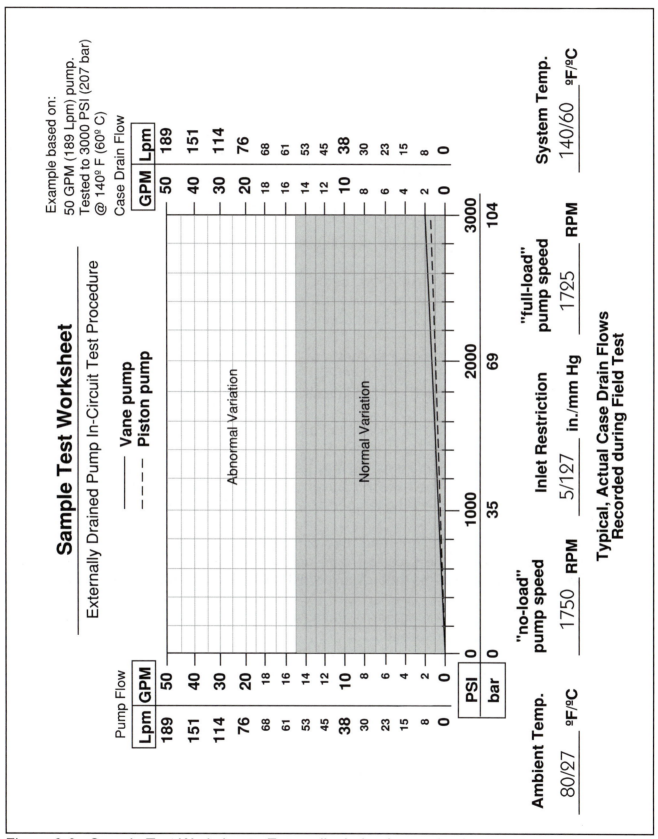

Figure 3-9   Sample Test Worksheet - Externally drained pump in-circuit test procedure

ANALYZING THE TEST RESULTS

1. **Diagnostic Observation: "full-load" cycle:** Flow meter (3) indicates a nominal increase in case drain flow as the pressure drop across the pump ports increases. The flow increase does not exceed the anticipated increase relative to the volumetric efficiency of the pump.

    The operating temperature of the fluid indicated on pyrometer (5) remains within design specification.

    Prime mover speed indicated on tachometer (2) remains within design specification.

    Inlet restriction indicated on vacuum gauge (4) remains within design specification.

    Diagnosis: The pump appears to be in satisfactory operating condition.

2. **Diagnostic Observation: "full-load" cycle:** Flow meter (3) indicates a moderate to high increase in case drain flow as the pressure drop across the pump ports increases. The flow increase does not exceed 30% of the theoretical or "no-load" flow (See **NOTE:** on page 3-7).

    Pyrometer (5) indicates a moderate increase in the operating temperature of the fluid.

    Prime mover speed indicated on tachometer (2) remains within design specification.

    Pump inlet restriction indicated on vacuum gauge (4) remains within design specification.

    Diagnosis: The condition of the pump may be marginal. If there is an abnormal variation (reduction) in actuator speed which causes an unacceptable production loss, and/or a marked increase in the operating temperature of the fluid, the pump should be replaced.

3. **Diagnostic Observation: "full-load" cycle:** Flow meter (3) indicates a progressive case drain flow increase as the pressure drop across the pump ports increases. The flow increase exceeds 30% of the theoretical or "no-load" flow (See **NOTE:** on page 3-7).

    Pyrometer (5) indicates a progressive increase in the operating temperature of the fluid which does not appear to "level-off."

    Prime mover speed indicated on tachometer (2) remains within design specification.

    Pump inlet restriction indicated on vacuum gauge (4) remains within design specification.

    Diagnosis: A case drain flow increase of this magnitude will invariably cause an unacceptable loss in actuator speed. It could also cause a marked increase in the temperature of the fluid. If neglected, elevated fluid temperatures could lead to the degradation of the entire hydraulic system.

    Replace the pump. Refer to chapter two for the "zero-fault" pump start-up procedures.

Pump Case Pressure
Direct-Access Test Procedure

What will this test procedure accomplish?

All hydraulic pumps have component parts which move in relation to one another separated by a small oil-filled clearance. These components are generally loaded toward one another by forces related to system pressure, surface area, and springs or seals.

Theoretically, a pump delivers an amount of fluid equal to its displacement during each revolution. The actual output is reduced because of internal leakage. As pressure increases, the leakage also increases causing a decrease in volumetric efficiency. Volumetric efficiency is equal to the actual output divided by the theoretical output, and is expressed as a percentage.

The amount of leakage is influenced by four factors:
1. Pressure difference across the clearances;
2. Oil viscosity;
3. Temperature; and
4. Size of clearance.

Internal leakage can either flow from the outlet port back to the inlet port (internal drain), or from the outlet port into the pump case (external drain). If a pump is externally drained, a separate port, which is generally smaller in diameter than the inlet and outlet ports, is provided in the pump case. A case drain-line is connected to the drain port to transport the leakage directly back to the fluid reservoir.

The port in the pump case and the external drain-line are designed for relatively low flow rates depending on the volumetric efficiency of the pump. The pump housing is designed for relatively low pressures: from 15 PSI (1 bar) to 25 PSI (1.7 bar) with lip-type shaft seals, and up to approximately 40 PSI (2.76 bar) with mechanical shaft seals.

Any increase in internal leakage will result in increased flow into the case drain-line. If case drain flow increases, case pressure will increase because of higher flow resistance through the drain port in the pump housing and the case drain-lines. Excessive case pressure could result in shaft seal failure and pump case seal leakage. If the pressure in the pump case is high enough, it can literally split the pump case apart.

The most common symptoms of problems associated with excessive pump case pressure are:
a. Shaft seal leakage.
b. Catastrophic shaft seal failure.
c. Pump case seal leakage.
d. Moderate to high increase in the operating temperature of the fluid.

This test procedure will determine:
- a. Pressure in the pump case.
- b. Pressure in the case drain line.
- c. If pump case pressure is excessive, why? Is it caused by:
    1. High internal leakage?
    2. Restriction in the case drain line?

 **-WARNING-** | Do not work on or around hydraulic systems without wearing safety glasses which conform to ANSI Z87.1-1989 standard.

YOU WILL NEED THE FOLLOWING DIAGNOSTIC EQUIPMENT TO CONDUCT THIS TEST PROCEDURE

CAUTION! *The pressure and flow ratings of the diagnostic equipment which will be used to conduct this test procedure must be equal to, or greater than, the pressure and flow ratings of the system being tested. Refer to the system schematic for recommended pressure and flow data.*

1.	Pressure gauge (system pressure)	4.	Needle valve
2.	Pressure gauge (case pressure)	5.	Pyrometer
3.	Pressure gauge (case pressure)		

PREPARATION

To conduct this test safely and accurately, refer to Figure 3-10 for correct placement of diagnostic equipment.

To record the test data, make a copy of the test worksheet on page 3-33 (Figure 3-12).

TEST PROCEDURE

Step 1. Shut the prime mover off.

Step 2. Lock the electrical system out or tag the keylock switch.

Step 3. Observe the system pressure gauge. Release any residual pressure trapped in the system by an accumulator, counterbalance or pilot-operated check valve, suspended load on an actuator, intensifier, or a pressurized reservoir.

Step 4. Install pressure gauge (2) in parallel with the connector at the case drain port of the pump.

WARNING! *DO NOT plug the case drain port in the pump housing. Plugging the case drain port could cause shaft seal or case seal failure. It could also cause catastrophic pump case failure.*

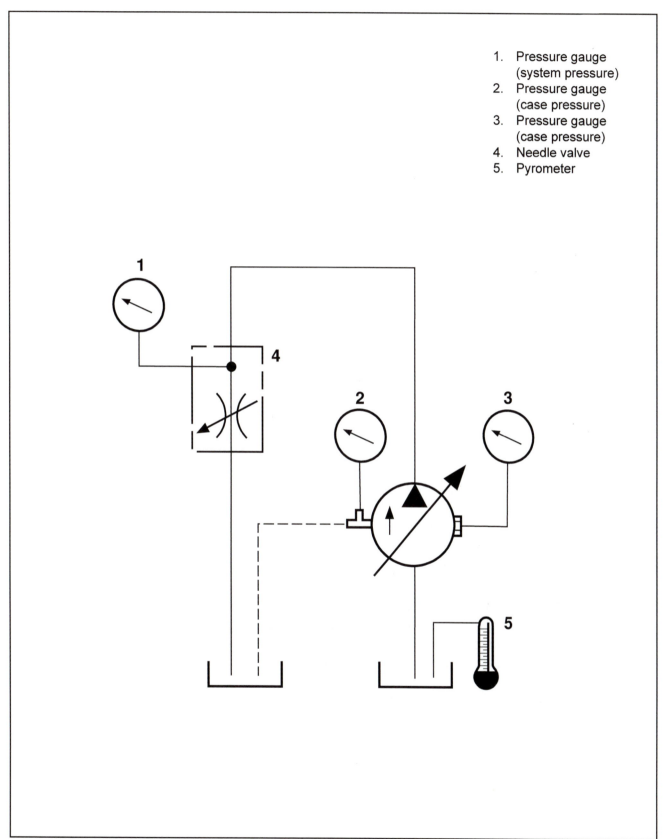

1. Pressure gauge (system pressure)
2. Pressure gauge (case pressure)
3. Pressure gauge (case pressure)
4. Needle valve
5. Pyrometer

Figure 3-10   Placement of Diagnostic Equipment

Step 5. Install pressure gauge (1) in parallel with the connector at the outlet port of the pump.

NOTE: *Look around the pump housing for a secondary access port to the pump case. The secondary access port will usually be plugged. If necessary, refer to the manufacturer's specifications for guidance.*

Not all hydraulic pumps have secondary pump case access ports. If the pump has no secondary case access port it is not possible to accurately measure case pressure. Substitute a case pressure test procedure for a case flow test procedure. Refer to the "Externally Drained" pump test procedures in this chapter.

Step 6. Remove the plug from the secondary case drain port. Install pressure gauge (3) in series with the drain port in the pump housing.

NOTE: *If the fluid drains from the pump case while installing pressure gauges (2) and/or (3), fill the case with clean fluid before starting.*

Step 7. Install needle valve (4) in series with the transmission line at the outlet port of the pump. Open the needle valve fully (turn counter-clockwise).

Step 8. The outlet port of needle valve (4) must be connected directly, to the reservoir. To do this it may be necessary to construct a length of hose. Fasten it securely to the reservoir.

Step 9. Start the prime mover. Inspect the diagnostic equipment connectors for leaks.

Step 10. Allow the system to warm up to approximately 130° F (54° C). Observe pyrometer (5).

Step 11. Record on the test worksheet:
    a. Fluid temperature indicated on pyrometer (5).
    b. Ambient temperature.

Step 12. Record on the test worksheet:
    a. "no-load" pressure indicated on pressure gauge (1).
    b. "no-load" pressure indicated on pressure gauge (2).
    c. "no-load" pressure indicated on pressure gauge (3).

NOTE: *If the pump is pressure compensated, conduct the case pressure test procedure while conducting a direct-access test procedure. Refer to page 3-15.*

Step 13. Gradually close needle valve (4) (turn clockwise). Stop when the value of the system's main pressure relief valve or pump compensator setting is reached.

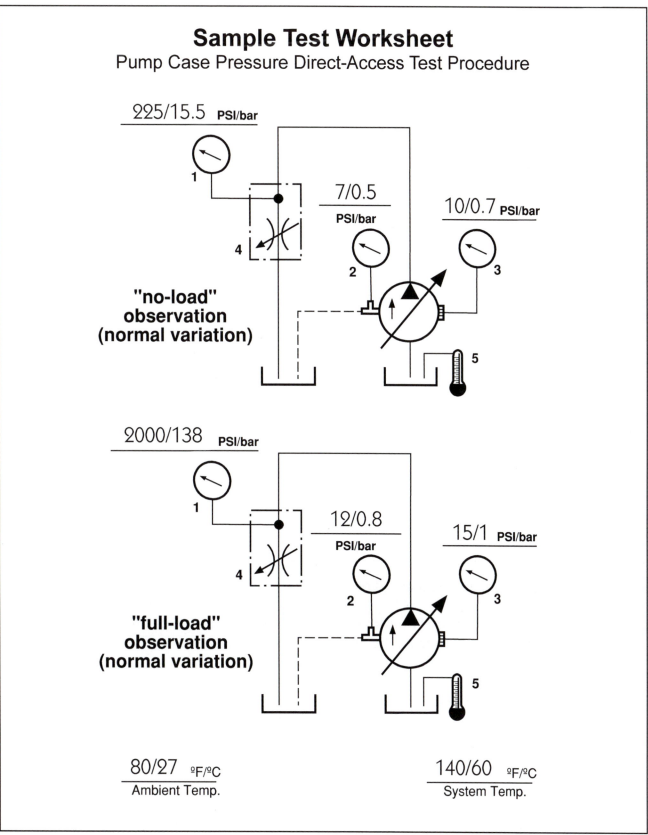

Sample Test Worksheet
Pump Case Pressure Direct-Access Test Procedure

225/15.5 **PSI/bar**

1

7/0.5
PSI/bar

10/0.7 **PSI/bar**

4

2

3

**"no-load"
observation
(normal variation)**

5

2000/138 **PSI/bar**

1

12/0.8
PSI/bar

15/1 **PSI/bar**

4

2

3

**"full-load"
observation
(normal variation)**

5

80/27 ºF/ºC
Ambient Temp.

140/60 ºF/ºC
System Temp.

Figure 3-11   Sample Test Worksheet - Pump case pressure direct-access test procedure

Step 14. Record on the test worksheet:
a. "full-load" pressure indicated on pressure gauge (1).
b. "full-load" pressure indicated on pressure gauge (2).
c. "full-load" pressure indicated on pressure gauge (3).

Step 15. Shut the prime mover off and analyze the test results.

Step 16. At the conclusion of this test procedure remove the diagnostic equipment. Reconnect the transmission lines, and tighten the connectors securely.

Step 17. Start the prime mover and inspect the connectors for leaks.

ANALYZING THE TEST RESULTS

1. **Diagnostic Observation: "No-load" cycle:** Pressure gauge (2) indicates nominal case drain-line pressure. Pressure gauge (3) indicates nominal case pressure. Pressure gauge (1) indicates nominal system pressure.

   The operating temperature of the fluid indicated on pyrometer (5) remains within design specification.

   Diagnosis: The pump appears to be in satisfactory operating condition.

2. **Diagnostic Observation: "Full-load" cycle:** Pressure gauge (2) indicates nominal pressure in the case drain-line. Pressure gauge (3) indicates nominal pressure in the pump case. Pressure gauge (1) indicates maximum system pressure (pressure relief valve or compensator setting).

   The operating temperature of the fluid indicated on pyrometer (5) remains within design specification.

   Diagnosis: The pump appears to be in satisfactory operating condition.

3. **Diagnostic Observation: "Full-load" cycle:** Case pressure indicated on pressure gauge (3) progressively increases as the system pressure increases. It exceeds the maximum case pressure recommended by the pump manufacturer.

   Case drain-line pressure indicated on pressure gauge (2) also increases progressively. However, it does not increase to the pressure level recorded on pressure gauge (3).

   Pyrometer (5) indicates a progressive increase in the operating temperature of the fluid which does not appear to "level-off."

   Diagnosis: There is evidence of excessive internal leakage in the pump. Excessive case pressure is usually caused by the reluctance of the internal leakage to flow through the port in the pump case and the case drain-line.

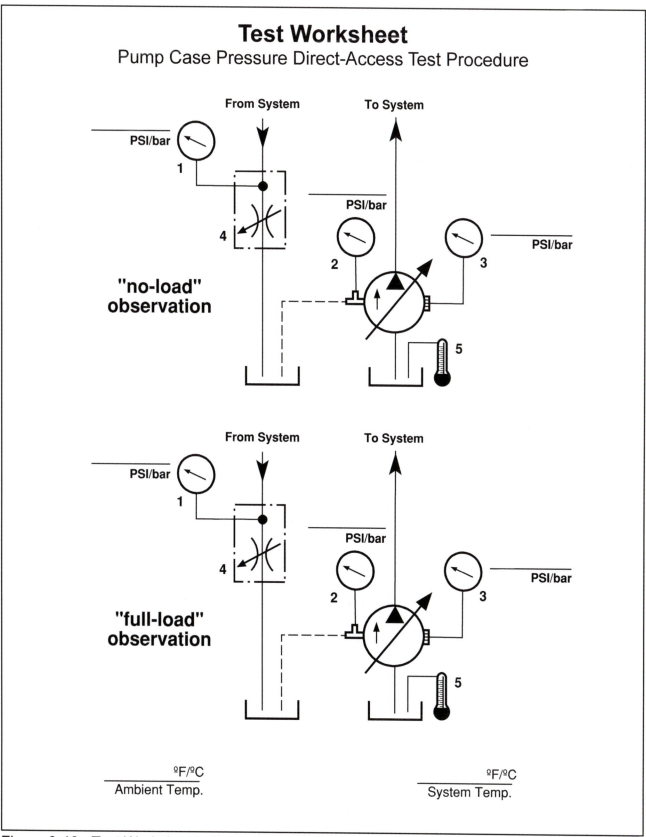

Figure 3-12   Test Worksheet - Pump case pressure direct-access test procedure

Since the case drain port is in series with the case drain-line, pressure caused by resistance to flow in the case drain-line is added to the pressure in the case.

If internal leakage is excessive the resistance offered by the case drain port could cause the pressure in the pump case to exceed design specification. This will usually result in shaft seal or housing seal leakage.

4. **Diagnostic Observation: "Full-load" cycle:** Case drain-line pressure indicated on pressure gauge (2) progressively increases as the system pressure increases. It increases to a level which exceeds the maximum case pressure recommended by the pump manufacturer.

    Case pressure indicated on pressure gauge (3) increases in proportion to the pressure increase indicated on pressure gauge (2). It too increases to a level which exceeds the maximum case pressure recommended by the pump manufacturer.

    Pressure gauge (1) indicates pressure relief valve or pump compensator setting.

    There is a moderate to high increase in the operating temperature of the fluid indicated on pyrometer (5).

    Diagnosis: There is evidence of excessive resistance in the case drain-line. Case drain-line pressure gauge (2) will generally record only the pressure caused by the resistance to flow in the case drain-line.

    To determine if excessive case drain-line pressure is caused by high internal leakage, excessive case drain-line resistance, or a combination of both, it may be necessary to conduct a case drain flow test.

    The following steps can be used to determine the root-cause with a pressure gauge:

Step 1. Install case drain-line pressure gauge (2) in parallel with the next downstream connector and repeat the "full-load" test procedure.

Step 2. If pressure gauge (2) indicates a reduction in pressure, the resistance is in the line upstream of the pressure gauge.

Step 3. If there is no change in the pressure, move pressure gauge (2) to the next downstream connector and repeat the test procedure.

Step 4. Continue moving pressure gauge (2) through the case drain-line progressively until the source of the high resistance is isolated.

NOTE: *If the high case drain-line resistance is caused by a return line filter, manifold, and/or cooler, the case drain-line will have to be re-routed directly back to the reservoir. Submit to application engineering for design review.*

Pump Case Pressure
In-Circuit Test Procedure

What will this test procedure accomplish?

All hydraulic pumps have component parts which move in relation to one another separated by a small oil-filled clearance. These components are generally loaded toward one another by forces related to system pressure, surface area, and springs or seals.

Theoretically, a pump delivers an amount of fluid equal to its displacement during each revolution. The actual output is reduced because of internal leakage. As pressure increases, the leakage also increases causing a decrease in volumetric efficiency. Volumetric efficiency is equal to the actual output divided by the theoretical output, and is expressed as a percentage.

The amount of leakage is influenced by four factors:
1. Pressure difference across the clearances;
2. Oil viscosity;
3. Temperature; and
4. Size of clearance.

Internal leakage can either flow from the outlet port back to the inlet port (internal drain), or from the outlet port into the pump case (external drain). If a pump is externally drained, a separate port, which is generally smaller in diameter than the inlet and outlet ports, is provided in the pump case. A case drain-line is connected to the drain port to transport the leakage directly back to the fluid reservoir.

The port in the pump case, and the external drain-line are designed for relatively low flow rates depending on the volumetric efficiency of the pump. The pump housing is designed for relatively low pressures: from 15 PSI (1 bar) to 25 PSI (1.7 bar) with lip-type shaft seals, and up to approximately 40 PSI (2.76 bar) with mechanical shaft seals.

Any increase in internal leakage will result in increased flow into the case drain-line. If case drain flow increases case pressure will increase because of higher flow resistance through the drain port in the pump housing, and the case drain-lines. Excessive case pressure could result in shaft seal failure and pump case seal leakage. If the pressure in the pump case is high enough, it can literally split the pump case apart.

The most common symptoms of problems associated with excessive pump case pressure are:
a. Shaft seal leakage.
b. Catastrophic shaft seal failure.
c. Pump case seal leakage.
d. Moderate to high increase in the operating temperature of the fluid.

This test procedure will determine:
   a. Pressure in the pump case.
   b. Pressure in the case drain-line.
   c. If pump case pressure is excessive, why? Is it caused by:
      1. High internal leakage?
      2. Restriction in the case drain-line?

-WARNING- | **Do not work on or around hydraulic systems without wearing safety glasses which conform to ANSI Z87.1-1989 standard.**

YOU WILL NEED THE FOLLOWING DIAGNOSTIC EQUIPMENT TO CONDUCT THIS TEST PROCEDURE

CAUTION! The pressure and flow ratings of the diagnostic equipment which will be used to conduct this test procedure must be equal to, or greater than, the pressure and flow ratings of the system being tested. Refer to the system schematic for recommended pressure and flow data.

1.	Pressure gauge (case pressure)	3.	Pressure gauge (case pressure)
2.	Pressure gauge (system pressure)	4.	Pyrometer

PREPARATION

To conduct this test safely and accurately, refer to Figure 3-13 for correct placement of diagnostic equipment.

To record the test data, make a copy of the test worksheet on page 3-41 (Figure 3-15).

TEST PROCEDURE

Step 1.   Shut the prime mover off.

Step 2.   Lock the electrical system out or tag the keylock switch.

Step 3.   Observe the system pressure gauge. Release any residual pressure trapped in the system by an accumulator, counterbalance or pilot-operated check valve, suspended load on an actuator, intensifier, or a pressurized reservoir.

Step 4.   Install pressure gauge (1) in parallel with the connector at the case drain port of the pump.

WARNING! DO NOT plug the case drain port in the pump housing. Plugging the case drain port could cause shaft seal or case seal failure. It could also cause catastrophic pump case failure.

---

1. Pressure gauge (case pressure)
2. Pressure gauge (system pressure)
3. Pressure gauge (case pressure)
4. Pyrometer

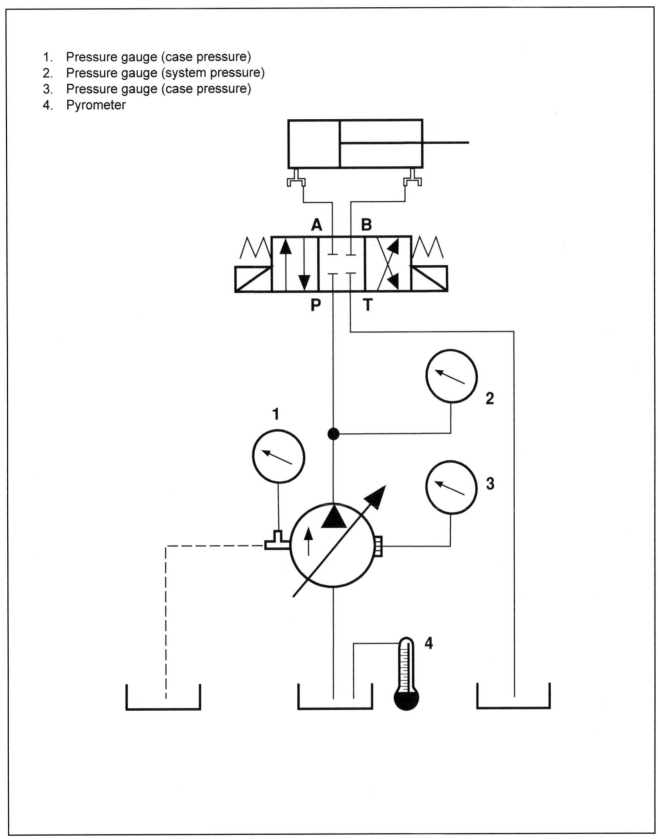

Figure 3-13   Placement of diagnostic equipment

NOTE: *Look around the pump housing for a secondary access port to the pump case. The secondary access port will usually be plugged. If necessary, refer to the manufacturer's specifications for guidance.*

Not all hydraulic pumps have secondary pump case access ports. If the pump has no secondary case access port it is not possible to accurately measure case pressure. Substitute a case pressure test procedure for a case flow test procedure. Refer to the "Externally Drained" pump test procedures in this chapter.

Step 5.   Remove the plug from the secondary case drain port. Install pressure gauge (3) in series with the drain port in the pump housing.

NOTE: *If the fluid drains from the pump case while installing pressure gauges (2) and/or (3), fill the pump case with fluid before starting the prime mover.*

Step 6.   Install pressure gauge (2) in parallel with the connector at the outlet port of the pump.

Step 7.   Start the prime mover. Inspect the diagnostic equipment connectors for leaks.

Step 8.   Allow the system to warm up to approximately 130° F. (54° C.). Observe pyrometer (4).

Step 9.   Record on the test worksheet the temperature indicated on pyrometer (4).

Step 10. Operate the machine through a "no-load" cycle. Record on the test worksheet:
   a. Pressure indicated on pressure gauge (1).
   b. Pressure indicated on pressure gauge (2).
   c. Pressure indicated on pressure gauge (3).

NOTE: *If the pump is pressure compensated, conduct the case pressure test procedure while conducting a direct-access test procedure. Refer to page 3-15.*

Step 11. Operate the machine through a "full-load" cycle. Record on the test worksheet:
   a. Pressure indicated on pressure gauge (1).
   b. Pressure indicated on pressure gauge (2).
   c. Pressure indicated on pressure gauge (3).

Step 12. Shut the prime mover off and analyze the test results.

Step 13. At the conclusion of this test procedure, remove the diagnostic equipment. Reconnect the transmission lines and tighten the connectors securely.

Step 14. Start the prime mover and inspect the connectors for leaks.

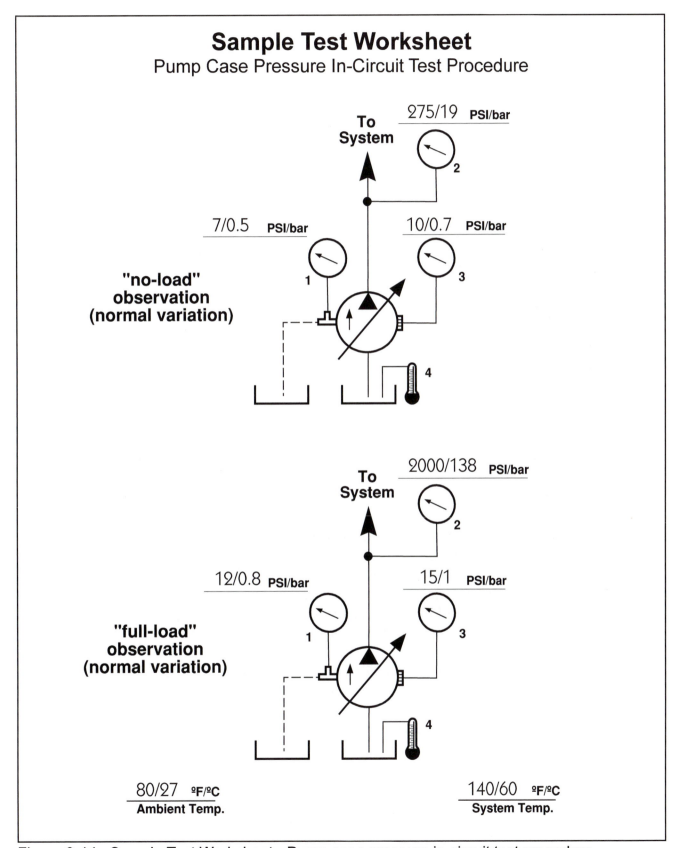

Figure 3-14   Sample Test Worksheet - Pump case pressure in-circuit test procedure

ANALYZING THE TEST RESULTS

1. **Diagnostic Observation: "No-load" cycle:** Pressure gauge (1) indicates nominal pressure in the case drain-line. Pressure gauge (3) indicates nominal case pressure. Pressure gauge (2) indicates nominal system pressure.

    The operating temperature of the fluid indicated on pyrometer (4) remains within design specification.

    Diagnosis: The pump appears to be in satisfactory operating condition.

2. **Diagnostic Observation: "Full-load" cycle:** Pressure gauge (1) indicates nominal pressure in the case drain-line. Pressure gauge (3) indicates nominal pressure in the pump case. Pressure gauge (2) indicates pressure relief valve or pump compensator setting.

    The operating temperature of the fluid indicated on pyrometer (4) remains within design specification.

    Diagnosis: The pump appears to be in satisfactory operating condition.

3. **Diagnostic Observation: "Full-load" cycle:** Case pressure indicated on pressure gauge (3) progressively increases as the system pressure increases. It exceeds the maximum case pressure recommended by the pump manufacturer.

    Case drain-line pressure indicated on pressure gauge (1) also increases progressively. However, it does not increase to the pressure level recorded on pressure gauge (3).

    Pyrometer (4) indicates a progressive increase in the operating temperature of the fluid which does not appear to "level-off."

    Diagnosis: There is evidence of excessive internal leakage in the pump. Excessive case pressure is usually caused by the reluctance of the internal leakage to flow through the port in the pump case and the case drain-line.

    Since the case drain port is in series with the case drain-line, pressure caused by resistance to flow in the case drain-line is added to the pressure in the case.

    If internal leakage is excessive, the resistance offered by the case drain port could cause the pressure in the pump case to exceed design specification. This will usually result in shaft seal or housing seal leakage.

4. **Diagnostic Observation: "Full-load" cycle:** Case drain-line pressure indicated on pressure gauge (1) progressively increases as the system pressure increases. It increases to a level which exceeds the maximum case pressure recommended by the pump manufacturer.

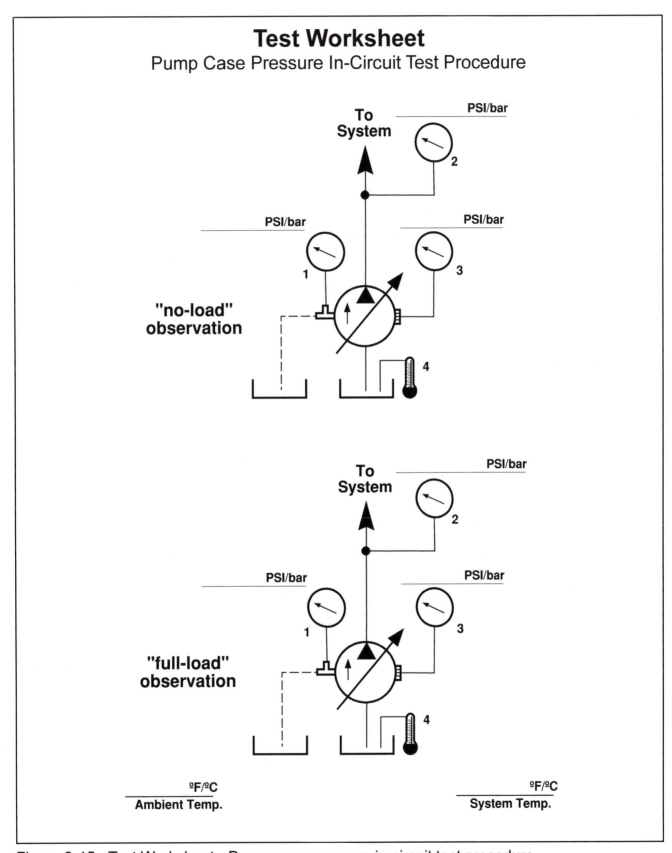

Figure 3-15   Test Worksheet - Pump case pressure in-circuit test procedure

Case pressure indicated on pressure gauge (3) increases in proportion to the pressure increase indicated on pressure gauge (1). It too increases to a level which exceeds the maximum case pressure recommended by the pump manufacturer.

Pressure gauge (2) indicates pressure relief valve or pump compensator setting.

There is a moderate to high increase in the operating temperature of the fluid indicated on pyrometer (4)

Diagnosis: There is evidence of excessive resistance in the case drain-line. Case drain-line pressure gauge (1) will generally record only the pressure caused by the resistance to flow in the case drain-line.

To determine if excessive case drain-line pressure is caused by high internal leakage, excessive case drain-line resistance, or a combination of both, it may be necessary to conduct a case drain flow test.

The following steps can be used to determine the root-cause with a pressure gauge:

Step 1.   Install case drain-line pressure gauge (1) in parallel with the next downstream connector and repeat the "full-load" test procedure.

Step 2.   If pressure gauge (1) indicates a reduction in pressure, the resistance is in the line upstream of the pressure gauge.

Step 3.   If there is no change in pressure, move pressure gauge (1) to the next downstream connector and repeat the test procedure.

Step 4.   Continue moving pressure gauge (1) through the case drain-line progressively until the source of the high resistance is isolated.

NOTE: *If the high case drain-line resistance is caused by a return line filter, manifold and/or cooler, the case drain-line will have to be re-routed directly back to the reservoir. Submit to application engineering for design review.*

Pump Cavitation
In-Circuit Test Procedure

What will this test procedure accomplish?

Cavitation by definition is: "The formation and collapse of gaseous cavities in a liquid."

Usually, the first indications of cavitation in a hydraulic system are: excessive noise from the pump, high operating temperatures, and erratic actuator operation.

Hydraulic fluid is made up of approximately 10% dissolved air. When the fluid is subjected to an absolute pressure (vacuum), the dissolved air is encouraged to come out of the fluid. As it does, the fluid stream fills with cavities.

When the cavities are exposed to pressure at the outlet port of the pump, they "implode" at tons of force per square inch against the components in the immediate vicinity of the pump outlet port. The "implosions" cause a high frequency noise. They also erode component surfaces.

The contaminants generated by erosion are transported throughout the system by the fluid. Further degradation of components is caused by the contaminants.

Interestingly, unlike air contamination of a liquid, the cavities "disappear" when the liquid is subjected to positive pressure at the outlet port of the pump.

Cavitation is often-times confused with aeration. If air is allowed to enter the inlet port of the pump, it will cause problems similar to cavitation. However, since aeration has little to do with the liquid's vapour pressure, aeration is referred to as "pseudo-cavitation."

Care must be taken when studying pump inlet restriction. It cannot be assumed that the "lower the inlet restriction the better" as is published in many texts. Without a point-of-reference the information can be extremely misleading.

It is of the utmost importance to learn what pump inlet restriction is under "normal" operating conditions. Once normal variations in pump inlet restriction are known, i.e. a point-of-reference has been established, an abnormal variation can be quickly determined.

For example, if a pump begins to show signs of cavitation-- excessive noise associated with high operating temperatures-- a vacuum gauge at the pump inlet port could be used to determine the root-cause. If the vaccum gauge indicates an inlet restriction which is well below the maximum inlet restriction recommended by the pump manufacturer, it would be easy to conclude that cavitation is not the root-cause of the problem.

However, if it is known that this pump normally operates with an inlet restriction of 5 in. Hg (127 mm Hg), at normal operating temperature, the conclusion could be very different. If the vacuum gauge indicates an inlet restriction of 2 in. Hg (51 mm Hg), there is a clear indication that air is entering the inlet port of the pump.

To be effective at troubleshooting hydraulic components it is essential to identify with normal variations in component performance so that abnormal variations can be easily identified.

Cavitation can be caused by:
      a. Restricting the oil flow at the pump inlet.
      b. Allowing air to enter the pump inlet port with the oil (aeration).

The most common symptoms of problems associated with pump cavitation are:
      a. Excessive noise from the pump.
      b. Moderate to high increase in the operating temperature of the fluid.
      c. Erratic actuator operation.
      d. Air contamination of the fluid (foaming).

This test procedure will determine:
      a. If the inlet restriction is **higher** than normal associated with no obvious signs of aeration (liquid foaming).
      b. If the inlet restriction is **higher** than normal associated with aeration.
      c. If the inlet restriction is **lower** than normal associated with aeration (liquid foaming).

-WARNING- | **Do not work on or around hydraulic systems without wearing safety glasses which conform to ANSI Z87.1-1989 standard.**

YOU WILL NEED THE FOLLOWING DIAGNOSTIC EQUIPMENT
TO CONDUCT THIS TEST PROCEDURE

CAUTION! *The pressure and flow ratings of the diagnostic equipment which will be used to conduct this test procedure must be equal to, or greater than, the pressure and flow ratings of the system being tested. Refer to the system schematic for recommended pressure and flow data.*

    1.   Vacuum gauge   3.   Tachometer
    2.   Pyrometer

PREPARATION

To conduct this test safely and accurately, refer to Figure 3-16 for correct placement of diagnostic equipment.

To record the test data, make a copy of the test worksheet on page 3-52 (Figure 3-18).

1. Vacuum gauge
2. Pyrometer
3. Tachometer

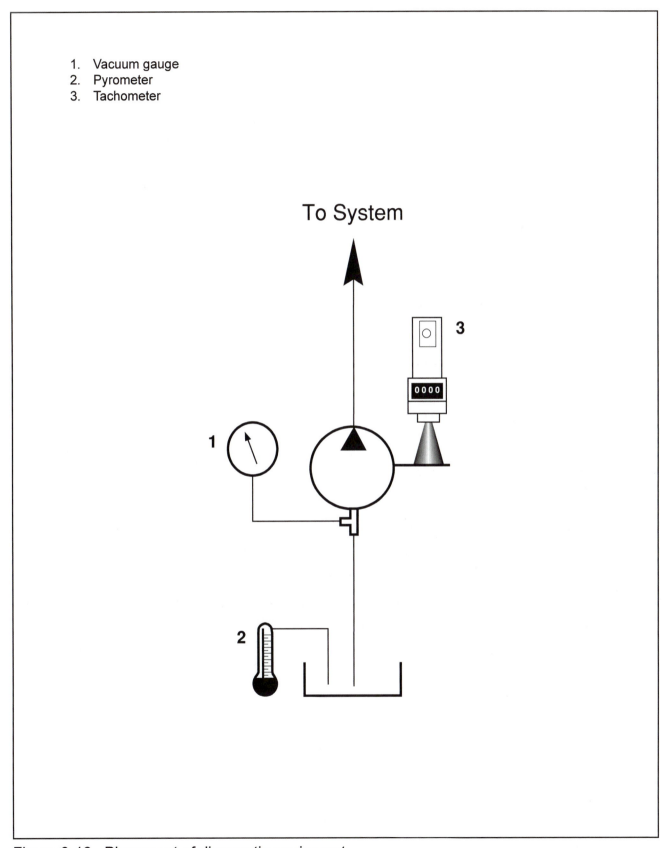

Figure 3-16   Placement of diagnostic equipment

TEST PROCEDURE

Step 1.   Shut the prime mover off.

Step 2.   Lock the electrical system out or tag the keylock switch.

Step 3.   Observe the system pressure gauge. Release any residual pressure trapped in the system by an accumulator, counterbalance or pilot-operated check valve, suspended load on an actuator, intensifier, or a pressurized reservoir.

Step 4.   Install vacuum gauge (1) in parallel with the connector at the inlet port of the pump.

Step 5.   Start the prime mover. Inspect the diagnostic equipment connectors for leaks.

Step 6.   Allow the system to warm up to approximately 130º F. (54º C.). Observe pyrometer (2).

Step 7.   Record on the test worksheet the temperature indicated on pyrometer (2).

Step 8.   Operate the prime mover at maximum governed speed. Record on the test worksheet the pump input shaft speed indicated on tachometer (3).

NOTE: *If the pump is engine or power-take-off (PTO) driven, refer to the manufacturer's specifications for gear ratios. The information can be used to calculate actual pump speed.*

Pump inlet restriction is directly affected by the speed at which a pump rotates. An increase in pump speed will proportionately increase pump flow. An increase in flow will increase resistance in the pump inlet transmission line, which could result in cavitation.

If necessary, adjust the prime mover to the recommended speed or change the drive gear ratios. Refer to the pump manufacturer's specifications for maximum and minimum speed recommendations.

Step 9.   Operate the prime mover at maximum governed speed . Record on the test worksheet the pump inlet restriction indicated on vacuum gauge (1).

NOTE: a. *If the pump inlet restriction exceeds pump manufacturer's specifications with no obvious signs of aeration, refer to the Root Cause Analysis chart on page 3-49.*
b. *If the pump inlet restriction exceeds pump manufacturer's specifications associated with aeration refer to the Root Cause Analysis chart on page 3-50.*
c. *If the pump inlet restriction is lower than pump manufacturer's specifications associated with aeration refer to the Root Cause Analysis chart on page 3-51.*

CAUTION! *If pump inlet restriction exceeds the maximum restriction recommended by the pump manufacturer, DO NOT operate the pump for longer than is absolutely necessary to conduct root-cause analysis.*

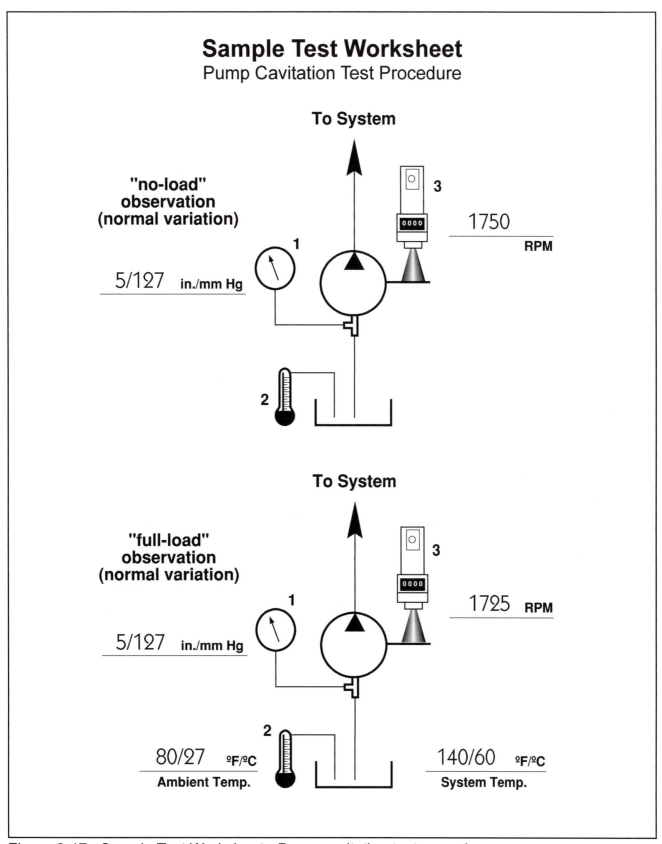

Figure 3-17   Sample Test Worksheet - Pump cavitation test procedure

Step 10.   Shut the prime mover off.

Step 11.   At the conclusion of this test procedure, remove the diagnostic equipment. Reconnect the transmission lines, and tighten the connectors securely.

Step 12.   Start the prime mover and inspect the connectors for leaks.

INITIAL EVALUATION

NOTE: *To determine the root-cause of high inlet restriction, every component between the pump and the fluid in the reservoir must be suspect. To isolate the source, progressively shift the vacuum gauge through the pump inlet transmission line. Install the vacuum gauge at each connection until the vacuum decreases.*

ANALYZING THE TEST RESULTS

CAVITATION - Generally, any restriction at the pump inlet which exceeds 10 in. Hg (254 mm Hg) at normal operating temperature is excessive. The maximum inlet restriction (unless otherwise specified by the manufacturer) for hydraulic pumps operating at normal temperature (approximately 140° F (60° C)) is:

Piston Pumps: 2 in. to 4 in. Hg (51 mm to 102 mm Hg)
Vane Pumps: 4 in. to 6 in. Hg (102 to 152 mm Hg)
Gear Pumps: 6 in. to 8 in. Hg (152 mm to 203 mm Hg)

AERATION - A lower than normal inlet restriction indicates a possible air leak (aeration) on the suction side of the pump.

CAUTION! *Cavitation should be regarded as a very serious condition which could lead to the rapid deterioration of the entire hydraulic system.*

There is a significant visual difference between what occurs when air enters the inlet side of the pump (low inlet restriction) and when the pump inlet port is restricted (high inlet restriction).

Aeration will cause the pump to make a high frequency noise which is pressure sensitive- i.e., the greater the pressure, the more intense the noise. Air pockets entering the inlet port of the pump do not disappear as they "implode" at the pump outlet port. Air is literally being pumped into the entire system.

If air entrainment is neglected, the entire lubrication system will become aerated. The machine will become erratic, overheat, and tend to vibrate. The oil becomes "spongy" and will lighten in color. Many times aeration is confused with water contamination. The entire contents of the reservoir will become contaminated with air resulting in oil-film separation between the moving parts in the system. Oil film separation will usually lead to adhesive wear and possible seizure.

Air entrainment is a very serious condition and could lead to the rapid degradation of the entire hydraulic system.

Pump Cavitation
Root Cause Analysis

OBSERVATION

High inlet restriction with no obvious signs of aeration.

Low ambient temperature	Causes oil viscosity to increase to a level which could severely restrict flow.
Pump speed	Prevents pumping chambers from filling adequately.
Oil viscosity	Incorrect oil resulting in high oil viscosity. Restricts oil flow into pump.
Suction screen	1) Incorrectly sized for pump flow. 2) Plugged due to contamination or lack of service.
Suction filter	Not recommended for most open-loop pump applications. Apply pump manufacturer's guidelines.
Suction line: Diameter	Not sized correctly for pump flow. Excessive oil velocity restricting flow.
Suction line: Length	Too long, causing high resistance to pump flow.
Suction line: Type	Incompatible with suction line application. Collapses or internal separation under vacuum.
Pump/Reservoir relationship	Pump inlet port too high above reservoir. Pump unable to "lift" oil from reservoir due to high resistance and weight.
Pump/Reservoir Isolator valve	Incorrectly sized or correctly sized but severely restricts flow due to internal design.
Vacuum in hydraulic reservoir	Reservoir unable to "breathe" to atmosphere. Inlet relief valve in pressure cap malfunction. Non-breather type cap installed on reservoir not compensating for contraction.
Pump inlet port	Incorrect port size in pump inlet port housing restricting oil flow into pumping chambers. This condition cannot be detected at the pump inlet with a vacuum gauge.
Reservoir not pressurized	In certain applications it is necessary to pressurize the reservoir to "boost" the pump inlet port -- high flow, open-loop pump applications.

Pump Cavitation
Root Cause Analysis

OBSERVATION

High inlet restriction associated with aeration.

Most internally drained pumps have low pressure shaft seals rated up to 25 PSI (1.7 bar). To minimize shaft seal stress, the shaft seal cavity is generally internally drained into the pump suction port. Consequently, the pump shaft seal is subjected to the same vacuum conditions as the pump suction port.

Abnormally high inlet restriction, caused by the conditions mentioned on page 3-49, can cause air to enter the suction side of the pump through the shaft seal, resulting in air contamination of the oil.

This condition can be corrected by following the Root Cause Analysis guidelines on page 3-49. Once the high inlet restriction has been reduced, air entrainment may stop if the shaft seal has not been damaged.

If aeration persists when inlet restriction is normalized replace the shaft seal.

Pump Cavitation
Root Cause Analysis

OBSERVATION

Low inlet restriction associated with aeration.

Low oil level -- Inlet port exposed to atmosphere	Causes air instead of oil to be pumped into hydraulic system.
Low oil level -- Inlet port not exposed to atmosphere	Causes a vortex to develop in the oil. Pumps oil and air into system.
Reservoir baffle design	Causes suction lines to become momentarily exposed to atmosphere due to machine movement.
Incline operation	Causes suction lines to become exposed to atmosphere due to angle of incline operation (mobile application).
Turbulence in reservoir	Caused by high return flow, relative to size and design of reservoir.
Turbulence in reservoir	Caused by oil return lines terminating above the oil level in reservoir.
Pump suction lines	Caused by loose or cracked connectors, cracked welds, loose clamps, overtightened connectors.
Pump rotation	Wrong direction of rotation causes air to enter through air bleed valves.
Accumulators	Caused by bladder failure or piston seal failure.
Pump/Reservoir Isolator valve	Caused by isolator valve seal leakage.
Pump shaft seal	Worn or damaged. Can cause air to enter pumping chambers through the shaft seal. Problem restricted to internally drained pumps.

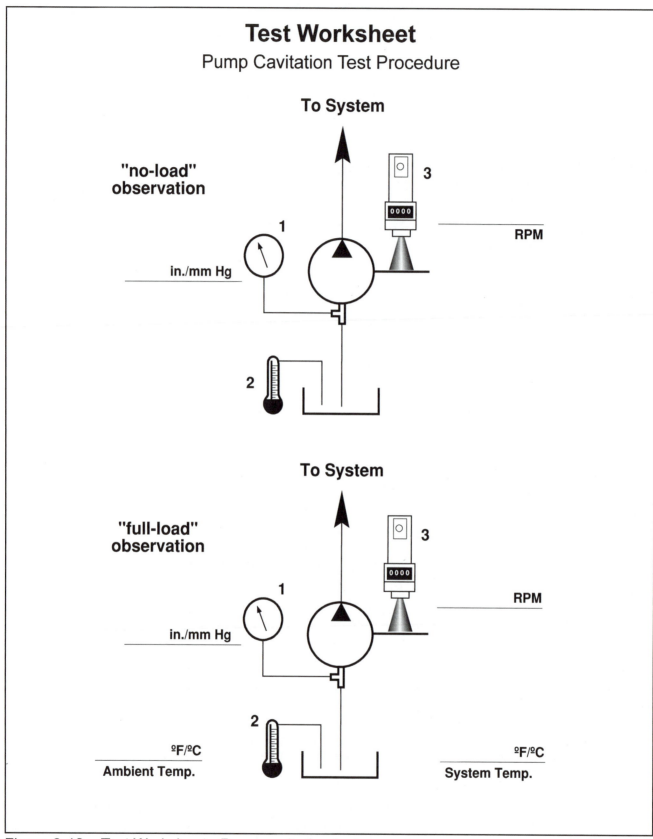

Figure 3-18   Test Worksheet - Pump cavitation test procedure

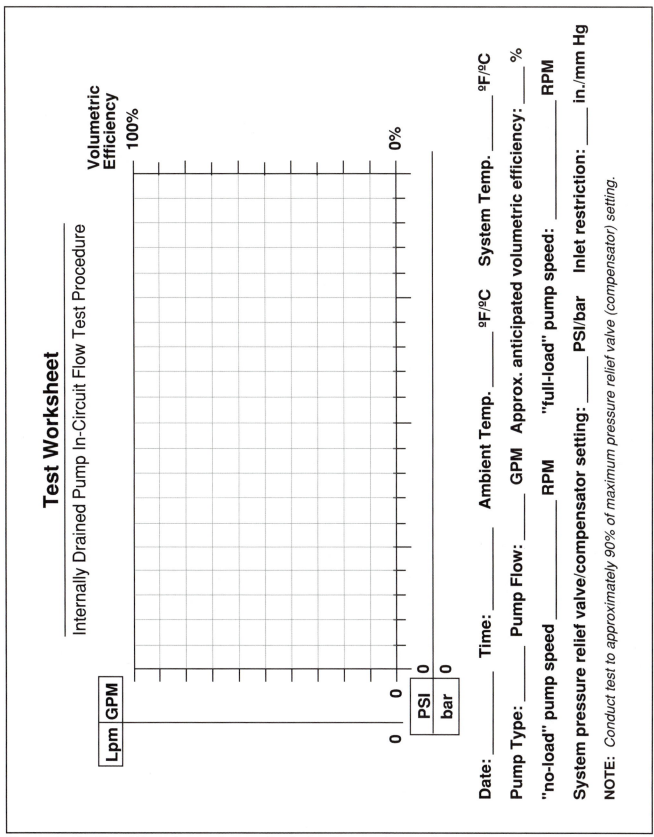

Figure 3-19   Test Worksheet - Internally drained pump in-circuit test procedure

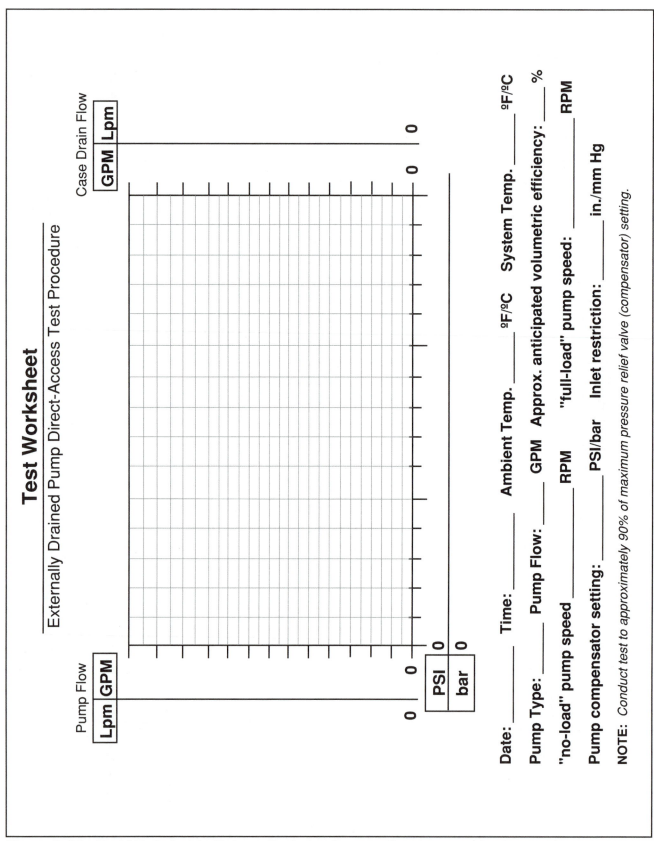

Figure 3-20   Test Worksheet - Externally drained pump direct-access test procedure

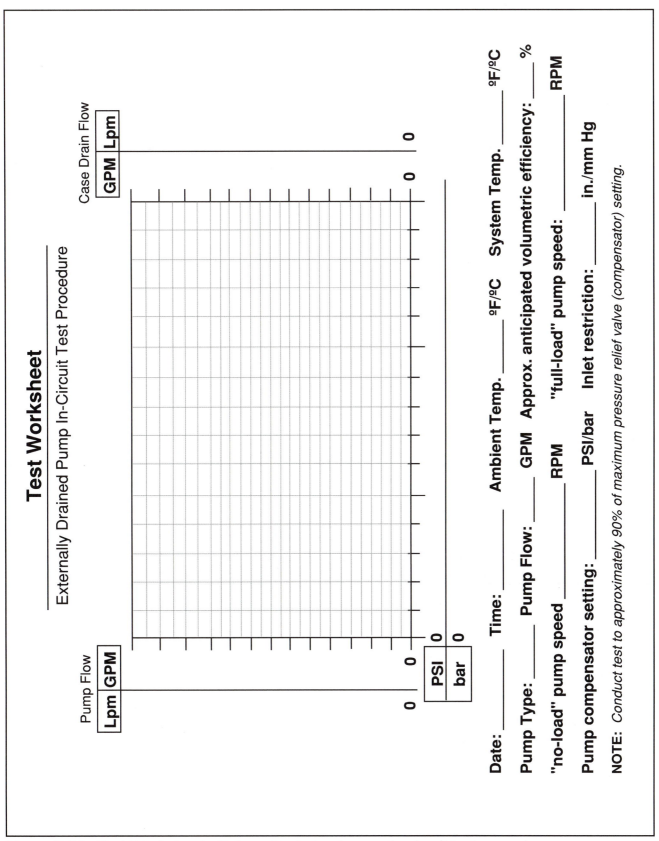

Figure 3-21   Test Worksheet - Externally drained pump in-circuit test procedure

notes

chapter four

Procedures for Testing Pressure Control Valves

Pilot-Operated Relief Valve
In-Circuit Test Procedure

What will this test procedure accomplish?

A pilot-operated relief valve consists of two direct-operated relief valves positioned in series within a single valve body. The valve body has an inlet and outlet port. The first valve in series is the main valve, which consists of a main spool and main spool bias spring. The second valve is the pilot valve which consists of a pilot valve spool, pilot valve bias spring, and screw adjustment.

An orifice, drilled in either the main spool or valve body, allows oil to flow into the pilot section of the valve. Pressure in the pilot section of the valve is determined by the pilot valve setting. Pilot pressure is applied to the top of the main spool which adds to the resistance offered by the main spool bias spring.

Since the areas on both sides of the main spool are equal, and the orifice allows pressure on both sides of the main spool to equalize, the force on top of the main spool is greater than the force on the bottom, by the value of the main spool bias spring - approximately 25 PSI (1.7 bar).

As the pressure in the system increases, to meet the value of the relief valve setting, the pilot valve opens first. The amount of oil flowing through the pilot valve is determined by the size of the orifice. The main spool will remain closed until system pressure increases an additional 25 PSI (1.7 bar) - the value of the main spool bias spring. When there is sufficient pressure in the system to overcome the pilot spool and main spool bias springs, both the pilot spool and main spool will open.

The most common symptoms of problems associated with pressure relief valves are:
    a. Actuator fails to respond when the directional control valve is activated.
    b. A loss of actuator speed which is "viscosity sensitive." As the fluid temperature increases, actuator speed progressively decreases.
    c. Moderate to high increase in the operating temperature of the fluid.
    d. Actuator slows as the load (pressure) increases.

This test procedure will determine the amount of leakage across the ports of a pilot-operated relief valve.

Since a pilot-operated relief valve is comprised of a number of moving parts, a cross-port leakage test will not isolate an internal leakage path. If the test reveals excessive leakage, the valve will have to be dismantled and each individual part visually inspected in order to determine the root-cause.

 **-WARNING-** | Do not work on or around hydraulic systems without wearing safety glasses which conform to ANSI Z87.1-1989 standard.

YOU WILL NEED THE FOLLOWING DIAGNOSTIC EQUIPMENT
TO CONDUCT THIS TEST PROCEDURE

CAUTION! *The pressure and flow ratings of the diagnostic equipment which will be used to conduct this test procedure must be equal to, or greater than, the pressure and flow ratings of the system being tested. Refer to the system schematic for recommended pressure and flow data.*

1.	Pressure gauge	3.	Needle valve
2.	Pyrometer	4.	Flow meter

PREPARATION

To conduct this test safely and accurately, refer to Figure 4-1 for correct placement of diagnostic equipment.

TEST PROCEDURE

Step 1.   Shut the prime mover off.

Step 2.   Lock the electrical system out or tag the keylock switch.

Step 3.   Observe the system pressure gauge. Release any residual pressure trapped in the system by an accumulator, counterbalance or pilot-operated check valve, suspended load on an actuator, intensifier, or a pressurized reservoir.

Step 4.   Install pressure gauge (1) in parallel with the connector at the outlet port of the pump.

Step 5.   Install flow meter (4) in series with the transmission line at the outlet port of the pilot operated relief valve.

Step 6.   Install needle valve (3) in parallel with the connector at the outlet port of the pump. Connect a length of hose to the outlet port of needle valve (3) and route it directly to the reservoir. Fasten it securely. The hose must be suitable for the pressure and flow rating of the system.

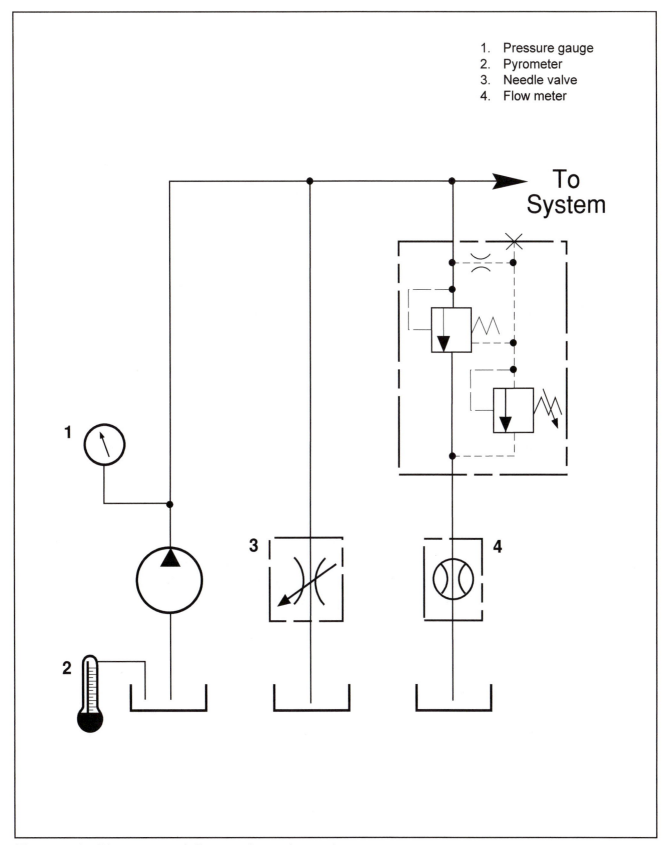

1. Pressure gauge
2. Pyrometer
3. Needle valve
4. Flow meter

To System

Figure 4-1 Placement of diagnostic equipment

NOTE: *It may be necessary to remove the reservoir inspection cover to provide access for the return line from the needle valve. If the reservoir cover is removed for this purpose, fasten the return line securely into the reservoir.*

Step 7.   Start the prime mover. Inspect the diagnostic equipment connectors for leaks.

Step 8.   Allow the system to warm up to approximately 130° F. (54° C.). Observe pyrometer (2).

Step 9.   Close needle valve (3) fully (turn clockwise).

Step 10.   Stall (dead-head) an actuator and verify that the pilot-operated relief valve is adjusted to manufacturer's specifications. Observe pressure gauge (1).

NOTE: *If the relief valve does not respond to pressure adjustment, either the pump is not providing flow to the valve, or the relief valve does not operate.*

To determine the root cause conduct a pump flow test procedure. If the pump test is positive, replace the relief valve.

CAUTION! *Certain variable speed prime mover applications (primarily mobile) require that the prime mover operate at a pre-determined speed when relief valve adjustment is made. Refer to the equipment service data for specific relief valve adjustment procedures.*

Step 11.   Open needle valve (3) fully (turn counter-clockwise).

Step 12.   Stall (dead-head) an actuator. Hold it in the "stalled" position.

NOTE: Full pump flow should be passing through open needle valve (3) and returning to the reservoir.

Step 13.   While observing flow meter (4), gradually close needle valve (3) (turn clockwise) and increase the pressure in 100 PSI (6.9 bar) increments. Observe pressure gauge (1).

NOTE: *There should be no sign of oil escaping from the drain port of the pilot operated relief valve at low pressure. Conduct the test procedure until the "cracking" pressure or pressure setting of the relief valve is reached.*

Step 14.   If the relief valve operates normally, adjust it to the recommended pressure setting.

Step 15.   Shut the prime mover off and analyze the test results.

Step 16.   At the conclusion of this test procedure, remove the diagnostic equipment. Reconnect the transmission lines and tighten the connectors securely.

Step 17.   Start the prime mover and inspect the connectors for leaks.

ANALYZING THE TEST RESULTS

1. **Diagnostic Observation:** The pilot-operated relief valve shows no signs of cross-port leakage until "cracking" pressure is reached.

    Diagnosis: The pressure relief valve appears to be in satisfactory operating condition.

    If necessary, refer to the valve manufacturer's specifications for maximum recommended leakage rates.

2. **Diagnostic Observation:** There is steady cross-port leakage at low pressure, which occurs well below the setting of the valve (below normal cracking pressure). Leakage appears to increase as system pressure increases.

    Diagnosis: There is evidence of cross-port leakage which could be caused by internal damage, wear, and/or, contamination between the spool and seat assemblies.

    If necessary, refer to the valve manufacturer's specifications for maximum recommended leakage rates.

3. **Diagnostic Observation:** Profuse cross-port leakage at low pressure which occurs well below the pressure setting of the valve (below normal cracking pressure). Leakage increases significantly as system pressure increases.

    Diagnosis: There is evidence of valve failure which could be caused by internal damage, wear, plugged control orifice, and/or contamination between the spool and seat assemblies.

    Repair or replace the pilot-operated relief valve.

CAUTION! *Component repair procedures must be conducted by trained, authorized personnel. Incorrect parts, or parts which are improperly installed, can cause catastrophic component malfunction.*

---

notes

Pilot-Operated Relief Valve
Direct-Access Test Procedure

What will this test procedure accomplish?

A pilot-operated relief valve consists of two direct-operated relief valves positioned in series within a single valve body. The valve body has an inlet and outlet port. The first valve in series is the main valve, which consists of a main spool and main spool bias spring. The second valve is the pilot valve which consists of a pilot valve spool, pilot valve bias spring, and screw adjustment.

An orifice, drilled in either the main spool or valve body, allows oil to flow into the pilot section of the valve. Pressure in the pilot section of the valve is determined by the pilot valve setting. Pilot pressure is applied to the top of the main spool which adds to the resistance offered by the main spool bias spring.

Since the areas on both sides of the main spool are equal, and the orifice allows pressure on both sides of the main spool to equalize, the force on top of the main spool is greater than the force on the bottom, by the value of the main spool bias spring - approximately 25 PSI (1.7 bar).

As the pressure in the system increases to meet the value of the relief valve setting, the pilot valve opens first. The amount of oil flowing through the pilot valve is determined by the size of the orifice. The main spool will remain closed until system pressure increases an additional 25 PSI (1.7 bar) - the value of the main spool bias spring. When there is sufficient pressure in the system to overcome the pilot spool and main spool bias springs, both the pilot spool and main spool will open.

The most common symptoms of problems associated with pressure relief valves are:
    a. Actuator fails to respond when the directional control valve is activated.
    b. A loss of actuator speed which is "viscosity sensitive." (As the fluid temperature increases, actuator speed progressively decreases).
    c. Moderate to high increase in the operating temperature of the fluid.
    d. Actuator slows as the load (pressure) increases.

This test procedure will determine the amount of leakage across the ports of a pilot-operated relief valve.

Since a pilot-operated relief valve is comprised of a number of moving parts, a cross-port leakage test will not isolate an internal leakage path. If the test reveals excessive leakage, the valve will have to be dismantled and each individual part visually inspected in order to determine the root-cause.

 **-WARNING-** | **Do not work on or around hydraulic systems without wearing safety glasses which conform to ANSI Z87.1-1989 standard.**

YOU WILL NEED THE FOLLOWING DIAGNOSTIC EQUIPMENT
TO CONDUCT THIS TEST PROCEDURE

CAUTION! *The pressure and flow ratings of the diagnostic equipment which will be used to conduct this test procedure must be equal to, or greater than, the pressure and flow ratings of the system being tested. Refer to the system schematic for recommended pressure and flow data.*

---

1.   MicroLeak analyzer   2.   Pressure gauge

PREPARATION

To conduct this test safely and accurately, refer to Figure 4-2 for correct placement of diagnostic equipment.

TEST PROCEDURE

Step 1.   Start the prime mover. Verify that the pilot-operated relief valve is adjusted to the manufacturer's specification.

NOTE: *If the relief valve does not respond to pressure adjustment, either the pump is not providing flow to the valve, or the relief valve does not operate.*

To determine the root cause conduct a pump flow test procedure. If the pump test is positive, replace the relief valve.

Step 2.   Shut the prime mover off.

Step 3.   Lock the electrical system out or tag the keylock switch.

Step 4.   Observe the system pressure gauge. Release any residual pressure trapped in the system by an accumulator, counterbalance or pilot-operated check valve, suspended load on an actuator, intensifier, or a pressurized reservoir.

NOTE: *The pilot-operated relief valve can be removed from the system for this test procedure.*

Step 5.   Connect MicroLeak analyzer (1) in series with the inlet port (pump port) of the pilot operated relief valve.

Step 6.   Install pressure gauge (2) in parallel with the connector at the outlet port of MicroLeak analyzer (1).

NOTE: *If the relief valve is not already set at a pre-determined pressure, turn the adjusting screw "in" a few turns and raise the pressure setting for the test procedure.*

Step 7.   Gradually pressurize the inlet port of the relief valve with MicroLeak analyzer (1). Stop and maintain pressure when the value of the pilot-operated relief valve setting is reached.

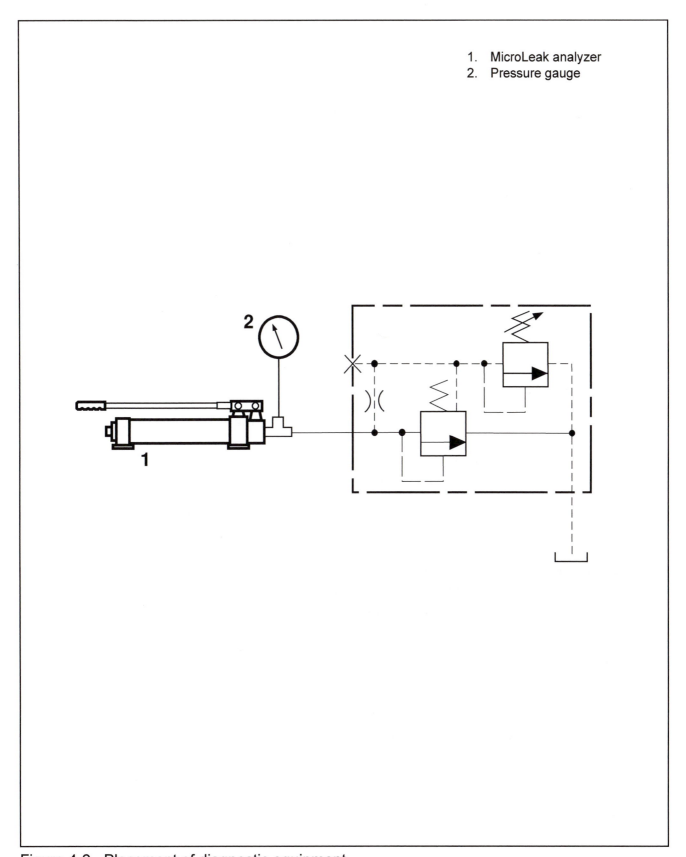

1. MicroLeak analyzer
2. Pressure gauge

Figure 4-2 Placement of diagnostic equipment

NOTE: *Oil should now be trapped between* MicroLeak analyzer *(1) and the inlet port of the pilot-operated relief valve.*

Step 8. Observe the leakage relative to the rate at which the pressure, indicated on pressure gauge (2), decreases. Refer to "Analyzing the Test Results," to determine the condition of the valve.

Step 9. Open the pressure release valve on the MicroLeak analyzer (1) and release the pressure between MicroLeak analyzer (1) and the relief valve.

ANALYZING THE TEST RESULTS

1. **Diagnostic Observation:** The pilot-operated relief valve shows no signs of cross-port leakage until "cracking" pressure is reached.

    Diagnosis: The pressure relief valve appears to be in satisfactory operating condition.

    If necessary, refer to the valve manufacturer's specifications for maximum recommended leakage rates.

2. **Diagnostic Observation:** The pressure attempts to increase as the MicroLeak analyzer is activated but rapidly declines. A steady stream of fluid flows from the outlet port of the valve, well below the pressure setting of the valve (below normal cracking pressure).

    Diagnosis: There is evidence of cross-port leakage which could be caused by internal damage, wear, and/or, contamination between the spool and seat assemblies.

    If necessary, refer to the valve manufacturer's specifications for maximum recommended leakage rates.

3. **Diagnostic Observation:** The MicroLeak analyzer fails to pressurize the inlet port of the valve while a steady stream of oil pours from the outlet port. Profuse cross-port leakage which occurs well below the setting of the valve (below normal cracking pressure).

    Diagnosis: There is evidence of valve failure which could be caused by internal damage, wear, plugged control orifice, and/or contamination between the spool and seat assemblies.

    Repair or replace the pilot-operated relief valve.

CAUTION! *Component repair procedures must be conducted by trained, authorized personnel. Incorrect parts, or parts which are improperly installed, can cause catastrophic component malfunction.*

WARNING! *Under no circumstances should a relief valve be adjusted with a hand-pump. Adjusting a relief valve in this manner could cause serious valve override. Refer to the manufacturer's recommendations for proper relief valve adjustment procedures.*

Direct-Operated Relief Valve
In-Circuit Test Procedure

What will this test procedure accomplish?

A direct-operated relief valve is a normally closed valve (flow path through the valve is blocked in the "inactive" position). It consists of a moveable, steel poppet located within a cast iron or aluminum body. The poppet is held against a seat in the valve body by an adjustable spring force. The seats of the poppet and body are ground to form a leak-tight seal to prevent leakage when the valve is closed.

Under normal operating conditions the poppet opens and closes frequently to dissipate pressure spikes caused by: initial actuator acceleration, changing actuator direction, actuator deceleration, and/or stalling the actuator.

The constant opening and closing of the poppet causes a seat "hammering" effect which can lead to seat deterioration. Seat deterioration could result in cross-port leakage.

Cross-port leakage can also be caused by fine particles of contamination, in a high velocity stream of oil, striking the poppet seat. This could eventually lead to seat erosion.

Premature leakage across the valve, during operation, will usually lead to a moderate to high operating temperature and/or a decline in actuator speed. Clearance-related problems are generally viscosity sensitive, i.e. increased leakage will be less obvious when the fluid temperature is low and will increase as the fluid temperature increases.

The most common symptoms of problems associated with pressure relief valves are:
    a. Actuator fails to respond when the directional control valve is activated.
    b. A loss of actuator speed which is "viscosity sensitive." As the fluid temperature increases, actuator speed progressively decreases.
    c. Moderate to high increase in the operating temperature of the fluid.
    d. Actuator slows as the load (pressure) increases.

This test procedure will determine the amount of leakage across the ports of a direct-operated relief valve.

-WARNING-   **Do not work on or around hydraulic systems without wearing safety glasses which conform to ANSI Z87.1-1989 standard.**

YOU WILL NEED THE FOLLOWING DIAGNOSTIC EQUIPMENT
TO CONDUCT THIS TEST PROCEDURE

CAUTION! *The pressure and flow ratings of the diagnostic equipment which will be used to conduct this test procedure must be equal to, or greater than, the pressure and flow ratings of the system being tested. Refer to the system schematic for recommended pressure and flow data.*

---

1. Pressure gauge
2. Pyrometer
3. Needle valve
4. Flow meter

PREPARATION

To conduct this test safely and accurately, refer to Figure 4-3 for correct placement of diagnostic equipment.

TEST PROCEDURE

Step 1. Shut the prime mover off.

Step 2. Lock the electrical system out or tag the keylock switch.

Step 3. Observe the system pressure gauge. Release any residual pressure trapped in the system by an accumulator, counterbalance or pilot-operated check valve, suspended load on an actuator, intensifier, or a pressurized reservoir.

Step 4. Install needle valve (3) in parallel with the connector at the outlet port of the pump. Connect a length of hose to the outlet port of needle valve (3) and route it directly to the reservoir. Fasten securely. The hose must be suitable for the pressure and flow ratings of the system.

NOTE: *It may be necessary to remove the reservoir inspection cover to provide access for the return line from the needle valve. If the reservoir cover is removed for this purpose, fasten the return line securely into the reservoir.*

Step 5. Install pressure gauge (1) in parallel with the connector at the outlet port of the pump.

Step 6. Install flow meter (4) in series with the transmission line at the outlet port of the relief valve.

Step 7. Start the prime mover. Inspect the diagnostic equipment connectors for leaks.

Step 8. Allow the system to warm up to approximately 130º F (54º C). Observe pyrometer (2).

Step 9. Close needle valve (3) fully (turn clockwise).

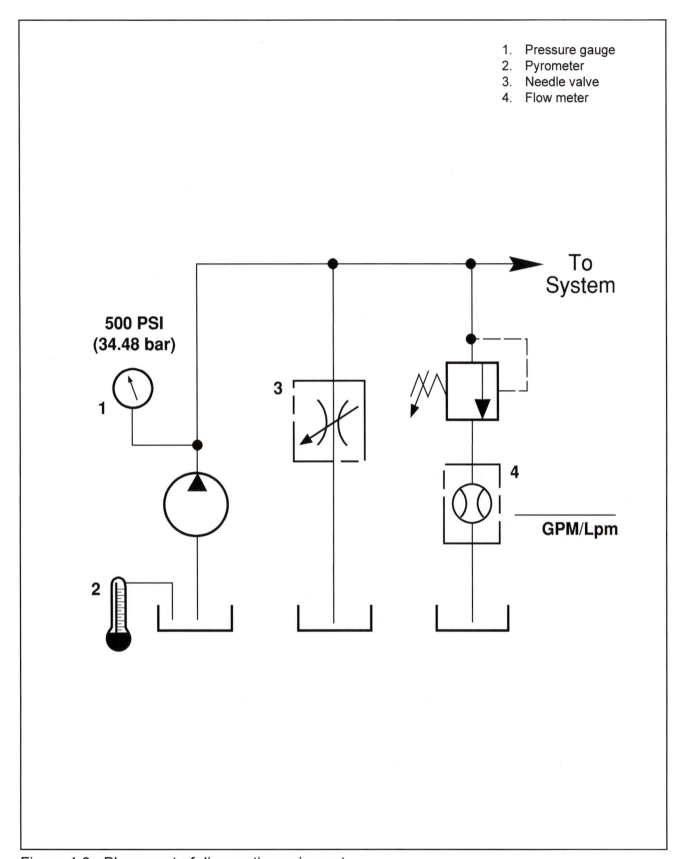

1. Pressure gauge
2. Pyrometer
3. Needle valve
4. Flow meter

**500 PSI
(34.48 bar)**

GPM/Lpm

To
System

Figure 4-3   Placement of diagnostic equipment

Step 10.   Stall (dead-head) an actuator and verify that the relief valve is adjusted to manufacturer's specifications. Observe pressure gauge (1).

NOTE: *If the relief valve does not respond to pressure adjustment, either the pump is not providing flow to the valve, or the relief valve does not operate.*

To determine the root-cause conduct a pump flow test procedure. If the pump test is positive, replace the relief valve.

CAUTION! *Certain variable speed prime mover applications (primarily mobile) require that the prime mover operate at a pre-determined speed when relief valve adjustment is made. Refer to the equipment service data for specific relief valve adjustment procedures.*

Step 11.   Open needle valve (3) fully (turn counter-clockwise).

Step 12.   Stall (dead-head) an actuator. Hold it in the "stalled" position.

NOTE: *Full pump flow should be passing through needle valve (3) and flowing to the reservoir.*

Step 13.   While observing flow meter (4), gradually close needle valve (3) (turn clockwise) and increase the pressure in 100 PSI (6.9 bar) increments. Observe pressure gauge (1). Conduct this test procedure up to the pressure setting of the relief valve.

NOTE: *There should be no sign of oil escaping from the drain port of the relief valve at low pressure. Conduct the test procedure until the "cracking" pressure or pressure setting of the relief valve is reached.*

Step 14.   If the relief valve operates normally, adjust it to the pressure recommended on the circuit diagram.

Step 15.   Shut the prime mover off and analyze the test results.

Step 16.   At the conclusion of this test procedure, remove the diagnostic equipment. Reconnect the transmission lines and tighten the connectors securely.

Step 17.   Start the prime mover and inspect the connectors for leaks.

NOTE: *It is normal for a direct-operated pressure relief valve to "crack" early (oil begins to exhaust from the valve before the full-flow pressure setting is reached). This is generally a functional characteristic of a direct-operated relief valve.*

ANALYZING THE TEST RESULTS

---

1.   **Diagnostic Observation:** The direct-operated relief valve shows no signs of cross-port leakage until "cracking" pressure is reached.

Diagnosis: The pressure relief valve appears to be in satisfactory operating condition.

If necessary, refer to the valve manufacturer's specifications for maximum recommended leakage rates.

2. **Diagnostic Observation:** There is steady cross-port leakage at low pressure, which occurs well below the setting of the valve (below normal cracking pressure). Leakage appears to increase as system pressure increases.

    Diagnosis: There is evidence of cross-port leakage which could be caused by internal damage, wear, and/or, contamination between the spool and seat assembly.

    If necessary, refer to the valve manufacturer's specifications for maximum recommended leakage rates.

3. **Diagnostic Observation:** Profuse cross-port leakage at low pressure which occurs well below the pressure setting of the valve (below normal cracking pressure). Leakage increases significantly as system pressure increases.

    Diagnosis: There is evidence of valve failure which could be caused by internal damage, wear, and/or contamination between the spool and seat assembly.

    Repair or replace the direct-operated relief valve.

CAUTION! *Component repair procedures must be conducted by trained, authorized personnel. Incorrect parts, or parts which are improperly installed, can cause catastrophic component malfunction.*

---

notes

Direct-Operated Relief Valve
Direct-Access Test Procedure

What will this test procedure accomplish?

A direct-operated relief valve is a normally closed valve (flow path through the valve is blocked in the "inactive" position). It consists of a moveable, steel poppet located within a cast iron or aluminum body. The poppet is held against a seat in the valve body by an adjustable spring force. The seats of the poppet and body are ground to form a leak-tight seal to prevent leakage when the valve is closed.

Under normal operating conditions the poppet opens and closes frequently to dissipate pressure spikes caused by: initial actuator acceleration, changing actuator direction, actuator deceleration, and/or stalling the actuator.

The constant opening and closing of the poppet causes a seat "hammering" effect which can lead to seat deterioration. Seat deterioration could result in cross-port leakage.

Cross-port leakage can also be caused by fine particles of contamination, in a high velocity stream of oil, striking the poppet seat. This could eventually lead to seat erosion.

Premature leakage across the valve, during operation, will usually lead to a moderate to high operating temperature and/or a decline in actuator speed. Clearance related problems are generally viscosity sensitive, i.e. increased leakage will be less obvious when the fluid temperature is low, and will increase as the fluid temperature increases.

The most common symptoms of problems associated with pressure relief valves are:
   a. Actuator fails to respond when the directional control valve is activated.
   b. A loss of actuator speed which is "viscosity sensitive." As the fluid temperature increases, actuator speed progressively decreases.
   c. Moderate to high increase in the operating temperature of the fluid.
   d. Actuator slows as the load (pressure) increases.

This test procedure will determine the amount of leakage across the ports of a direct-operated relief valve.

 **-WARNING-** | Do not work on or around hydraulic systems without wearing safety glasses which conform to ANSI Z87.1-1989 standard.

YOU WILL NEED THE FOLLOWING DIAGNOSTIC EQUIPMENT
TO CONDUCT THIS TEST PROCEDURE

CAUTION! *The pressure and flow ratings of the diagnostic equipment which will be used to conduct this test procedure must be equal to, or greater than, the pressure and flow ratings of the system being tested. Refer to the system schematic for recommended pressure and flow data.*

1.   MicroLeak analyzer   2.   Pressure gauge

PREPARATION

To conduct this test safely and accurately, refer to Figure 4-4 for correct placement of diagnostic equipment.

TEST PROCEDURE

Step 1.   Start the prime mover. Verify that the direct-operated relief valve is adjusted to the manufacturer's specification.

NOTE: *If the relief valve does not respond to pressure adjustment, either the pump is not providing flow to the valve, or the relief valve does not operate.*

To determine the root-cause conduct a pump flow test procedure. If the pump test is positive, replace the relief valve.

Step 2.   Shut the prime mover off.

Step 3.   Lock the electrical system out or tag the keylock switch.

Step 4.   Observe the system pressure gauge. Release any residual pressure trapped in the system by an accumulator, counterbalance or pilot-operated check valve, suspended load on an actuator, intensifier, or a pressurized reservoir.

NOTE: *The relief valve can be removed from the system for this test procedure.*

Step 5.   Connect MicroLeak analyzer (1) in series with the inlet port (pump port) of the relief valve.

Step 6.   Install pressure gauge (2) in parallel with the connector at the outlet port of MicroLeak analyzer (1).

NOTE: *If the relief valve is not already set at a pre-determined pressure, turn the adjusting screw "in" a few turns and raise the pressure setting for the test procedure.*

Step 7.   Gradually pressurize the inlet port of the relief valve with MicroLeak analyzer (1). Stop and maintain pressure when the value of the direct-operated relief valve setting is reached.

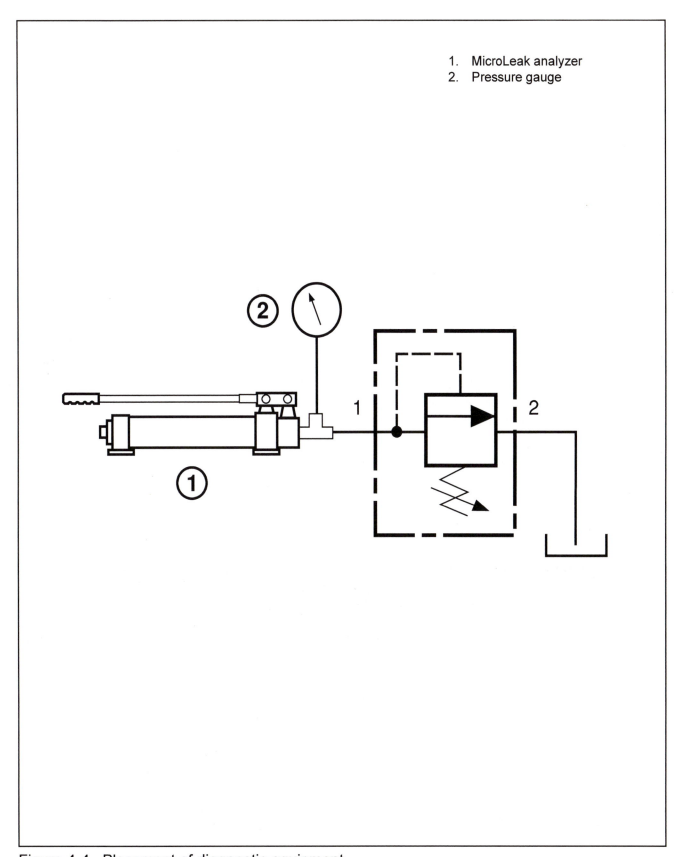

1. MicroLeak analyzer
2. Pressure gauge

Figure 4-4   Placement of diagnostic equipment

NOTE: *Oil should now be trapped between MicroLeak analyzer (1) and the inlet port of the relief valve.*

Step 8. Observe the leakage relative to the rate at which the pressure, indicated on pressure gauge (2), decreases. Refer to "Analyzing the Test Results," to determine the condition of the valve.

Step 9. Open the pressure release valve on the MicroLeak analyzer (1). Release the pressure between MicroLeak analyzer (1) and the relief valve.

ANALYZING THE TEST RESULTS

1. **Diagnostic Observation:** The direct-operated relief valve shows no signs of cross-port leakage until "cracking" pressure is reached.

    Diagnosis: The relief valve appears to be in satisfactory operating condition.

    If necessary, refer to the valve manufacturer's specifications for maximum recommended leakage rates.

2. **Diagnostic Observation:** The pressure attempts to increase as the MicroLeak analyzer is activated but rapidly declines. A steady stream of fluid flows from the outlet port of the valve, well below the pressure setting of the valve (below normal cracking pressure).

    Diagnosis: There is evidence of cross-port leakage which could be caused by internal damage, wear, and/or, contamination between the spool and seat assembly.

    If necessary, refer to the valve manufacturer's specifications for maximum recommended leakage rates.

3. **Diagnostic Observation:** The MicroLeak analyzer fails to pressurize the inlet port of the valve while a steady stream of oil pours from the outlet port. Profuse cross-port leakage which occurs well below the setting of the valve (below normal cracking pressure).

    Diagnosis: There is evidence of valve failure which could be caused by internal damage, wear, and/or contamination between the spool and seat assembly.

    Repair or replace the direct-operated relief valve.

WARNING! *Under no circumstances should a relief valve be adjusted with a hand-pump. Adjusting a relief valve in this manner could cause serious valve override. Refer to the manufacturer's recommendations for proper relief valve adjustment procedures.*

CAUTION! *Component repair procedures must be conducted by trained, authorized personnel. Incorrect parts, or parts which are improperly installed, can cause catastrophic component malfunction.*

Series Restriction Pressure Test Procedure

What will this test procedure accomplish?

A pressure relief valve is a normally closed valve (flow path through the valve blocked in the "inactive" position). It consists of a moveable, steel poppet located within a cast iron or aluminum body. The poppet is held against a seat in the valve body by an adjustable spring force. The seats of the poppet and body are ground to form a leak-tight connection to prevent leakage when the valve is closed.

This test procedure addresses the end of the spool opposite the fluid - the side which interfaces with the mechanical spring force.

Leakage which occurs across the poppet in the relief valve will accumulate above the poppet (in the spring cavity) and prevent the spool from shifting. This condition is commonly referred to as "hydraulic-lock." Since the relief valve is a normally-closed valve, the valve will malfunction (lock) to the closed position.

To prevent hydraulic-lock, the spring cavity is drained through an internal drain passage which is connected to the outlet port of the valve.

The outlet or reservoir port of a relief valve is generally connected directly to the reservoir. The pressure in the reservoir return-line is usually nominal. Any pressure increase in the reservoir return-line will be sensed in the spring cavity above the poppet. The increased pressure will act against the poppet with the spring force, increasing the valve's pressure setting proportionately.

A sudden increase in pressure relief valve setting could be caused by added resistance in series with the outlet port of the valve rather than by a mechanical malfunction.

Restriction in return lines caused by return-line filters, coolers, and/or manifolds will indirectly increase pressure relief valve settings. Damage to the inside or outside diameter of the return-line could increase return-line resistance and proportionately increase relief valve setting.

Do not decrease relief valve adjustment to compensate for return-line restriction. Reducing the pressure relief valve setting to compensate for return-line resistance could indirectly reduce the force or torque output of the system.

The most common symptoms of problems associated with resistance in series are:
    a. A sudden increase in pressure relief valve setting.
    b. A noticeable reduction in actuator force or torque output.
    c. Seal failure or damage to an upstream component (power beyond).
    d. Erratic pressure relief valve operation.
    e. Total relief valve malfunction (hydraulic-lock to "closed" position).

This test procedure will:
- a. Determine the pressure at the outlet port of a pressure relief valve.
- b. Determine the root-cause of excessive return-line resistance.

 **-WARNING-** **Do not work on or around hydraulic systems without wearing safety glasses which conform to ANSI Z87.1-1989 standard.**

YOU WILL NEED THE FOLLOWING DIAGNOSTIC EQUIPMENT TO CONDUCT THIS TEST PROCEDURE

CAUTION! *The pressure and flow ratings of the diagnostic equipment which will be used to conduct this test procedure must be equal to, or greater than, the pressure and flow ratings of the system being tested. Refer to the system schematic for recommended pressure and flow data.*

1. Pressure gauge (system pressure)	3. Pyrometer
2. Pressure gauge (system pressure)	

PREPARATION

To conduct this test safely and accurately, refer to Figure 4-5 for correct placement of diagnostic equipment.

TEST PROCEDURE

Step 1.   Shut the prime mover off.

Step 2.   Lock the electrical system out or tag the keylock switch.

Step 3.   Observe the system pressure gauge. Release any residual pressure trapped in the system by an accumulator, counterbalance or pilot-operated check valve, suspended load on an actuator, intensifier, or a pressurized reservoir.

Step 4.   Install pressure gauge (1) in parallel with the connector at the inlet port of the relief valve.

Step 5.   Install pressure gauge (2) in parallel with the connector at the outlet port of the relief valve.

Step 6.   Start the prime mover. Inspect the diagnostic equipment connectors for leaks.

Step 7.   Allow the system to warm up to approximately 130ºF (54ºC). Observe pyrometer (3).

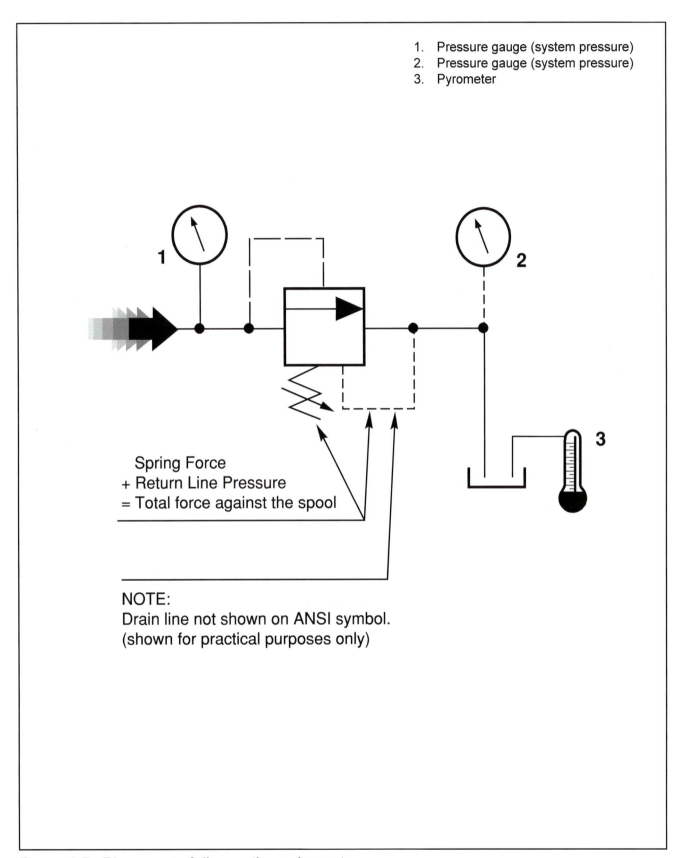

1. Pressure gauge (system pressure)
2. Pressure gauge (system pressure)
3. Pyrometer

Spring Force
+ Return Line Pressure
= Total force against the spool

NOTE:
Drain line not shown on ANSI symbol.
(shown for practical purposes only)

Figure 4-5   Placement of diagnostic equipment

Step 8.   Stall an actuator. Record the system relief valve setting indicated on pressure gauge (1).

NOTE: *If pressure gauge (1) indicates a pressure which is higher than the pressure recommended by the manufacturer, do not reduce the relief valve setting until the test procedure has been completed.*

Step 9.   Operate the machine through a number of normal load cycles.

Step 10.   Record the highest pressure in the return-line indicated on pressure gauge (2) during the load test cycle.

Step 11.   If there is an abnormal variation in return-line pressure, install pressure gauge (2) in the next downstream connector and repeat the test procedure.

NOTE: *Progressively move the pressure gauge through the return-line to determine the source of the restriction which may be causing the abnormal variation in return-line pressure.*

Step 12.   Remove the restriction and repeat the test procedure.

Step 13.   Adjust the pressure relief valve to the recommended pressure. Refer to the equipment manufacturer's specifications for the recommended pressure relief valve setting.

Step 14.   At the conclusion of this test procedure, remove the diagnostic equipment. Reconnect the transmission lines and tighten the connectors securely.

Step 15.   Start the prime mover and inspect the connectors for leaks.

CAUTION! *Certain variable speed prime mover applications (primarily mobile) require that the prime mover operate at a pre-determined speed when relief valve adjustment is made. Refer to the equipment service data for specific relief valve adjustment procedures.*

ANALYZING THE TEST RESULTS

1.   **Diagnostic Observation:** Pressure gauge (1) indicates recommended pressure relief valve setting. Pressure gauge (2) indicates nominal pressure.

   Diagnosis: The pressure relief valve appears to be in satisfactory operating condition.

2.   **Diagnostic Observation:** Pressure gauge (1) indicates a pressure which is higher than the recommended pressure relief valve setting.

Pressure gauge (2) indicates a pressure which appears to be the difference between what the pressure relief valve should be adjusted to and what pressure gauge (1) indicates.

Diagnosis: There is evidence of a restriction or resistance in the pressure relief valve's drain-line.

NOTE: *This condition typically occurs when a directional control valve with integral relief valve is connected in series. Refer to the manufacturer's specifications for "power-beyond" directional control valve applications.*

If there is resistance at the outlet port of a directional control valve, which is equipped with an integral pressure relief valve, it will have a direct and adverse affect on the force or torque output of the actuator.

If it is not possible to reduce the resistance at the outlet port of a relief valve, it can be substituted for a relief valve with an external spool drain (sequence valve). This will prevent drain-line resistance and/or pressure surges from affecting the pressure setting of the valve.

WARNING! *If a component in series with the outlet port of the pressure relief valve can completely block the flow path, it could cause unpredictable catastrophic failure. This type of failure could cause injury or death to persons working on or around the machine.*

---

notes

Sequence Valve
External Drain Back-Pressure Test Procedure

What will this test procedure accomplish?

A sequence valve is a normally closed (flow path closed in the "inactive" position) valve. It is generally used in a hydraulic system to establish sequential actuator operation.

A sequence valve consists of a moveable steel poppet located within a cast iron or aluminum body. The poppet is held against a seat in the valve body by an adjustable spring force. It is normal for a small amount of fluid to leak across the poppet into the bias spring cavity.

If fluid accumulates in the spring cavity above the poppet, it will cause the poppet to "hydraulically-lock". Since the sequence valve is a normally-closed valve, it will hydraulically-lock in the closed position.

Although a sequence valve and a pressure relief valve are identical in terms of their operation, there is one distinct difference between the two - the outlet port of a pressure relief valve always returns to the reservoir at nominal pressure, while the outlet port of a sequence valve is in series with an actuator.

Consequently, it will encounter system pressure at both the inlet and outlet ports simultaneously. For this reason, a sequence valve **must** be externally drained to prevent pressure from building up in the spring cavity on the top of the spool.

A pressure relief valve is affected by return (drain) line resistance because it has an internal drain passage which drains the spring cavity to the outlet port of the valve. Since the sequence valve is positioned in series with an actuator, it is necessary to drain the top of the poppet or spring cavity externally to prevent system pressure from acting against the top of the poppet with the bias spring force.

The most common symptoms of problems associated with excessive, external drain-line pressure are:
    a. Actuators do not operate sequentially.
    b. Actuators in series with the sequence valve:
        1. Fail to respond (lock up) when the directional control valve is actuated.
        2. Fail to allow the actuator to move in the opposite direction (reverse-flow).
        3. Operate erratically.

This test procedure will determine:
    a. The amount of back-pressure in the external drain-line of a sequence valve.
    b. If back-pressure is preventing the valve from operating normally.

 **-WARNING-** | Do not work on or around hydraulic systems without wearing safety glasses which conform to ANSI Z87.1-1989 standard.

YOU WILL NEED THE FOLLOWING DIAGNOSTIC EQUIPMENT TO CONDUCT THIS TEST PROCEDURE

CAUTION! *The pressure and flow ratings of the diagnostic equipment which will be used to conduct this test procedure must be equal to, or greater than, the pressure and flow ratings of the system being tested. Refer to the system schematic for recommended pressure and flow data.*

1. Pressure gauge (system pressure)
2. Pressure gauge (low pressure - <100 PSI (6.9 bar))
3. Pyrometer

PREPARATION

To conduct this test safely and accurately, refer to Figure 4-6 for correct placement of diagnostic equipment.

TEST PROCEDURE

Step 1. Shut the prime mover off.

Step 2. Lock the electrical system out or tag the keylock switch.

Step 3. Observe the system pressure gauge. Release any residual pressure trapped in the system by an accumulator, counterbalance or pilot-operated check valve, suspended load on an actuator, intensifier, or a pressurized reservoir.

Step 4. Install pressure gauge (1) in parallel with the connector at the inlet port of the sequence valve.

Step 5. Install pressure gauge (2) in parallel with the connector at the external drain port of the sequence valve.

Step 6. Start the prime mover. Inspect the diagnostic equipment connectors for leaks.

Step 7. Allow the system to warm up to approximately 130ºF. (54ºC.). Observe pyrometer (2).

Step 8. Operate the machine through a series of normal "full-load" cycles, and record the highest pressures registered on pressure gauges (1) and (2).

Step 9. Shut the prime mover off and analyze the test results.

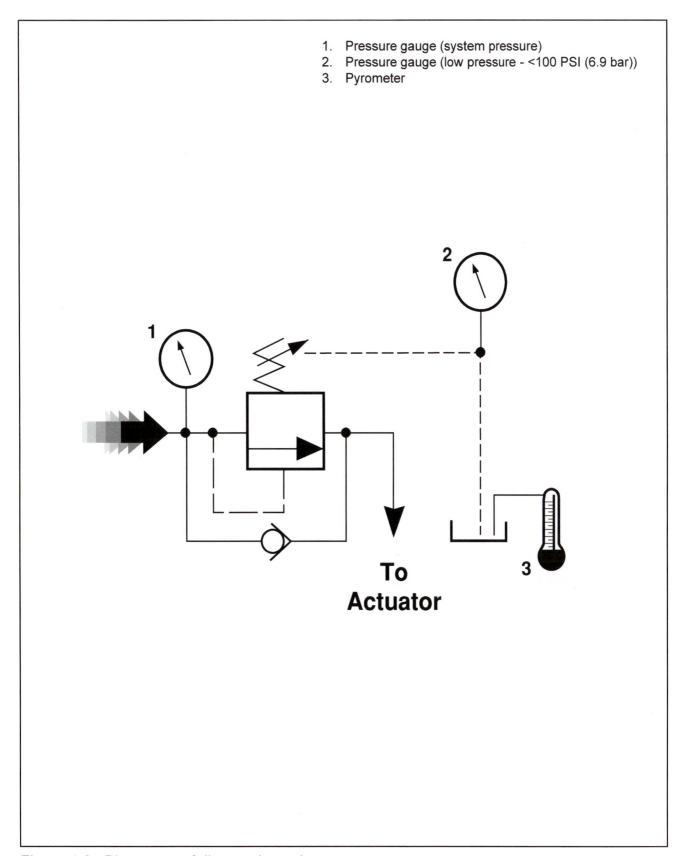

1. Pressure gauge (system pressure)
2. Pressure gauge (low pressure - <100 PSI (6.9 bar))
3. Pyrometer

To
Actuator

Figure 4-6   Placement of diagnostic equipment

Step 10. At the conclusion of this test procedure, remove the diagnostic equipment. Reconnect the transmission lines and tighten the connectors securely.

Step 11. Start the prime mover and inspect the connectors for leaks.

ANALYZING THE TEST RESULTS

1. **Diagnostic Observation:** Pressure gauge (2) indicates nominal pressure. Pressure gauge (1) indicates the setting of the sequence valve.

   Diagnosis: The sequence valve appears to be in satisfactory operating condition.

2. **Diagnostic Observation:** Pressure gauge (2) indicates pressure fluctuations as the actuator operates which appear to cause erratic actuator operation.

   Pressure gauge (1) indicates pressure fluctuations which appear to equal the pressure setting of the valve plus the pressure fluctuations indicated on pressure gauge (2).

   Diagnosis: There is an indication of excessive resistance and/or pressure fluctuations in the sequence valve's external drain-line.

   This problem is typically caused by resistance in multiple-connection return-line manifolds and/or return-line filters and heat exchangers.

NOTE: *Components (typically linear actuators) operating in parallel with sequence circuits have the tendency to cause flow surges. The flow surges can be rather significant relative to the cylinder's rod-to-bore ratio.*

   Flow surges may cause pressure transients in the drain-lines of pressure sensitive valves causing them to operate erratically.

   To prevent return-line resistance from affecting the operation of pressure sensitive, externally drained valves, route external drain-lines directly back to the reservoir independently.

3. **Diagnostic Observation:** Pressure gauge (2) indicates system relief valve pressure. Pressure gauge (1) indicates system relief valve pressure.

   Diagnosis: There is evidence that the spool in the sequence valve is "hydraulically-locked" in the closed position.

   In this case the spool is most likely locked because of a plugged external drain-line. This is determined by the pressure indicated on pressure gauge (1). If internal leakag restricted from leaving the sequence valve, it will cause the spool to "hydraulically-lock" in the closed position.

   Spool jamming can also be caused by contamination in the fluid, valve distortion from sub-standard mounting procedures and over-tightened, tapered pipe connectors. However, none of these conditions are likely to cause the pressure in the external drain-line to reach the system's main pressure relief valve setting.

Sequence Valve
Direct-Access Test Procedure

What will this test procedure accomplish?

A sequence valve is a normally-closed (flow path closed in the "inactive" position) valve. It is generally used in a hydraulic system to establish sequential actuator operation.

A sequence valve consists of a moveable steel poppet located within a cast iron or aluminum body. The poppet is held against a seat in the valve body by an adjustable spring force. It is normal for a small amount of fluid to leak across the poppet into the bias spring cavity.

If fluid accumulates in the spring cavity above the poppet, it will cause the poppet to "hydraulically-lock." Since the sequence valve is a normally-closed valve, it will hydraulically-lock in the closed position.

Although a sequence valve and a pressure relief valve are identical in terms of their operation, there is one distinct difference between the two-- the outlet port of a pressure relief valve always returns to the reservoir at nominal pressure, while the outlet port of a sequence valve is in series with an actuator.

Consequently, it will encounter system pressure at both the inlet and outlet ports simultaneously. For this reason, a sequence valve **must** be externally drained to prevent pressure from building up in the spring cavity on the top of the spool.

A pressure relief valve is affected by return (drain) line resistance because it has an internal drain passage which drains the spring cavity to the outlet port of the valve. Since the sequence valve is positioned in series with an actuator, it is necessary to drain the top of the poppet or spring cavity externally to prevent system pressure from acting against the top of the poppet with the bias spring force.

The most common symptoms of problems associated with sequence valves are:
- a. Actuators do not operate sequentially.
- b. Actuators in series with the sequence valve:
    1. Fail to respond (lock up) when the directional control valve is actuated.
    2. Fail to allow the actuator to move in the opposite direction (reverse-flow).
    3. Operate erratically.

This test procedure will determine the amount of leakage across the ports of a sequence valve.

 **-WARNING-** | **Do not work on or around hydraulic systems without wearing safety glasses which conform to ANSI Z87.1-1989 standard.**

YOU WILL NEED THE FOLLOWING DIAGNOSTIC EQUIPMENT
TO CONDUCT THIS TEST PROCEDURE

CAUTION! *The pressure and flow ratings of the diagnostic equipment which will be used to conduct this test procedure must be equal to, or greater than, the pressure and flow ratings of the system being tested. Refer to the system schematic for recommended pressure and flow data.*

---

1.   MicroLeak analyzer   2.   Pressure gauge

PREPARATION

To conduct this test safely and accurately, refer to Figure 4-7 for correct placement of diagnostic equipment.

TEST PROCEDURE

Step 1.   Start the prime mover and operate the machine through a normal load cycle. Verify that the sequence valve is adjusted to the recommended pressure.

Step 2.   Shut the prime mover off.

Step 3.   Lock the electrical system out or tag the keylock switch.

Step 4.   Observe the system pressure gauge. Release any residual pressure trapped in the system by an accumulator, counterbalance or pilot-operated check valve, suspended load on an actuator, intensifier, or a pressurized reservoir.

NOTE: *The sequence valve can be removed from the system for this test procedure. However, if it is removed, the external drain-line must remain open during the test procedure.*

Step 5.   Disconnect the transmission lines from the inlet, outlet, and drain ports of the sequence valve.

Step 6.   Connect MicroLeak analyzer (1) in series with the transmission line at the inlet port of the sequence valve.

Step 7.   Install pressure gauge (2) in parallel with the connector at the inlet port of the sequence valve.

Step 8.   Gradually pressurize the inlet port of the sequence valve with MicroLeak analyzer (1). Stop and maintain pressure when the value of the sequence valve setting is reached.

NOTE: *Oil should now be trapped between MicroLeak analyzer (1) and the inlet port of the sequence valve.*

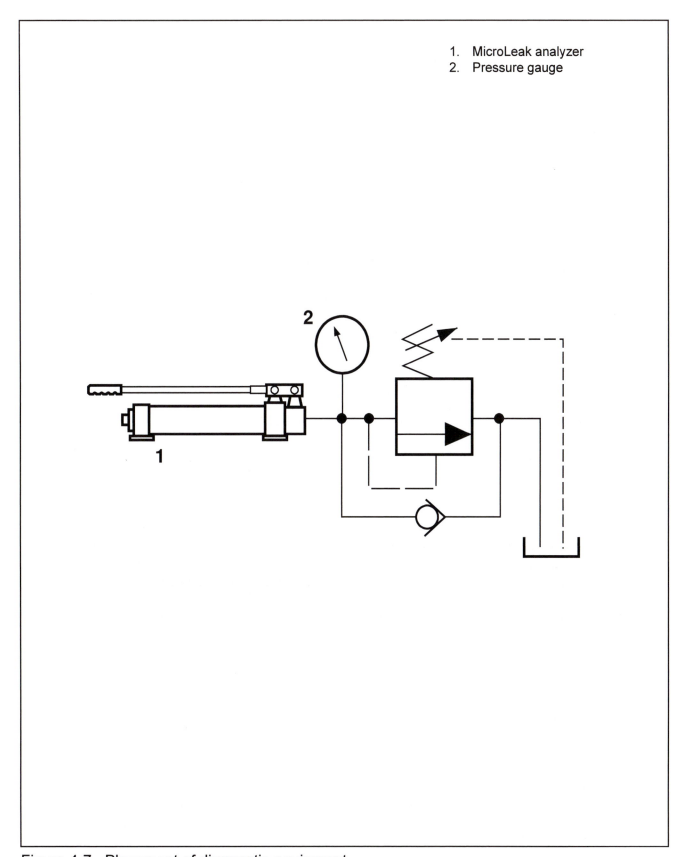

1. MicroLeak analyzer
2. Pressure gauge

Figure 4-7   Placement of diagnostic equipment

Step 9.   Compare the leakage at the outlet port of the valve with the rate at which the pressure indicated on pressure gauge (2) decreases.

Step 10.   Open the pressure release valve on MicroLeak analyzer (1). Release the pressure between MicroLeak analyzer (1) and the sequence valve.

Step 11.   Analyze the test results.

> ## ANALYZING THE TEST RESULTS

1.   **Diagnostic Observation:** The sequence valve shows no signs of cross-port leakage until "cracking" pressure is reached.

Diagnosis: The sequence valve appears to be in satisfactory operating condition.

If necessary, refer to the valve manufacturer's specifications for maximum recommended leakage rates.

2.   **Diagnostic Observation:** Pressure attempts to increase as the portable Mini-Tester is activated but rapidly declines. A steady stream of fluid flows from the outlet port of the valve, well below the pressure setting of the valve (below normal cracking pressure).

Diagnosis: There is evidence of cross-port leakage which could be caused by internal damage, reverse-flow check valve malfunction, wear, and/or contamination between the spool and seat assembly.

If necessary, refer to the valve manufacturer's specifications for maximum recommended leakage rates.

3.   **Diagnostic Observation:** The MicroLeak analyzer fails to pressurize the inlet port of the sequence valve while a steady stream of oil pours from the outlet port. Profuse cross-port leakage which occurs well below the setting of the valve (below normal cracking pressure).

Diagnosis: There is evidence of valve failure which could be caused by internal damage, reverse-flow check valve malfunction, wear, and/or contamination between the spool and seat assembly.

Repair or replace the sequence valve.

WARNING! *Under no circumstances should a relief valve be adjusted with a hand-pump. Adjusting a relief valve in this manner could cause serious valve override. Refer to the manufacturer's recommendations for proper relief valve adjustment procedures.*

CAUTION! *Component repair procedures must be conducted by trained, authorized personnel. Incorrect parts, or parts which are improperly installed, can cause catastrophic component malfunction.*

Pressure Reducing Valve
External Drain Back-Pressure Test Procedure

What will this test procedure accomplish?

A pressure reducing valve is a normally-open (flow path open in the "inactive" position) valve designed to reduce the pressure in a "leg" of a hydraulic system to a lower value than the system's main pressure relief valve setting. It consists of a moveable piston located within a cast iron body, or a cartridge valve assembly. The valve body has a primary (inlet) port and a secondary (reduced pressure) port.

Since a pressure reducing valve is generally positioned in series with a load, it is necessary to drain the bias spring cavity, on the top of the spool, externally. A third port is located in the valve body for this purpose.

Contamination can cause the spool to bind, which could cause the valve to malfunction to the "open" position. Pressure surges in the external drain-line could cause erratic operation. A plugged drain-line will cause internal leakage to accumulate in the bias spring cavity, "hydraulically locking" the spool and preventing it from shifting.

Since a pressure reducing valve is a normally-open valve, it will malfunction to the open position allowing full system pressure (system relief valve pressure setting) at the secondary or reduced pressure port.

Sensitive components downstream of the pressure reducing valve may suffer catastrophic failure should full system pressure reach the secondary port of a pressure reducing valve.

The most common symptoms of problems associated with excessive back-pressure in the external drain-line of a pressure reducing valve are:
    a. Catastrophic failure to components downstream of the pressure reducing valve.
    b. Erratic operation of actuators in series with the pressure reducing valve.
    c. Damage to parts in process due to excessive actuator force or torque output.

This test procedure will determine:
    a. The pressure in the external drain-line.
    b. If there is a restriction in the external drain-line.
    c. If there are pressure surges in the external drain-line.

 **-WARNING-** Do not work on or around hydraulic systems without wearing safety glasses which conform to ANSI Z87.1-1989 standard.

YOU WILL NEED THE FOLLOWING DIAGNOSTIC EQUIPMENT TO CONDUCT THIS TEST PROCEDURE

CAUTION! *The pressure and flow ratings of the diagnostic equipment which will be used to conduct this test procedure must be equal to, or greater than, the pressure and flow ratings of the system being tested. Refer to the system schematic for recommended pressure and flow data.*

1. Pressure gauge (system pressure)
2. Pressure gauge (low pressure)
3. Pressure gauge (system pressure)
4. Pyrometer

PREPARATION

To conduct this test safely and accurately, refer to Figure 4-8 for correct placement of diagnostic equipment.
To record the test data, make a copy of the test worksheet on page 4-41 (Figure 4-10).

TEST PROCEDURE

Step 1.   Start the prime mover and verify that the main pressure relief valve is adjusted to the manufacturer's specifications.

Step 2.   Shut the prime mover off.

Step 3.   Lock the electrical system out or tag the keylock switch.

Step 4.   Observe the system pressure gauge. Release any residual pressure trapped in the system by an accumulator, counterbalance or pilot-operated check valve, suspended load on an actuator, intensifier, or a pressurized reservoir.

Step 5.   Install pressure gauge (1) in parallel with the connector at the inlet port of the pressure reducing valve.

Step 6.   Install pressure gauge (2) in parallel with the connector at the drain port of the pressure reducing valve.

Step 7.   Install pressure gauge (3) in parallel with the connector at the outlet port of the pressure reducing valve.

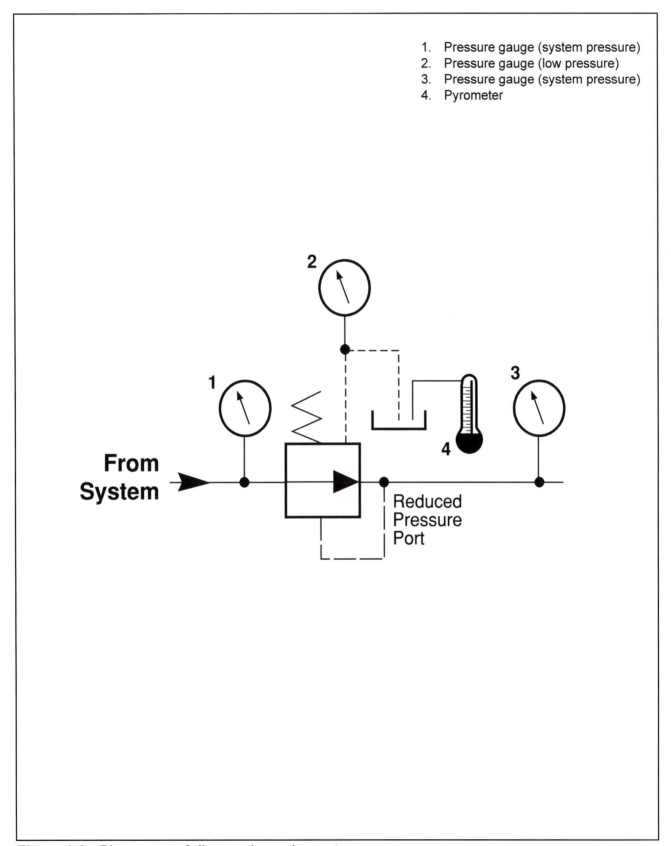

1. Pressure gauge (system pressure)
2. Pressure gauge (low pressure)
3. Pressure gauge (system pressure)
4. Pyrometer

Figure 4-8   Placement of diagnostic equipment

NOTE: *If a component on the "reduced pressure" side of the pressure reducing valve has failed catastrophically, DO NOT attempt to troubleshoot the valve while the secondary port is connected to the system. Disconnect the transmission line from the secondary port and install a suitable pressure gauge in series with the port. The pressure gauge must be rated for the main pressure relief valve setting. Refer to the circuit diagram for recommended main pressure relief valve and pressure reducing valve settings.*

Step 8.   Start the prime mover. Inspect the diagnostic equipment connectors for leaks.

Step 9.   Allow the system to warm up to approximately 130°F (54°C). Observe pyrometer (4).

Step 10.   Operate the machine through a series of normal "full-load" cycles. Record on the test worksheet:
   a. The highest pressure indicated on pressure gauge (1).
   b. The highest pressure indicated on pressure gauge (2).
   c. The highest pressure indicated on pressure gauge (3).

Step 11.   Shut the prime mover off and analyze the test results.

Step 12.   At the conclusion of this test procedure, remove the diagnostic equipment. Reconnect the transmission lines and tighten the connectors securely.

Step 13.   Start the prime mover and inspect the connectors for leaks.

ANALYZING THE TEST RESULTS

1.   **Diagnostic Observation:** Pressure gauge (1) indicates the system's main pressure relief valve setting. Pressure gauge (2) indicates nominal pressure. Presssure gauge (3) indicates the recommended pressure setting of the pressure reducing valve.

   Diagnosis: The pressure reducing valve appears to be in satisfactory operating condition.

2.   **Diagnostic Observation:** Pressure gauge (1) indicates the system's main pressure relief valve setting.

   Pressure gauge (2) indicates pressure fluctuations in the external drain-line as the actuator operates which causes pressure fluctuations in the reduced pressure port of the pressure reducing valve.

   Pressure gauge (3) indicates pressure fluctuations in the reduced pressure port of the valve which appear to track the pressure fluctuations indicated on pressure gauge (2).

   Diagnosis: There is an indication of excessive resistance and/or pressure fluctuations in the pressure reducing valve's external drain-line.

Sample Test Worksheet

Pressure Reducing Valve External Drain Back-Pressure Test Procedure

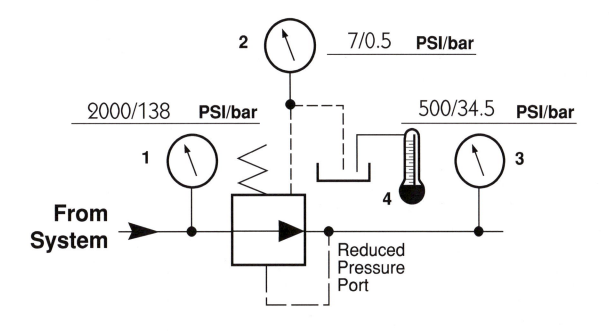

Figure 4-9   Sample Test Worksheet - Pressure reducing valve external drain back-pressure test procedure

This problem is typically caused by resistance in multiple-connection return-line manifolds and/or return-line filters and heat exchangers.

NOTE: *Components (typically linear actuators) operating in parallel with pressure reducing valves have the tendency to cause flow surges. The flow surges can be rather significant relative to the cylinder's rod-to-bore ratio.*

Flow surges may cause pressure transients in the drain-line of a pressure sensitive valve causing it to operate erratically.

To prevent return-line resistance from affecting the operation of a pressure sensitive, externally drained valve, route the external drain-line directly back to the resevoir independently.

3. **Diagnostic Observation:** Pressure gauges (1), (2), and (3), indicate the system's main pressure relief valve setting.

   Diagnosis: There is evidence that the spool is "hydraulically-locked" in the open (inactive) position.

   In this case the spool is most likely locked because of a plugged external drain-line. This is determined by the pressure indicated on pressure gauge (2). If internal leakage is restricted from leaving the pressure reducing valve, it will cause the spool to "hydraulically-lock" in the open position.

   Spool locking can also be caused by contamination in the fluid, valve distortion from sub-standard mounting procedures, and over-tightened, tapered pipe connectors. However, none of these conditions are likely to cause the pressure in the external drain-line to reach the system's main pressure relief valve setting.

Test Worksheet

Pressure Reducing Valve External Drain Back-Pressure Test Procedure

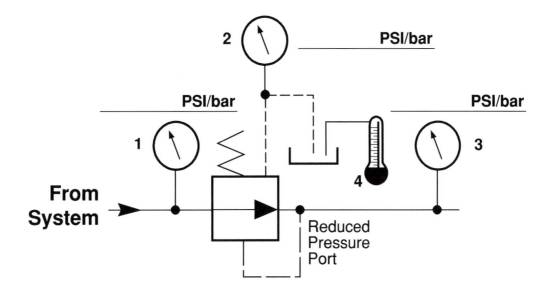

Figure 4-10   Test Worksheet - Pressure reducing valve external drain back-pressure
        test procedure

notes

Pressure Reducing Valve
Direct-Access Pressure Test Procedure

What will this test procedure accomplish?

A pressure reducing valve is a normally open (flow path open in the "inactive" position) valve designed to reduce the pressure in a "leg" of a hydraulic system to a lower value than the system's main pressure relief valve setting. It consists of a moveable piston located within a cast iron body, or a cartridge valve assembly. The valve body has a primary (inlet) port, and a secondary (reduced pressure) port.

Since a pressure reducing valve is generally positioned in series with a load, it is necessary to drain the bias spring cavity, on the top of the spool, externally. A third port is located in the valve body for this purpose.

Contamination can cause the spool to bind, which could cause the valve to malfunction to the "open" position. Pressure surges in the external drain line could cause erratic operation. A plugged drain line will cause internal leakage to accumulate in the bias spring cavity, "hydraulically locking" the spool and preventing it from shifting.

Since a pressure reducing valve is a normally open valve, it will malfunction to the "open" position allowing full system pressure (system pressure relief valve setting) at the secondary or reduced pressure port.

Sensitive components downstream of the pressure reducing valve may suffer catastrophic failure should full system pressure reach the secondary port of a pressure reducing valve.

The most common symptoms of problems associated with pressure reducing valves are:
     a. Catastrophic failure to components downstream of the pressure reducing valve.
     b. Erratic operation of actuators in series with the pressure reducing valve.
     c. Damage to parts in manufacturing operation due to excessive actuator force or torque output.

This test procedure will determine:
     a. If the valve is operating correctly.
     b. What pressure the pressure reducing valve is adjusted to.

-WARNING- | **Do not work on or around hydraulic systems without wearing safety glasses which conform to ANSI Z87.1-1989 standard.**

YOU WILL NEED THE FOLLOWING DIAGNOSTIC EQUIPMENT TO CONDUCT THIS TEST PROCEDURE

CAUTION! *The pressure and flow ratings of the diagnostic equipment which will be used to conduct this test procedure must be equal to, or greater than, the pressure and flow ratings of the system being tested. Refer to the system schematic for recommended pressure and flow data.*

1. Pressure gauge (system pressure)
2. Pressure gauge (system pressure)
3. Pyrometer

PREPARATION

To conduct this test safely and accurately, refer to Figure 4-11 for correct placement of diagnostic equipment.

TEST PROCEDURE

Step 1. Start the prime mover and verify that the main pressure relief valve is adjusted to the manufacturer's specifications.

Step 2. Shut the prime mover off.

Step 3. Lock the electrical system out or tag the keylock switch.

Step 4. Observe the system pressure gauge. Release any residual pressure trapped in the system by an accumulator, counterbalance or pilot-operated check valve, suspended load on an actuator, intensifier, or a pressurized reservoir.

Step 5. Install pressure gauge (1) in parallel with the connector at the primary port (inlet port) of the pressure reducing valve.

Step 6. Install pressure gauge (2) in series with the connector at the secondary (reduced pressure) port of the pressure reducing valve.

Step 7. Start the prime mover. Inspect the diagnostic equipment connectors for leaks.

Step 8. Allow the system to warm up to approximately 130ºF (54ºC). Observe pyrometer (3).

1. Pressure gauge (system pressure)
2. Pressure gauge (system pressure)
3. Pyrometer

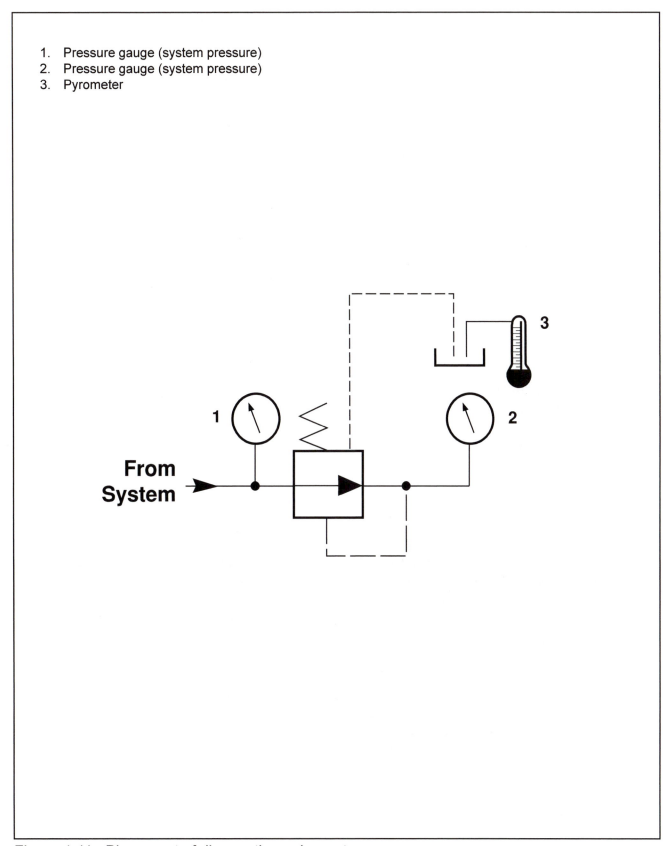

Figure 4-11 Placement of diagnostic equipment

Step 9.   If the pressure reducing valve is an integral part of the system, record the pressures indicated on pressure gauges (1) and (2). If not, activate the valve which directs flow to the pressure reducing valve and record the pressures indicated on pressure gauges (1) and (2).

Step 10.   Shut the prime mover off and analyze the test results.

Step 11.   At the conclusion of this test procedure, remove the diagnostic equipment. Reconnect the transmission lines and tighten the connectors securely.

Step 12.   Start the prime mover and inspect the connectors for leaks.

ANALYZING THE TEST RESULTS

1.   **Diagnostic Observation:** Pressure gauge (1) indicates main pressure relief valve setting.

Pressure gauge (2) indicates the recommended pressure reducing valve setting.

Diagnosis: The pressure reducing valve appears to be in satisfactory operating condition.

2.   **Diagnostic Observation:** Pressure gauge (1) indicates recommended main pressure relief valve setting.

Pressure gauge (2) indicates main pressure relief valve setting.

Diagnosis: There is evidence that the pressure reducing valve may require adjustment or has failed for some reason.

First attempt to adjust the pressure by turning the adjusting screw. If the valve fails to respond to the pressure adjustment, there is probably internal damage, or restriction in the external drain-line.

Internal damage can be caused by contamination in the fluid, valve distortion from sub-standard mounting procedures, over-tightened, tapered pipe connectors, and/or a plugged external drain-line.

To determine if the problem is caused by restriction in the external drain-line, conduct an external drain-line test procedure (Page 4-35). If the pressure in the external drain-line is within specification, repair or replace the pressure reducing valve.

chapter five

Procedures for Testing Industrial Directional Control Valves

"A" Port Passage
Direct-Access Test Procedure

What will this test procedure accomplish?

Generally, directional control valves have radial clearances between the valve bore and spool of between 5 to 15 micrometers.

The production of perfectly round and straight bores is exceptionally difficult to achieve, so it is highly unlikely that any spool will lie precisely concentric within its bore.

Valve manufacturers strive to develop casting and machining methods that will improve bore and spool production thereby reducing cross-port leakage. However, since the spool moves in relation to the bore, some leakage is expected.

If the fluid is heavily contaminated with particles which are equal in size to the clearances between the spool and bore, it will cause abrasive wear. Abrasive wear will gradually increase the radial and diametrical clearances, resulting in increased cross-port leakage.

High leakage will cause system overheating and/or a decline in actuator speed. It can also cause increased cylinder rod drifting in pressure manifold applications.

Clearance-related problems are generally viscosity sensitive, i.e. increased leakage will be less obvious when the fluid temperature is low and will increase as the fluid temperature increases.

This test procedure is applicable to directional control valves which have no sandwich mounted circuit modules. If one or more sandwich circuit modules are located between the directional control valve and sub-plate or manifold, refer to the "Sandwich Circuit Module Cross-port Leakage" test procedure located in this chapter.

The most common symptoms of problems associated with excessive cross-port leakage in directional control valves are:
  a. Linear actuator drifting in the rod extend position while the directional control valve. is in the neutral (hold) position.
  b. Linear actuator vertical load "down-drift" while the directional control valve is in the neutral (hold) position.

   c. Increased rotary actuator drifting while the directional control valve is in the neutral (hold) position.

   d. Moderate to high increase in the operating temperature of the fluid.

   e. Loss of precise position control.

This test procedure will determine the amount of leakage across the "A" port passage of a directional control valve.

-WARNING- | **Do not work on or around hydraulic systems without wearing safety glasses which conform to ANSI Z87.1-1989 standard.**

YOU WILL NEED THE FOLLOWING DIAGNOSTIC EQUIPMENT TO CONDUCT THIS TEST PROCEDURE

CAUTION! *The pressure and flow ratings of the diagnostic equipment which will be used to conduct this test procedure must be equal to, or greater than, the pressure and flow ratings of the system being tested. Refer to the system schematic for recommended pressure and flow data.*

   1. MicroLeak analyzer   3. Stopwatch (not shown)
   2. Pressure gauge

PREPARATION

To conduct this test safely and accurately, refer to Figure 5-1 for correct placement of diagnostic equipment.

TEST PROCEDURE

Step 1.   Shut the prime mover off.

Step 2.   Lock the electrical system out or tag the keylock switch.

Step 3.   Observe the system pressure gauge. Release any residual pressure trapped in the system by an accumulator, counterbalance or pilot-operated check valve, suspended load on an actuator, intensifier, or a pressurized reservoir.

NOTE: *It is not necessary to remove the "P" or "T" port transmission lines from the valve ports to conduct this test procedure.*

Step 4.   Remove the transmission line from the "A" port of the directional control valve sub-plate.

Step 5.   Install MicroLeak analyzer (1) in series with the "A" port of the directional control valve sub-plate.

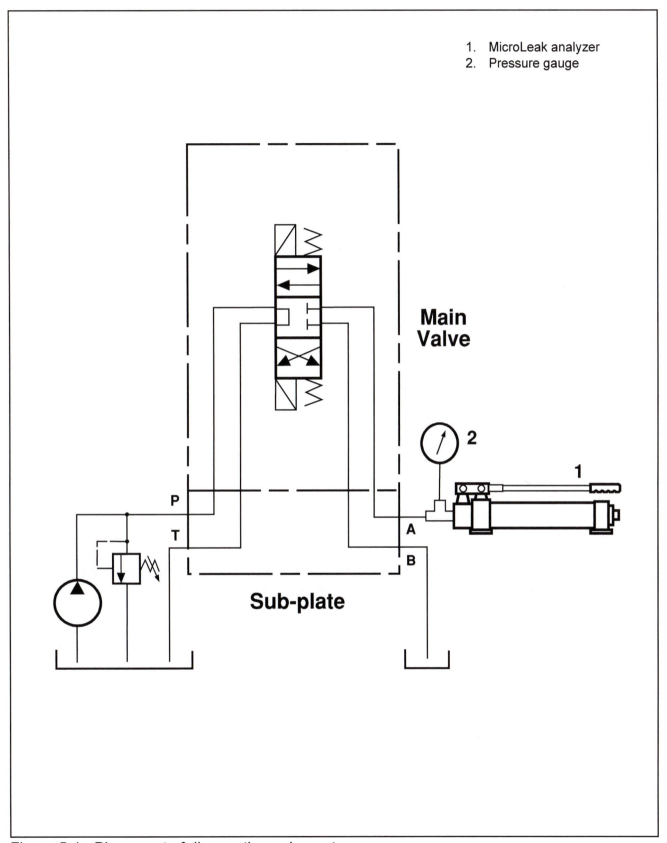

1. MicroLeak analyzer
2. Pressure gauge

Main Valve

Sub-plate

P

T

A

B

Figure 5-1   Placement of diagnostic equipment

Step 6.   Install pressure gauge (2) in parallel with the connector at the outlet port of MicroLeak analyzer (1).

Step 7.   Remove the transmission line from the "B" port of the directional control valve sub-plate.

Step 8.   Gradually pressurize the "A" port passage with MicroLeak analyzer (1). Stop when the value of the system's main pressure relief valve or compensator setting is reached.

NOTE: *Oil is now trapped between the MicroLeak analyzer and the spool in the "A" port passage of the directional control valve.*

Step 9.   Observe the needle on pressure gauge (2). Using a stopwatch, time how long it takes for the pressure to drop to 200 PSI (13.8 bar).

NOTE: *The condition of the "A" port passage is determined by timing how long it takes for the pressure to drop from the value of the system's main pressure relief valve setting down to approximately 200 PSI (13.8 bar).*

Step 10.   Repeat the test procedure in the "B" port passage and compare the leakage rates.

ANALYZING THE TEST RESULTS

1.   **Diagnostic Observation:** The pressure indicated on pressure gauge (2) immediately begins to decrease when the pumping action of the MicroLeak analyzer stops.

    Initially, the pressure drops fairly rapidly due to the high pressure differential across the ports. The pressure gauge indicates that the leakage across the ports decreases progressively as the pressure differential narrows.

    It takes a "reasonable" length of time for the pressure to drop from the value of the system's main pressure relief valve setting down to approximately 200 PSI (13.8 bar).

NOTE: *There is no general rule-of-thumb regarding the volumetric efficiency of directional control valves. Refer to the manufacturer's specifications for maximum recommended leakage rates.*

    Diagnosis: Leakage across the "A" port passage appears to be within design specification.

2.   **Diagnostic Observation:** The pressure indicated on pressure gauge (2) attempts to increase when the MicroLeak analyzer is stroked. However, between pumping strokes the pressure drops rapidly.

    If the "P," "T," and "B" ports are open, a steady stream of oil can be seen pouring from the open ports.

The MicroLeak analyzer fails to pressurize the "A" port passage to the value of the system's main pressure relief valve setting.

Diagnosis: Leakage across the "A" port passage appears to be excessive. Although this amount of leakage may appear to be insignificant, it could cause actuator drifting and/or a marked increase in the operating temperature of the fluid.

If it is not possible to determine if the cross-port leakage is excessive, conduct an actuator leakage path analysis test procedure. If actuator leakage is within specification, replace the directional control valve.

If necessary, refer to the manufacturer's specifications for recommended leakage data.

3.   **Diagnostic Observation:** The MicroLeak analyzer fails to pressurize the "A" port passage. Oil leaks profusely from the open ports in the valve.

Diagnosis: Leakage across the "A" port passage is excessive. Replace the directional control valve.

NOTE: *Before removing the valve check to see if excessive leakage is caused by the spool either not centering or binding.*

The spool may not be centering because there is an accumulation of contamination between either the spool and bore, or the end-caps.

Spool binding could also be caused by valve body misalignment due to substandard mounting procedures, overtightened mounting bolts, or tapered pipe connectors.

CAUTION! *Component repair procedures must be conducted by trained, authorized personnel. Incorrect parts, or parts which are improperly installed, can cause catastrophic component malfunction.*

---

notes

"B" Port Passage
Direct-Access Test Procedure

What will this test procedure accomplish?

Generally, directional control valves have radial clearances between the valve bore and spool of between 5 to 15 micrometers.

The production of perfectly round and straight bores is exceptionally difficult to achieve, so it is highly unlikely that any spool will lie precisely concentric within its bore.

Valve manufacturers strive to develop casting and machining methods that will improve bore and spool production thereby reducing cross-port leakage. However, since the spool moves in relation to the bore, some leakage is expected.

If the fluid is heavily contaminated with particles which are equal in size to the clearances between the spool and bore, it will cause abrasive wear. Abrasive wear will gradually increase the radial and diametrical clearances, resulting in increased cross-port leakage.

High leakage will cause system overheating and/or a decline in actuator speed. It can also cause increased cylinder rod drifting in pressure manifold applications.

Clearance-related problems are generally viscosity sensitive, i.e. increased leakage will be less obvious when the fluid temperature is low and will increase as the fluid temperature increases.

This test procedure is applicable to directional control valves which have no sandwich mounted circuit modules. If one or more sandwich circuit modules are located between the directional control valve and sub-plate or manifold, refer to the "Sandwich Circuit Module Cross-port Leakage" test procedure located in this chapter.

The most common symptoms of problems associated with excessive cross-port leakage in directional control valves are:
    a. Linear actuator drifting in the rod extend position while the directional control valve is in the neutral (hold) position.
    b. Linear actuator vertical load "down-drift" while the directional control valve is in the neutral (hold) position.
    c. Increased rotary actuator drifting while the directional control valve is in the neutral (hold) position.
    d. Moderate to high increase in the operating temperature of the fluid.
    e. Loss of precise position control.

This test procedure will determine the amount of leakage across the "B" port passage of a directional control valve.

-WARNING- | **Do not work on or around hydraulic systems without wearing safety glasses which conform to ANSI Z87.1-1989 standard.**

YOU WILL NEED THE FOLLOWING DIAGNOSTIC EQUIPMENT TO CONDUCT THIS TEST PROCEDURE

CAUTION! *The pressure and flow ratings of the diagnostic equipment which will be used to conduct this test procedure must be equal to, or greater than, the pressure and flow ratings of the system being tested. Refer to the system schematic for recommended pressure and flow data.*

1. MicroLeak analyzer
2. Pressure gauge
3. Stopwatch (not shown)

PREPARATION

To conduct this test safely and accurately, refer to Figure 5-2 for correct placement of diagnostic equipment.

TEST PROCEDURE

Step 1. Shut the prime mover off.

Step 2. Lock the electrical system out or tag the keylock switch.

Step 3. Observe the system pressure gauge. Release any residual pressure trapped in the system by an accumulator, counterbalance or pilot-operated check valve, suspended load on an actuator, intensifier, or a pressurized reservoir.

NOTE: *It is not necessary to remove the "P" or "T" port transmission lines from the valve ports to conduct this test procedure.*

Step 4. Remove the transmission line from the "B" port of the directional control valve sub-plate.

Step 5. Install MicroLeak analyzer (1) in series with the "B" port of the directional control valve sub-plate.

Step 6. Install pressure gauge (2) in parallel with the connector at the outlet port of MicroLeak analyzer (1).

Step 7. Remove the transmission line from the "A" port of the directional control valve sub-plate.

Step 8. Gradually pressurize the "B" port passage with MicroLeak analyzer (1). Stop when the value of the system's main pressure relief valve is reached.

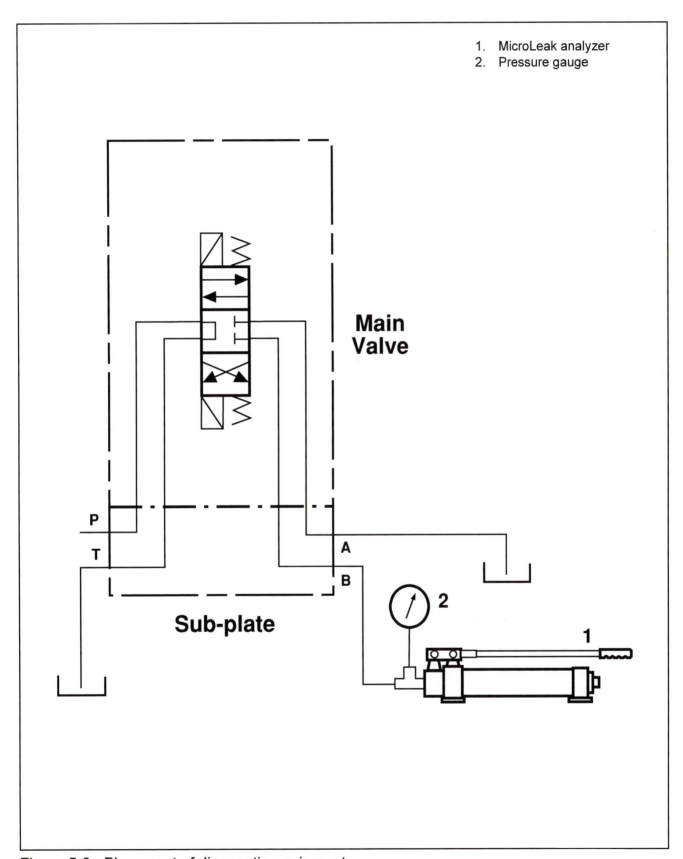

1. MicroLeak analyzer
2. Pressure gauge

Main
Valve

P

T

A

B

Sub-plate

Figure 5-2   Placement of diagnostic equipment

NOTE: *Oil is now trapped between the MicroLeak analyzer and the spool in the "B" port passage of the directional control valve.*

Step 9.   Observe the needle on pressure gauge (2). Using a stopwatch, time how long it takes for the pressure to drop to 200 PSI (13.8 bar).

NOTE: *The condition of the "B" port passage is determined by timing how long it takes for the pressure to drop from the value of the system's main pressure relief valve setting down to approximately 200 PSI (13.8 bar).*

Step 10.   Repeat the test procedure in the "A" port passage and compare the leakage rates.

ANALYZING THE TEST RESULTS

1.   **Diagnostic Observation:** The pressure indicated on pressure gauge (2) immediately begins to decrease when the pumping action of the MicroLeak analyzer stops.

Initially, the pressure drops fairly rapidly due to the high pressure differential across the ports. The pressure gauge indicates that the leakage across the ports decreases progressively as the pressure differential narrows.

It takes a "reasonable" length of time for the pressure to drop from the value of the system's main pressure relief valve setting down to approximately 200 PSI (13.8 bar)

NOTE: *There is no general rule-of-thumb regarding the volumetric efficiency of directional control valves. If it is necessary to determine accurate leakage rates, refer to the valve manufacturer's specifications.*

Diagnosis: Leakage across the "B" port passage appears to be within design specification.

2.   **Diagnostic Observation:** The pressure, indicated on pressure gauge (2), attempts to increase when the MicroLeak analyzer is stroked. However, between pumping strokes the pressure drops rapidly.

If the "P," "T," and "A" ports are open, a steady stream of oil can be seen pouring from the valve.

The MicroLeak analyzer fails to pressurize the "B" port passage to the value of the system's main pressure relief valve setting.

Diagnosis: Leakage across the "B" port passage appears to be excessive. Although this amount of leakage may appear to be insignificant, it could cause actuator drifting and/or a marked increase in the operating temperature of the fluid.

If it is not possible to determine if cross-port leakage is excessive, conduct an actuator leakage path analysis test procedure. If actuator leakage is within specification, replace the directional control valve.

If necessary, refer to the manufacturer's specifications for recommended leakage data.

3.   **Diagnostic Observation:** The MicroLeak analyzer fails to pressurize the "B" port passage. Oil leaks profusely from the open ports in the valve.

Diagnosis: Leakage across the "B" port passage is excessive. Replace the directional control valve.

NOTE: *Before removing the valve, check to see if excessive leakage is caused by the spool either not centering or binding.*

The spool may not be centering because there is an accumulation of contamination between either the spool and bore, or the end-caps.

Spool binding could also be caused by valve body misalignment due to substandard mounting procedures, overtightened mounting bolts, or tapered pipe connectors.

notes

"P" Port to "A" Port Passage
Direct-Access Test Procedure

What will this test procedure accomplish?

Generally, directional control valves have radial clearances between the valve bore and spool of between 5 to 15 micrometers.

The production of perfectly round and straight bores is exceptionally difficult to achieve, so it is highly unlikely that any spool will lie precisely concentric within its bore.

Valve manufacturers strive to develop casting and machining methods that will improve bore and spool production thereby reducing cross-port leakage. However, since the spool moves in relation to the bore, some leakage is expected.

If the fluid is heavily contaminated with particles which are equal in size to the clearances between the spool and bore, it will cause abrasive wear. Abrasive wear will gradually increase the radial and diametrical clearances, resulting in increased cross-port leakage.

High leakage will cause system overheating and/or a decline in actuator speed. It can also cause increased cylinder rod drifting in pressure manifold applications.

Clearance-related problems are generally viscosity sensitive, i.e. increased leakage will be less obvious when the fluid temperature is low and will increase as the fluid temperature increases.

This test procedure is applicable to directional control valves which have no sandwich mounted circuit modules. If one or more sandwich circuit modules are located between the directional control valve and sub-plate or manifold, refer to the "Sandwich Circuit Module Cross-port Leakage" test procedure located in this chapter.

The most common symptoms of problems associated with excessive cross-port leakage in directional control valves are:
   a. Linear actuator drifting in the rod extend position while the directional control valve is in the neutral (hold) position.
   b. Linear actuator vertical load "down-drift" while the directional control valve is in the neutral (hold) position.
   c. Increased rotary actuator drifting while the directional control valve is in the neutral (hold) position.
   d. Moderate to high increase in the operating temperature of the fluid.
   e. Loss of precise position control.

This test procedure will determine the amount of leakage across the "P" port to "A" port passage of a directional control valve.

 **-WARNING-** | Do not work on or around hydraulic systems without wearing safety glasses which conform to ANSI Z87.1-1989 standard.

YOU WILL NEED THE FOLLOWING DIAGNOSTIC EQUIPMENT TO CONDUCT THIS TEST PROCEDURE

CAUTION! *The pressure and flow ratings of the diagnostic equipment which will be used to conduct this test procedure must be equal to, or greater than, the pressure and flow ratings of the system being tested. Refer to the system schematic for recommended pressure and flow data.*

1. MicroLeak analyzer
2. Pressure gauge
3. Stopwatch (not shown)

PREPARATION

To conduct this test safely and accurately, refer to Figure 5-3 for correct placement of diagnostic equipment.

TEST PROCEDURE

Step 1.   Shut the prime mover off.

Step 2.   Lock the electrical system out or tag the keylock switch.

Step 3.   Observe the system pressure gauge. Release any residual pressure trapped in the system by an accumulator, counterbalance or pilot-operated check valve, suspended load on an actuator, intensifier, or a pressurized reservoir.

Step 4.   Remove the transmission line from the "P" port of the directional control valve sub-plate.

Step 5.   Install MicroLeak analyzer (1) in series with the "P" port of the directional control valve sub-plate.

Step 6.   Install pressure gauge (2) in parallel with the connector at the outlet port of MicroLeak analyzer (1).

Step 7.   Remove the transmission line from the "A" port of the directional control valve sub-plate.

Step 8.   Plug the "A" port of the directional control valve sub-plate.

Step 9.   Activate the directional control valve spool and align the "P" and "A" ports. Hold this position.

1. MicroLeak analyzer
2. Pressure gauge

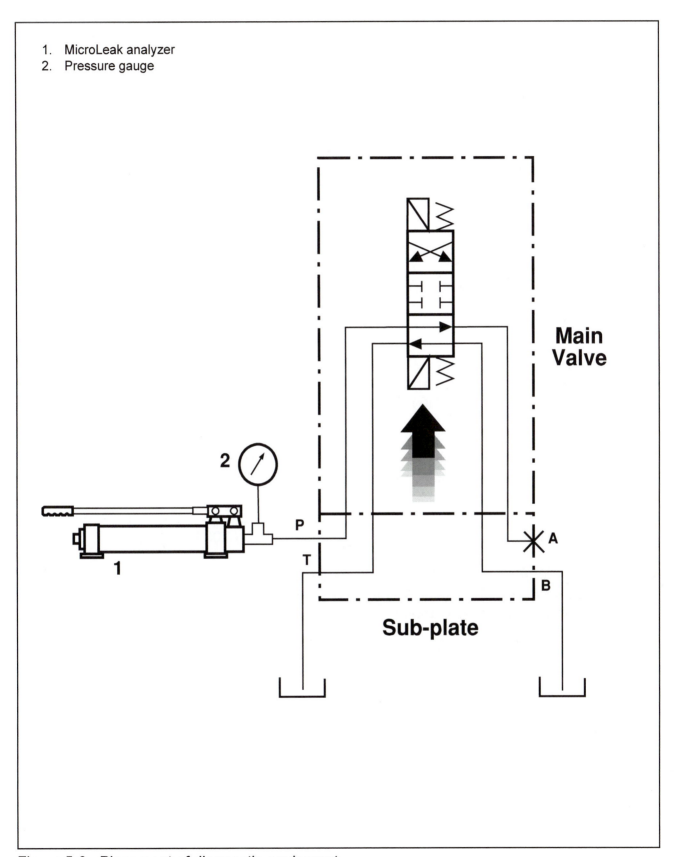

Figure 5-3   Placement of diagnostic equipment

Step 10. Gradually pressurize the "P" port to "A" port passage with MicroLeak analyzer (1). Stop when the value of the system's main pressure relief valve or compensator setting is reached.

NOTE: *Oil is now trapped between the MicroLeak analyzer and the plugged "P" port to "A" port passage.*

Step 11. Observe the needle on pressure gauge (2). Using a stopwatch, time how long it takes the pressure to drop to 200 PSI (13.8 bar).

NOTE: *The condition of the "P" port to "A" port passage is determined by timing how long it takes for the pressure to drop from the value of the system's main pressure relief valve setting down to approximately 200 PSI (13.8 bar).*

Step 12. Repeat this test procedure in the "P" port to "B" port passage and compare the leakage rates.

ANALYZING THE TEST RESULTS

1. **Diagnostic Observation:** The pressure indicated on pressure gauge (2) immediately begins to decrease when the pumping action of the MicroLeak analyzer stops.

    Initially, the pressure drops fairly rapidly due to the high pressure differential across the ports. The pressure gauge indicates that the leakage across the ports decreases progressively as the pressure differential narrows.

    It takes a "reasonable" length of time for the pressure to drop from the value of the system's main pressure relief valve setting down to approximately 200 PSI (13.8 bar)

NOTE: *There is no general rule-of-thumb regarding the volumetric efficiency of directional control valves. If it is necessary to determine accurate leakage rates, refer to the valve manufacturer's specifications.*

    Diagnosis: Leakage across the "P" port to "A" port passage appears to be within design specification.

2. **Diagnostic Observation:** The pressure indicated on pressure gauge (2) attempts to increase when the MicroLeak analyzer is stroked. However, between pumping strokes the pressure drops rapidly.

    If the "T" and "B" ports are open, a steady stream of oil can be seen pouring from the valve.

    The MicroLeak analyzer fails to pressurize the "P" port to "A" port passage to the value of the system's main pressure relief valve or compensator setting.

Diagnosis: Leakage across the "P" port to "A" port passage appears to be excessive. Although this amount of leakage may appear to be insignificant, it could cause actuator drifting and/or a marked increase in the operating temperature of the fluid.

If it is not possible to determine if the cross-port leakage is excessive, conduct an actuator leakage path analysis test procedure. If actuator leakage is within specification, replace the directional control valve.

If necessary, refer to the manufacturer's specifications for recommended leakage data.

3. **Diagnostic Observation:** The MicroLeak analyzer fails to pressurize the "P" port to "A" port passage. Oil leaks profusely from the open ports in the valve.

   Diagnosis: Leakage across the "P" port to "A" port passage is excessive. Replace the directional control valve.

NOTE: *Before removing the valve check to see if excessive leakage is caused by the spool either not centering or binding.*

The spool may not be centering because there is an accumulation of contamination between either the spool and bore, or the end-caps.

Spool binding could also be caused by valve body mis-alignment due to sub-standard mounting procedures, over-tightened mounting bolts, or tapered pipe connectors.

notes

"P" Port to "B" Port Passage Direct-Access Test Procedure

What will this test procedure accomplish?

Generally, directional control valves have radial clearances between the valve bore and spool of between 5 to 15 micrometers.

The production of perfectly round and straight bores is exceptionally difficult to achieve, so it is highly unlikely that any spool will lie precisely concentric within its bore.

Valve manufacturers strive to develop casting and machining methods that will improve bore and spool production thereby reducing cross-port leakage. However, since the spool moves in relation to the bore, some leakage is expected.

If the fluid is heavily contaminated with particles which are equal in size to the clearances between the spool and bore, it will cause abrasive wear. Abrasive wear will gradually increase the radial and diametrical clearances, resulting in increased cross-port leakage.

High lakage will cause system overheating and/or a decline in actuator speed. It can also cause increased cylinder rod drifting in pressure manifold applications.

Clearance-related problems are generally viscosity sensitive, i.e. increased leakage will be less obvious when the fluid temperature is low and will increase as the fluid temperature increases.

This test procedure is applicable to directional control valves which have no sandwich mounted circuit modules. If one or more sandwich circuit modules are located between the directional control valve and sub-plate or manifold, refer to the "Sandwich Circuit Module Cross-port Leakage" test procedure located in this chapter.

The most common symptoms of problems associated with excessive cross-port leakage in directional control valves are:
  a. Linear actuator drifting in the rod extend position while the directional control valve is in the neutral (hold) position.
  b. Linear actuator vertical load "down-drift" while the directional control valve is in the neutral (hold) position.
  c. Increased rotary actuator drifting while the directional control valve is in the neutral (hold) position.
  d. Moderate to high increase in the operating temperature of the fluid.
  e. Loss of precise position control.

This test procedure will determine the amount of leakage across the "P" port to "B" port passage of a directional control valve.

 **-WARNING-** Do not work on or around hydraulic systems without wearing safety glasses which conform to ANSI Z87.1-1989 standard.

YOU WILL NEED THE FOLLOWING DIAGNOSTIC EQUIPMENT TO CONDUCT THIS TEST PROCEDURE

CAUTION! *The pressure and flow ratings of the diagnostic equipment which will be used to conduct this test procedure must be equal to, or greater than, the pressure and flow ratings of the system being tested. Refer to the system schematic for recommended pressure and flow data.*

1. MicroLeak analyzer   2.   Pressure gauge   3.   Stopwatch (not shown)

PREPARATION

To conduct this test safely and accurately, refer to Figure 5-4 for correct placement of diagnostic equipment.

TEST PROCEDURE

Step 1.   Shut the prime mover off.

Step 2.   Lock the electrical system out or tag the keylock switch.

Step 3.   Observe the system pressure gauge. Release any residual pressure trapped in the system by accumulators, counterbalance or pilot-operated check valves, suspended loads on cylinders, or intensifiers, or pressurized reservoirs.

Step 4.   Remove the transmission line from the "P" port of the directional control valve sub-plate.

Step 5.   Install MicroLeak analyzer (1) in series with the "P" port of the directional control valve sub-plate.

Step 6.   Install pressure gauge (2) in parallel with the connector at the outlet port of MicroLeak analyzer (1).

Step 7.   Remove the transmission line from the "B" port of the directional control valve sub-plate.

Step 8.   Plug the "B" port of the directional control valve sub-plate.

Step 9.   Activate the directional control valve spool and align the "P" and "B" ports. Hold this position.

1. MicroLeak analyzer
2. Pressure gauge

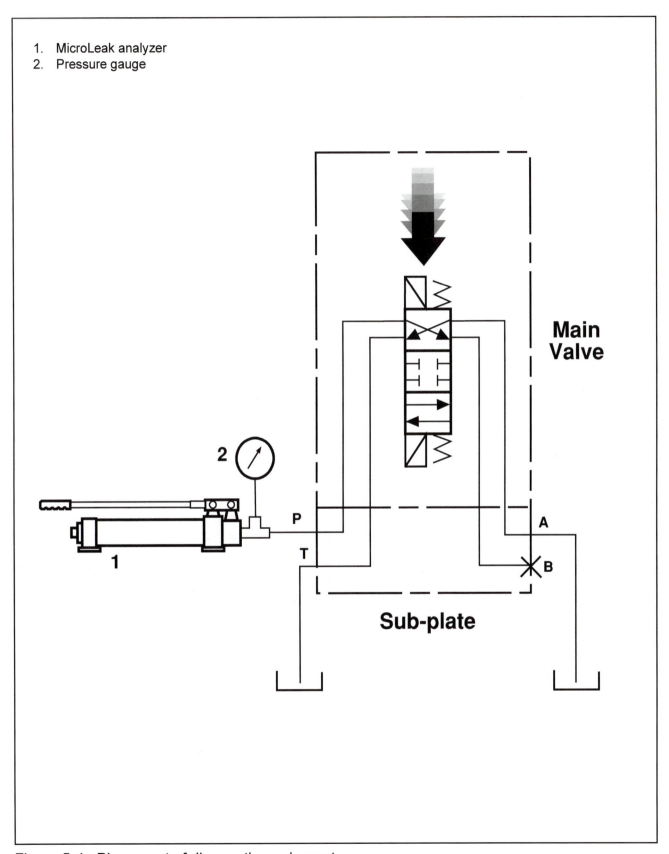

Figure 5-4   Placement of diagnostic equipment

Step 10. Gradually pressurize the "P" port to "B" port passage with MicroLeak analyzer (1). Stop when the value of the system's main pressure relief valve or compensator setting is reached.

NOTE: *Oil is now trapped between the MicroLeak analyzer and the plugged "P" port to "B" port passage.*

Step 11. Observe the needle on pressure gauge (2). Using a stopwatch, time how long it takes for the pressure to drop to 200 PSI (13.8 bar).

NOTE: *The condition of the "P" port to "B" port passage is determined by timing how long it takes for the pressure to drop from the value of the system's main pressure relief valve setting down to approximately 200 PSI (13.8 bar).*

Step 12. Repeat this test procedure in the "P" port to "A" port passage and compare the leakage rates.

ANALYZING THE TEST RESULTS

1. **Diagnostic Observation:** The pressure indicated on pressure gauge (2) immediately begins to decrease when the pumping action of the MicroLeak analyzer stops.

   Initially, the pressure drops fairly rapidly due to the high pressure differential across the ports. The pressure gauge indicates that the leakage across the ports decreases progressively as the pressure differential narrows.

   It takes a "reasonable" length of time for the pressure to drop from the value of the system's main pressure relief valve setting down to approximately 200 PSI (13.8 bar)

NOTE: *There is no general rule-of-thumb regarding the volumetric efficiency of directional control valves. If it is necessary to determine accurate leakage rates, refer to the valve manufacturer's specifications.*

   Diagnosis: Leakage across the "P" port to "B" port passage appears to be within design specification.

2. **Diagnostic Observation:** The pressure indicated on pressure gauge (2) attempts to increase when the MicroLeak analyzer is stroked. However, between pumping strokes the pressure drops rapidly.

   If the "T" and "A" ports are open, a steady stream of oil can be seen pouring from the valve.

   The MicroLeak analyzer fails to pressurize the "P" port to "B" port passage to the value of the system's main pressure relief valve or compensator setting.

Diagnosis: Leakage across the "P" port to "B" port passage appears to be excessive. Although this amount of leakage may appear to be insignificant, it could cause actuator drifting and/or a marked increase in the operating temperature of the fluid.

If it is not possible to determine if the cross-port leakage is excessive, conduct an actuator leakage path analysis test procedure. If actuator leakage is within specification, replace the directional control valve.

If necessary, refer to the manufacturer's specifications for recommended leakage data.

3.   **Diagnostic Observation:** The MicroLeak analyzer fails to pressurize the "P" port to "B" port passage. Oil leaks profusely from the open ports in the valve.

   Diagnosis: Leakage across the "P" port to "B" port passage is excessive. Replace the directional control valve.

NOTE: *Before removing the valve check to see if excessive leakage is caused by the spool either not centering or binding.*

   The spool may not be centering because there is an accumulation of contamination between either the spool and bore, or the end-caps.

   Spool binding could also be caused by valve body mis-alignment due to sub-standard mounting procedures, over-tightened mounting bolts, or tapered pipe connectors.

notes

Sandwich Circuit Module - "A" Port Passage Direct-Access Test Procedure

What will this test procedure accomplish?

To simplify installation, eliminate resonance sources, minimize external transmission lines, and reduce cost, an unlimited combination of simple and complex circuits can be easily constructed in a single "stack" using sub-plates, sandwich circuit modules, and industrial solenoid directional control valves.

Sandwich valves make it simple to modify existing circuits. Additional control functions can be added without disturbing existing transmission lines and connectors.

As many as three or four sandwich circuit modules can be stacked under a single industrial directional control valve to achieve any number of control functions. The four internal oil flow passages are separated by valve cartridges, spools, and "O"-ring seals.

Since system control valves are generally integrated in one or more circuit modules, the source of a cross-port leakage path is usually difficult to detect. The most common causes of cross-port leakage problems related to stack valves are:
   a. Contamination in the fluid causing valve malfunction or erratic operation.
   b. Wear in the directional control valve or circuit module.
   c. "O"-ring leakage.

Cross-port leakage could cause actuator drifting, system overheating, poor single or bi-directional actuator performance, or a total loss of actuator control in both directions.

The most common symptoms of problems associated with excessive cross-port leakage in directional control valves are:
   a. Linear actuator drifting in the rod extend position while the directional control valve is in the neutral (hold) position.
   b. Linear actuator vertical load "down-drift" while the directional control valve is in the neutral (hold) position.
   c. Increased rotary actuator drifting while the directional control valve is in the neutral (hold) position.
   d. Moderate to high increase in the operating temperature of the fluid.
   e. Loss of precise position control.

This test procedure will determine:
     a. The amount of leakage across the "A" port passage of a directional control valve.
     b. If there is a cross-port leakage path in a circuit module.
     c. How to isolate a leakage path in a valve stack that has one or more circuit modules.

 **-WARNING-** | **Do not work on or around hydraulic systems without wearing safety glasses which conform to ANSI Z87.1-1989 standard.**

YOU WILL NEED THE FOLLOWING DIAGNOSTIC EQUIPMENT
TO CONDUCT THIS TEST PROCEDURE

CAUTION! *The pressure and flow ratings of the diagnostic equipment which will be used to conduct this test procedure must be equal to, or greater than, the pressure and flow ratings of the system being tested. Refer to the system schematic for recommended pressure and flow data.*

1.	MicroLeak analyzer	4.	Sandwich Circuit Module Bolt Kit/s (not shown)
2.	Pressure Gauge	5.	Sandwich Circuit Module Sub-plate
3.	Stopwatch (not shown)		

PREPARATION

To conduct this test safely and accurately, refer to Figure 5-5 for correct placement of diagnostic equipment.

TEST PROCEDURE

Step 1.   Shut the prime mover off.

Step 2.   Lock the electrical system out or tag the keylock switch.

Step 3.   Observe the system pressure gauge. Release any residual pressure trapped in the system by an accumulator, counterbalance or pilot-operated check valve, suspended load on an actuator, intensifier, or a pressurized reservoir.

Step 4.   Remove the sandwich circuit module(s) and directional control valve from the sub-plate.

NOTE: *If it is safe and practical to disconnect the "A" port and "B" port transmission lines from the existing sub-plate, this test procedure can be conducted with the valves in position on the machine.*

Step 5.   Install the valve assembly, as it was installed on its original sub-plate, on the test sub-plate. Tighten the retaining bolts to the recommended torque values.

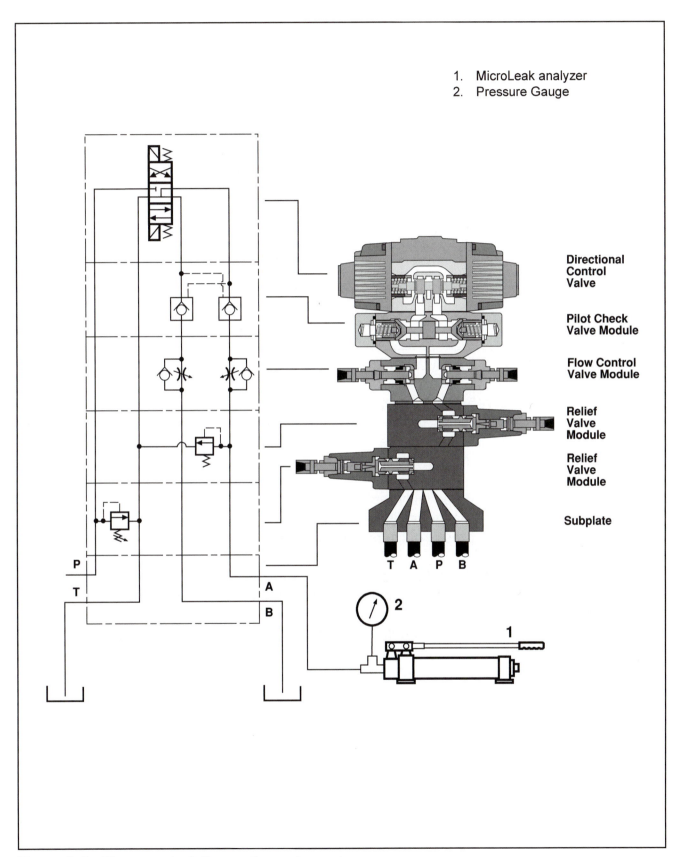

1. MicroLeak analyzer
2. Pressure Gauge

Directional Control Valve

Pilot Check Valve Module

Flow Control Valve Module

Relief Valve Module

Relief Valve Module

Subplate

Figure 5-5   Placement of diagnostic equipment

Step 6.   Connect MicroLeak analyzer (1) in series with the "A" port of the test sub-plate.

Step 7.   Install pressure gauge (2) in parallel with the connector at the outlet port of MicroLeak analyzer (1).

Step 8.   Attempt to pressurize the "A" port passage with MicroLeak analyzer (1). Stop when the value of the system's main pressure relief valve or compensator setting is reached.

NOTE: *Oil is now trapped between the MicroLeak analyzer and the "A" port passage in the sandwich circuit module test sub-plate.*

Step 9.   Observe the needle on pressure gauge (2). Using a stopwatch, time how long it takes for the pressure to drop to 200 PSI (13.8 bar).

NOTE: *The condition of the directional control valve and circuit module(s) is determined by timing how long it takes for the pressure to drop from the value of the system's main pressure relief valve setting down to approximately 200 PSI (13.8 bar).*

ANALYZING THE TEST RESULTS

1.   **Diagnostic Observation**: Closed-center directional control valve without a load-check or counter-balance valve circuit module (Direct access to spool through the "A" port passage): The pressure indicated on pressure gauge (2) immediately begins to decrease when the pumping action of the MicroLeak analyzer stops.

Initially, the pressure drops fairly rapidly due to the high pressure differential across the ports. The pressure gauge indicates that the leakage across the ports decreases progres-sively as the pressure differential narrows.

It takes a "reasonable" length of time for the pressure to drop from the value of the system's main pressure relief valve setting down to approximately 200 PSI (13.8 bar).

NOTE: *There is no general rule-of-thumb regarding the volumetric efficiency of directional control valves. If it is necessary to determine accurate leakage rates, refer to the valve manufacturer's specifications.*

Diagnosis: Leakage across the "A" port passage appears to be within design specification.

2.   **Diagnostic Observation:** Closed-center directional control valve without a load-check or counter-balance valve circuit module (Direct access to spool through the "A" port passage): The pressure indicated on pressure gauge (2) attempts to increase when the MicroLeak analyzer is stroked. However, between pumping strokes the pressure drops rapidly.

If the "P," "T," and "B" ports are open, a steady stream of oil can be seen pouring from the valve.

The MicroLeak analyzer fails to pressurize the "A" port passage to the value of the system's main pressure relief valve or compensator setting.

Diagnosis: Leakage across the "A" port passage appears to be excessive. Although this amount of leakage may appear to be insignificant, it could cause actuator drifting and/or a marked increase in the operating temperature of the fluid.

If it is not possible to determine if cross-port leakage is excessive, conduct an actuator leakage path analysis test procedure. If actuator leakage is within specification, replace the directional control valve.

NOTE: *One or more circuit modules may be mounted in parallel with the directional control valve and the sub-plate. The leakage path could be caused by any one of the circuit modules or the directional control valve. Use the following steps to determine the root-cause:*

Step 1.　Remove the bolts fastening the sandwich circuit module(s) and directional control valve to the test sub-plate.

Step 2.　Remove the sandwich circuit module closest to the test sub-plate.

Step 3.　Install the remaining module(s) and directional control valve on the test sub-plate. Tighten the retaining bolts to the recommended torque values.

NOTE: *Use a shorter set of retaining bolts to re-install the valve assembly.*

Step 4.　Once again, attempt to pressurize the "A" port passage with MicroLeak analyzer (1). Stop when the value of the system's main pressure relief valve setting is reached.

NOTE: *If it is now possible to temporarily pressurize the "A" port passage, the cross-port leakage path is isolated to the first sandwich circuit module.*

Step 5.　If it is still not possible to pressurize the "A" port passage, continue removing sandwich circuit modules one at a time until the cross-port leakage path has been isolated.

Step 6.　Repeat this test procedure in the "B" port passage.

Step 7.　Once the cross-port leakage path has been isolated, repair or replace the defective component, and install it on its original sub-plate. Tighten the retaining bolts to the recommended torque values.

If necessary, refer to the manufacturer's specifications for recommended leakage data.

3.　　**Diagnostic Observation:** The MicroLeak analyzer fails to pressurize the "A" port passage. Oil leaks profusely from the open ports in the valve.

Diagnosis: Leakage across the "A" port passage is excessive.

NOTE: *One or more circuit modules may be mounted in parallel with the directional control valve and the sub-plate. The leakage path could be caused by any one of the circuit modules or the directional control valve. Use the following steps to determine the root-cause:*

Step 1.   Remove the bolts fastening the sandwich circuit module(s) and directional control valve to the test sub-plate.

Step 2.   Remove the sandwich circuit module closest to the test sub-plate.

Step 3.   Install the remaining module(s) and directional control valve on the test sub-plate and tighten to the recommended torque values.

NOTE: *Use a shorter set of retaining bolts to re-install the valve assembly.*

Step 4.   Once again, attempt to pressurize the "A" port passage with MicroLeak analyzer (1). Stop when the value of the system's main pressure relief valve or compensator setting is reached.

NOTE: *If it is now possible to temporarily pressurize the "A" port passage, the cross-port leakage path is isolated to the first sandwich circuit module.*

Step 5.   If it is still not possible to pressurize the "A" port passage, continue removing sandwich circuit modules one at a time until the cross-port leakage path has been isolated.

Step 6.   Repeat this test procedure in the "B" port passage.

Step 7.   Once the cross-port leakage path has been isolated, repair or replace the defective component, and install it on its original sub-plate. Tighten the retaining bolts to the recommended torque values.

NOTE: *Before removing the valve check to see if excessive leakage is caused by the spool either not centering or binding.*

The spool may not be centering because there is an accumulation of contamination between either the spool and bore, or the end-caps.

Spool binding could also be caused by valve body mis-alignment due to sub-standard mounting procedures, over-tightened mounting bolts, or tapered pipe connectors.

4.   **Diagnostic Observation:** Closed-center, or open-center, directional control valve with a load-check or counter-balance valve circuit module: The pressure indicated on pressure gauge (2) holds steady at the value of the system's main pressure relief valve or compensator setting.

There may be some minor leakage across a circuit module indicated by a very gradual pressure loss.

Diagnosis: Leakage across the "A" port passage appears to be within design specification.

5.   **Diagnostic Observation:** Closed-center, or open-center, directional control valve with a load-check or counter-balance valve circuit module: The pressure indicated on pressure gauge (2) attempts to increase when the MicroLeak analyzer is stroked. However, between pumping strokes the pressure drops rapidly.

If the "P," "T," and "B" ports are open, a steady stream of oil can be seen pouring from the valve.

The MicroLeak analyzer fails to pressurize the "A" port passage to the value of the system's main pressure relief valve or compensator setting.

Diagnosis: Leakage across the "A" port passage appears to be excessive. The leakage path is caused by one or more of the circuit modules. Although this amount of leakage may appear to be insignificant, it could cause actuator drifting and/or a marked increase in the operating temperature of the fluid.

NOTE: *One or more circuit modules may be mounted in parallel with the directional control valve and the sub-plate. The leakage path could be caused by any one of the circuit modules.*

Use the following steps to determine the root-cause:

Step 1.   Remove the bolts fastening the sandwich circuit module(s) and directional control valve to the test sub-plate.

Step 2.   Remove the sandwich circuit module closest to the test sub-plate.

Step 3.   Install the remaining module(s) and directional control valve on the test sub-plate. Tighten the retaining bolts to the recommended torque values.

NOTE: *Use a shorter set of retaining bolts to re-install the valve assembly.*

Step 4.   Once again, attempt to pressurize the "A" port passage with MicroLeak analyzer (1). Stop when the value of the system's main pressure relief valve or compensator setting is reached.

NOTE: *If it is now possible to temporarily pressurize the "A" port passage, the cross-port leakage path is isolated to the first sandwich circuit module.*

Step 5.   If it is still not possible to pressurize the "A" port passage, continue removing sandwich circuit modules one at a time until the cross-port leakage path has been isolated.

Step 6.   Repeat this test procedure in the "B" port passage.

Step 7.   Once the cross-port leakage path has been isolated, repair or replace the defective component, and install it on its original sub-plate. Tighten the retaining bolts to the recommended torque values.

          If necessary, refer to the manufacturer's specifications for recommended leakage data.

6.   **Diagnostic Observation:** Closed-center, or open-center, directional control valve with a load-check or counter-balance valve circuit module: The MicroLeak analyzer fails to pressurize the "A" port passage. Oil leaks profusely from the open ports in the valve.

          Diagnosis: Leakage across the "A" port passage is excessive.

NOTE: *One or more circuit modules may be mounted in parallel with the directional control valve and the sub-plate. The leakage path could be caused by any one of the circuit modules. Use the following steps to determine the root-cause:*

Step 1.   Remove the bolts fastening the sandwich circuit module(s) and directional control valve to the test sub-plate.

Step 2.   Remove the sandwich circuit module closest to the test sub-plate.

Step 3.   Install the remaining module(s) and directional control valve on the test sub-plate. Tighten the retaining bolts to the recommended torque values.

NOTE: *Use a shorter set of retaining bolts to re-install the valve assembly.*

Step 4.   Once again, attempt to pressurize the "A" port passage with MicroLeak analyzer (1). Stop when the value of the system's main pressure relief valve or compensator setting is reached.

NOTE: *If it is now possible to temporarily pressurize the "A" port passage, the cross-port leakage path is isolated to the first sandwich circuit module.*

Step 5.   If it is still not possible to pressurize the "A" port passage, continue removing sandwich circuit modules one at a time until the cross-port leakage path has been isolated.

Step 6.   Repeat this test procedure in the "B" port passage.

Step 7.   Once the cross-port leakage path has been isolated, repair or replace the defective component, and install it on its original sub-plate. Tighten the retaining bolts to the recommended torque values.

Sandwich Circuit Module - "B" Port Passage Direct-Access Test Procedure

What will this test procedure accomplish?

To simplify installation, eliminate resonance sources, minimize external transmission lines, and reduce cost, an unlimited combination of simple and complex circuits can be easily constructed in a single "stack" using sub-plates, sandwich circuit modules, and industrial solenoid directional control valves.

Sandwich valves make it simple to modify existing circuits. Additional control functions can be added without disturbing existing transmission lines and connectors.

As many as three or four sandwich circuit modules can be stacked under a single industrial directional control valve to achieve any number of control functions. The four internal oil flow passages are separated by valve cartridges, spools, and "O"-ring seals.

Since system control valves are generally integrated in one or more circuit modules, the source of a cross-port leakage path is usually difficult to detect. The most common causes of cross-port leakage problems related to stack valves are:
    a. Contamination in the fluid causing valve malfunction or erratic operation.
    b. Wear in the directional control valve or circuit module.
    c. "O"-ring leakage.

Cross-port leakage could cause actuator drifting, system overheating, poor single or bi-directional actuator performance, or a total loss of actuator control in both directions.

The most common symptoms of problems associated with excessive cross-port leakage in directional control valves are:
    a. Linear actuator drifting in the rod extend position while the directional control valve is in the neutral (hold) position.
    b. Linear actuator vertical load "down-drift" while the directional control valve is in the neutral (hold) position.
    c. Increased rotary actuator drifting while the directional control valve is in the neutral (hold) position.
    d. Moderate to high increase in the operating temperature of the fluid.
    e. Loss of precise position control.

This test procedure will determine:

    a. The amount of leakage across the "B" port passage of a directional control valve.

    b. If there is a cross-port leakage path in a circuit module.

    c. How to isolate a leakage path in a valve stack that has one or more circuit modules.

-WARNING- | **Do not work on or around hydraulic systems without wearing safety glasses which conform to ANSI Z87.1-1989 standard.**

YOU WILL NEED THE FOLLOWING DIAGNOSTIC EQUIPMENT TO CONDUCT THIS TEST PROCEDURE

CAUTION! *The pressure and flow ratings of the diagnostic equipment which will be used to conduct this test procedure must be equal to, or greater than, the pressure and flow ratings of the system being tested. Refer to the system schematic for recommended pressure and flow data.*

1. MicroLeak analyzer
2. Pressure gauge
3. Stopwatch (not shown)
4. Sandwich Circuit Module Bolt Kit/s (not shown)

PREPARATION

To conduct this test safely and accurately, refer to Figure 5-6 for correct placement of diagnostic equipment.

TEST PROCEDURE

Step 1. Shut the prime mover off.

Step 2. Lock the electrical system out or tag the keylock switch.

Step 3. Observe the system pressure gauge. Release any residual pressure trapped in the system by an accumulator, counterbalance or pilot-operated check valve, suspended load on an actuator, intensifier, or a pressurized reservoir.

Step 4. Remove the sandwich circuit module(s) and directional control valve from the sub-plate.

NOTE: *If it is safe and practical to disconnect the "A" port and "B" port transmission line from the existing sub-plate, this test procedure can be conducted with the valves in position on the machine.*

Step 5. Install the valve assembly, as it was installed on its original sub-plate, on the test sub-plate. Tighten the retaining bolts to the recommended torque values.

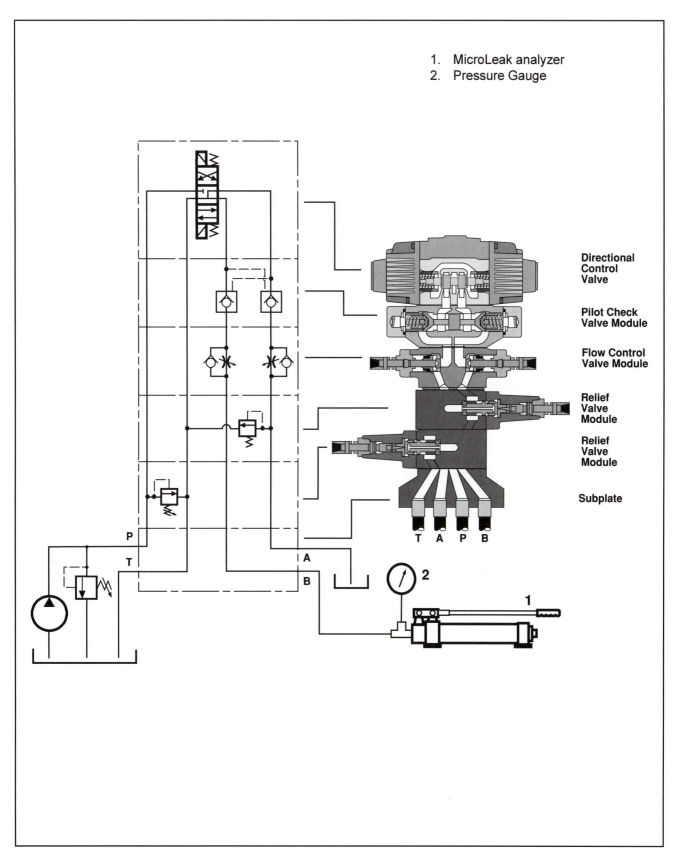

1. MicroLeak analyzer
2. Pressure Gauge

Directional Control Valve

Pilot Check Valve Module

Flow Control Valve Module

Relief Valve Module

Relief Valve Module

Subplate

Figure 5-6 Placement of diagnostic equipment

Step 6.   Connect MicroLeak analyzer (1) in series with the "B" port of the valve sub-plate.

Step 7.   Install pressure gauge (2) in parallel with the connector at the outlet port of MicroLeak analyzer (1).

Step 8.   Attempt to pressurize the "B" port passage with MicroLeak analyzer (1). Stop when the value of the system's main pressure relief valve or compensator setting is reached.

NOTE: *Oil is now trapped between the MicroLeak analyzer and the "B" port passage in the sandwich circuit module.*

Step 9.   Observe the needle on pressure gauge (2). Using a stopwatch, time how long it takes for the pressure to drop to 200 PSI (13.8 bar).

NOTE: *The condition of the directional control valve and circuit module(s) is determined by timing how long it takes for the pressure to drop from the value of the system's main pressure relief valve setting down to approximately 200 PSI (13.8 bar).*

ANALYZING THE TEST RESULTS
---

1.   **Diagnostic Observation:** Closed-center directional control valve without a load-check or counter-balance valve circuit module (Direct access to spool through the "B" port passage): The pressure indicated on pressure gauge (2) immediately begins to decrease when the pumping action of the MicroLeak analyzer stops.

Initially, the pressure drops fairly rapidly due to the high pressure differential across the ports. The pressure gauge indicates that the leakage across the ports decreases progressively as the pressure differential narrows.

It takes a "reasonable" length of time for the pressure to drop from the value of the system's main pressure relief valve setting down to approximately 200 PSI (13.8 bar)

NOTE: *There is no general rule-of-thumb regarding the volumetric efficiency of directional control valves. If it is necessary to determine accurate leakage rates, refer to the valve manufacturer's specifications.*

Diagnosis: Leakage across the "B" port passage appears to be within design specification.

2.   **Diagnostic Observation:** Closed-center directional control valve without a load-check or counter-balance valve circuit module (Direct access to spool through the "B" port passage): The pressure, indicated on pressure gauge (2), attempts to increase when the MicroLeak analyzer is stroked. However, between pumping strokes the pressure drops rapidly.

If the "P," "T," and "A" ports are open, a steady stream of oil can be seen pouring from the valve.

The MicroLeak analyzer fails to pressurize the "B" port passage to the value of the system's main pressure relief valve or compensator setting.

Diagnosis: Leakage across the "B" port passage appears to be excessive. Although this amount of leakage may appear to be insignificant, it could cause actuator drifting and/or a marked increase in the operating temperature of the fluid.

If it is not possible to determine if the cross-port leakage is excessive, conduct an actuator leakage path analysis test procedure. If actuator leakage is within specification, replace the directional control valve.

NOTE: *One or more circuit modules may be mounted in parallel with the directional control valve and the sub-plate. The leakage path could be caused by any one of the circuit modules. Use the following steps to determine the root-cause:*

Step 1.   Remove the bolts fastening the sandwich circuit module(s) and directional control valve to the test sub-plate.

Step 2.   Remove the sandwich circuit module closest to the test sub-plate.

Step 3.   Install the remaining module(s) and directional control valve on the test sub-plate. Tighten the retaining bolts to the recommended torque values.

NOTE: *Use a shorter set of retaining bolts to re-install the valve assembly.*

Step 4.   Once again, attempt to pressurize the "B" port passage with MicroLeak analyzer (1). Stop when the value of the system's main pressure relief valve or compensator setting is reached.

NOTE: *If it is now possible to temporarily pressurize the "B" port passage, the cross-port leakage path is isolated to the first sandwich circuit module.*

Step 5.   If it is still not possible to pressurize the "B" port passage, continue removing sandwich circuit modules one at a time until the cross-port leakage path has been isolated.

Step 6.   Repeat this test procedure in the "A" port passage.

Step 7.   Once the cross-port leakage path has been isolated, repair or replace the defective component, and install it on its original sub-plate. Tighten the retaining bolts to the recommended torque values.

If necessary, refer to the manufacturer's specifications for recommended leakage data.

3.   Diagnostic Observation: The MicroLeak analyzer fails to pressurize the "B" port passage. Oil leaks profusely from the open ports in the valve.

Diagnosis: Leakage across the "B" port passage is excessive.

NOTE: *One or more circuit modules may be mounted in parallel with the directional control valve and the sub-plate. The leakage path could be caused by any one of the circuit modules. Use the following steps to determine the root-cause:*

Step 1.   Remove the bolts fastening the sandwich circuit module(s) and directional control valve to the test sub-plate.

Step 2.   Remove the sandwich circuit module closest to the test sub-plate.

Step 3.   Install the remaining module(s) and directional control valve on the test sub-plate. Tighten the retaining bolts to the recommended torque values.

NOTE: *Use a shorter set of retaining bolts to re-install the valve assembly.*

Step 4.   Once again, attempt to pressurize the "B" port passage with MicroLeak analyzer (1). Stop when the value of the system's main pressure relief valve is reached.

NOTE: *If it is now possible to temporarily pressurize the "B" port passage, the cross-port leakage path is isolated to the first sandwich circuit module.*

Step 5.   If it is still not possible to pressurize the "B" port passage, continue removing sandwich circuit modules one at a time until the cross-port leakage path has been isolated.

Step 6.   Repeat this test procedure in the "A" port passage.

Step 7.   Once the cross-port leakage path has been isolated, repair or replace the defective component, and install it on its original sub-plate. Tighten the retaining bolts to the recommended torque values.

NOTE: *Before removing the valve check to see if excessive leakage is caused by the spool either not centering or binding.*

The spool may not be centering because there is an accumulation of contamination between either the spool and bore, or the end-caps.

Spool binding could also be caused by valve body mis-alignment due to sub-standard mounting procedures, over-tightened mounting bolts, or tapered pipe connectors.

4.   Diagnostic Observation: Closed-center, or open-center, directional control valve with a load-check or counter-balance valve circuit module: The pressure indicated on pressure gauge (2) holds steady at the value of the system's main pressure relief valve or compensator setting.

There may be some minor leakage indicated by a very gradual pressure loss.

Diagnosis: Leakage across the "B" port passage appears to be within design specification.

5.   **Diagnostic Observation:** Closed-center, or open-center, directional control valve with a load-check or counter-balance valve circuit module: The pressure indicated on pressure gauge (2) attempts to increase when the MicroLeak analyzer is stroked. However, between pumping strokes the pressure drops rapidly.

If the "P," "T," and "A" ports are open, a steady stream of oil can be seen pouring from the valve.

The MicroLeak analyzer fails to pressurize the "B" port passage to the value of the system's main pressure relief valve setting.

Diagnosis: Leakage across the "B" port passage appears to be excessive. Although this amount of leakage may appear to be insignificant, it could cause actuator drifting and/or a marked increase in the operating temperature of the fluid.

If it is not possible to determine if cross-port leakage is excessive, conduct an actuator leakage path analysis test procedure. If actuator leakage is within specification, replace the directional control valve.

NOTE: *One or more circuit modules may be mounted in parallel with the directional control valve and the sub-plate. The leakage path could be caused by any one of the circuit modules. Use the following steps to determine the root-cause:*

Step 1.   Remove the bolts fastening the sandwich circuit module(s) and directional control valve to the test sub-plate.

Step 2.   Remove the sandwich circuit module closest to the test sub-plate.

Step 3.   Install the remaining module(s) and directional control valve on the test sub-plate. Tighten the retaining bolts to the recommended torque values.

NOTE: *Use a shorter set of retaining bolts to re-install the valve assembly.*

Step 4.   Once again, attempt to pressurize the "B" port passage with MicroLeak analyzer (1). Stop when the value of the system's main pressure relief valve or compensator setting is reached.

NOTE: *If it is now possible to temporarily pressurize the "B" port passage, the cross-port leakage path is isolated to the first sandwich circuit module.*

Step 5.   If it is still not possible to pressurize the "B" port passage, continue removing sandwich circuit modules one at a time until the cross-port leakage path has been isolated.

Step 6.   Repeat this test procedure in the "A" port passage.

Step 7.   Once the cross-port leakage path has been isolated, repair or replace the defective component, and install it on its original sub-plate. Tighten the retaining bolts to the recommended torque values.

If necessary, refer to the manufacturer's specifications for recommended leakage data.

6.   Diagnostic Observation: The MicroLeak analyzer fails to pressurize the "B" port passage. Oil leaks profusely from the open ports in the valve.

Diagnosis: Leakage across the "B" port passage is excessive.

NOTE: *One or more circuit modules may be mounted in parallel with the directional control valve and the sub-plate. The leakage path could be caused by any one of the circuit modules. Use the following steps to determine the root-cause:*

Step 1.   Remove the bolts fastening the sandwich circuit module(s) and directional control valve to the test sub-plate.

Step 2.   Remove the sandwich circuit module closest to the test sub-plate.

Step 3.   Install the remaining module(s) and directional control valve on the test sub-plate. Tighten the retaining bolts to the recommended torque values.

NOTE: *Use a shorter set of retaining bolts to re-install the valve assembly.*

Step 4.   Once again, attempt to pressurize the "B" port passage with MicroLeak analyzer (1). Stop when the value of the system's main pressure relief valve is reached.

NOTE: *If it is now possible to temporarily pressurize the "B" port passage, the cross-port leakage path is isolated to the first sandwich circuit module.*

Step 5.   If it is still not possible to pressurize the "B" port passage, continue removing sandwich circuit modules one at a time until the cross-port leakage path has been isolated.

Step 6.   Repeat this test procedure in the "A" port passage.

Step 7.   Once the cross-port leakage path has been isolated, repair or replace the defective component, and install it on its original sub-plate. Tighten the retaining bolts to the recommended torque values.

Internal Pilot Pressure
Test Procedure

What will this test procedure accomplish?

A solenoid-controlled, pilot-operated, directional control valve is made up of a smaller, direct-operated valve (pilot valve) mounted on top of a larger valve (main valve). The spool of the pilot valve is shifted with a solenoid, while the spool of the main valve is shifted with pilot pressure.

There are two internal oil passages in the main valve; the pilot pressure supply passage and the pilot drain passage. Pilot pressure can be supplied internally or externally.

When internally piloted, the pilot pressure passage is connected, in parallel, from the main valve inlet port "P" through the inlet pilot passage to the "P" port of the pilot valve (Figure 5-7).

When the pilot valve solenoid is energized, the pilot spool shifts, directing pressurized fluid from the "P" port to the "A" or "B" ports which are connected through internal passages to both ends of the main valve spool. When pressurized fluid enters the end-caps of the main spool it forces the spool to shift, directing fluid from the "P" port of the main valve to the "A" or "B" actuator ports.

Generally, a minimum pressure difference of 70 PSI (4.8 bar) is required across the "P" and "T" ports of the main valve to shift the main spool. Any surges in the reservoir return-line could cause the pressure on both sides of the main spool to equalize. This would cause the spring-centered main spool to momentarily move to the neutral position resulting in erratic actuator operation.

The most common symptoms of problems associated with pilot pressure related problems are:
    a. Directional control valve does not respond when solenoid is activated.
    b. Erratic actuator operation.

This test procedure will determine:
    a. Pilot pressure on new system start-up.
    b. If erratic actuator operation is caused by an abnormal variation in pilot pressure.
    c. If the directional control valve is correctly configured for internal pilot operation.

-WARNING- | **Do not work on or around hydraulic systems without wearing safety glasses which conform to ANSI Z87.1-1989 standard.**

YOU WILL NEED THE FOLLOWING DIAGNOSTIC EQUIPMENT
TO CONDUCT THIS TEST PROCEDURE

CAUTION! *The pressure and flow ratings of the diagnostic equipment which will be used to conduct this test procedure must be equal to, or greater than, the pressure and flow ratings of the system being tested. Refer to the system schematic for recommended pressure and flow data.*

---

1. Pressure gauge
2. Pressure gauge
3. Pyrometer

PREPARATION

To conduct this test safely and accurately, refer to Figure 5-7 for correct placement of diagnostic equipment.
To record the test data, make a copy of the test worksheet on page 5-46 (Figure 5-8).

TEST PROCEDURE

Step 1.   Shut the prime mover off.

Step 2.   Lock the electrical system out or tag the keylock switch.

Step 3.   Observe the system pressure gauge. Release any residual pressure trapped in the system by an accumulator, counterbalance or pilot-operated check valve, suspended load on an actuator, intensifier, or a pressurized reservoir.

Step 4.   Install pressure gauge (1) in series with the pilot pressure test port ("X") in the valve subplate. An alternative port is provided in the main valve housing. Refer to the manufacturer's specification for guidance.

Step 5.   Install pressure gauge (2) in series with the drain pressure test port ("Y") in the sub-plate. An alternative port is provided in the main valve housing. Refer to the manufacturer's specification for guidance.

Step 6.   Start the prime mover. Inspect the diagnostic equipment connectors for leaks.

Step 7.   Allow the system to warm up to approximately 130° F. (54° C.). Observe pyrometer (3).

Step 8.   With the directional control valve in the neutral position, record on the test worksheet:
a. The pressure indicated on pressure gauge (1).
b. The pressure indicated on pressure gauge (2).

NOTE: *If the pressure indicated on gauge (1) is not at least 70 PSI (4.8 bar) higher than the pressure indicated on gauge (2), refer to "ANALYZING THE TEST RESULTS" at the end of this test.*

1. Pressure gauge
2. Pressure gauge
3. Pyrometer

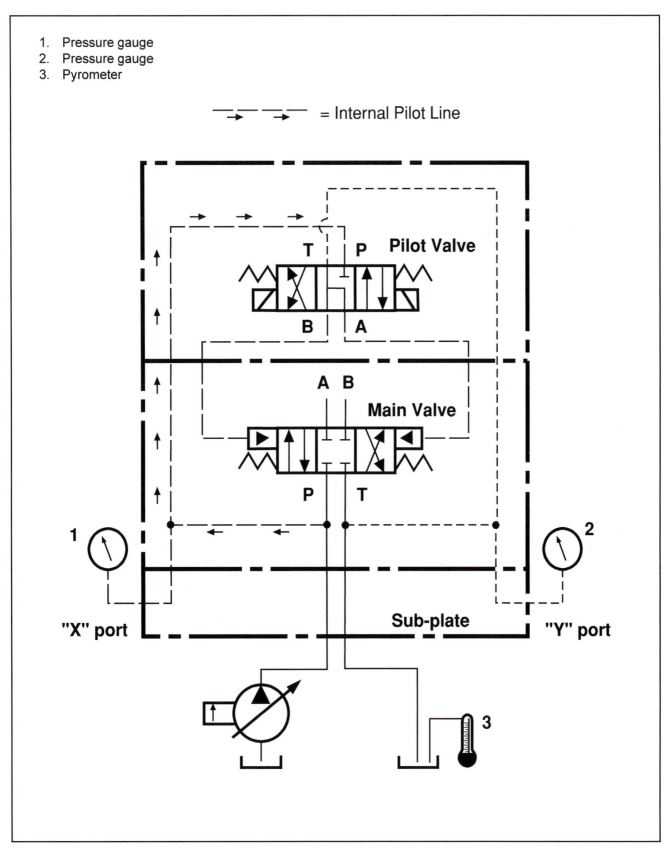

Figure 5-7   Placement of diagnostic equipment

Step 9.   Operate the machine through a number of normal load cycles. While the machine is in operation, record on the test worksheet:
    a. The highest pressure indicated on pressure gauge (1).
    b. The highest pressure indicated on pressure gauge (2).

Step 10.   Shut the prime mover off and analyze the test results.

Step 11.   At the conclusion of this test procedure, remove the diagnostic equipment. Reconnect the transmission lines and tighten the connectors securely.

Step 12.   Start the prime mover and inspect the connectors for leaks.

ANALYZING THE TEST RESULTS

1.   **Diagnostic Observation:** The pressures indicated on gauges (1) and (2) are equal or almost equal.

    Diagnosis: If the problem coincides with a new system start-up, the valve is either incorrectly configured, or lacks a back-pressure check valve.

    Refer to chapter six for directional control valve conversion procedures. Back-pressure check valve test procedures can be found on page 5-63 in this chapter.

2.   **Diagnostic Observation:** The pressure indicated on pressure gauge (1) is not at least 70 PSI (4.8 bar) higher than the pressure indicated on pressure gauge (2).

    Diagnosis: If an open-center or tandem-center valve is used, the pressure difference across the "P" and "T" ports may be insufficient to generate pilot pressure (at least approximately 70 PSI (4.8 bar)) to shift the main spool.

    Normally, an internal or external back-pressure check valve is used to generate pilot pressure. It is usually installed in the "P" port of the directional control valve. However, it is sometimes installed in series with the tank return transmission line.

    If it is installed in series with the tank return transmission line, the directional control valve must be configured for external drain.

    If the directional control valve is correctly configured for internal pilot, and the problem coincides with a new system start-up, check if the internal back-pressure check valve is correctly installed. Refer to page 5-63 for back-pressure check valve test procedures.

3.   **Diagnostic Observation:** Pressure in the internal drain passage indicated on pressure gauge (2) momentarily equals the pressure in the internal pilot passage indicated on pressure gauge (1).

Diagnosis: If the pressure in the internal drain passage momentarily equals the pressure in the pilot pressure passage, it will cause the spring centered main spool to momentarily shift to the center (neutral) position. This will cause erratic actuator operation.

Since the internal drain passage of the pilot valve is in parallel with the directional control valve's tank return-line, pressure surges in the tank return-line will directly affect the pressure in the internal drain passage.

Return-line pressure surges can be caused by:
      a. Flow intensification in linear actuator applications.
      b. Load induced actuator overspeed conditions.
      c. Return-line filters and/or heat exchangers
      d. Return-line manifolds.

If return-line pressure surges, which appear to be an operational characteristic of the system, cause the main spool to operate erratically, it may be necessary to convert the directional control valve to external drain configuration.

Refer to chapter six for external drain conversion procedures.

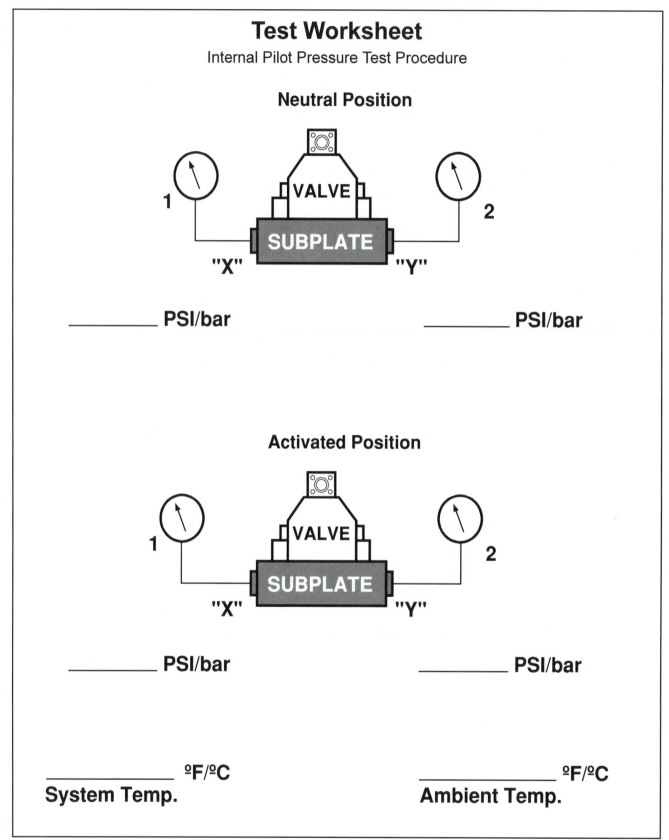

Figure 5-8   Test Worksheet - Internal pilot pressure test procedure

External Pilot Pressure
Test Procedure

What will this test procedure accomplish?

A solenoid-controlled, pilot-operated, directional control valve is made up of a smaller, direct-operated valve (pilot valve) mounted on top of a larger valve (main valve). The spool of the pilot valve is shifted with a solenoid, while the spool of the main valve is shifted with pilot pressure.

There are two internal oil passages in the main valve; the pilot pressure supply passage, and the pilot drain passage. Pilot pressure can be supplied internally or externally.

When externally piloted, the internal pilot pressure passage is blocked off from the main valves inlet port "P" (Figure 5-9). An independent transmission line, which is connected to a port in the main valve housing, is used to provide pilot pressure to the main valve from an external pilot pressure source.

When the pilot valve solenoid is energized, the pilot spool shifts, directing external pilot pressure to the pilot valve's "A" or "B" ports which are connected through internal passages to both ends of the main valve spool. When pressurized fluid enters the end-caps of the main spool, it forces the spool to shift, directing pump flow to the "A" or "B" actuator ports.

Generally, a minimum pressure difference of 70 PSI (4.8 bar) is required across the "P" and "T" ports of the main valve to allow the main spool to shift. Any surges in the tank return-line could cause the pressure on both sides of the main spool to equalize. This would cause the spring-centered main spool to momentarily move to the neutral position resulting in erratic actuator operation.

The most common symptoms of problems associated with pilot pressure related problems are:
    a. Main directional control valve does not respond when the pilot valve is activated.
    b. Erratic actuator operation.

This test procedure will determine:
    a. Pilot pressure on new system start-up.
    b. If erratic actuator operation is caused by an abnormal variation in pilot pressure.
    c. If the directional control valve is correctly configured for external pilot operation.

-WARNING- | **Do not work on or around hydraulic systems without wearing safety glasses which conform to ANSI Z87.1-1989 standard.**

YOU WILL NEED THE FOLLOWING DIAGNOSTIC EQUIPMENT TO CONDUCT THIS TEST PROCEDURE

CAUTION! *The pressure and flow ratings of the diagnostic equipment which will be used to conduct this test procedure must be equal to, or greater than, the pressure and flow ratings of the system being tested. Refer to the system schematic for recommended pressure and flow data.*

1.   Pressure gauge   2.   Pyrometer

PREPARATION

To conduct this test safely and accurately, refer to Figure 5-9 for correct placement of diagnostic equipment.

TEST PROCEDURE

Step 1.   Shut the prime mover off.

Step 2.   Lock the electrical system out or tag the keylock switch.

Step 3.   Observe the system pressure gauge. Release any residual pressure trapped in the system by an accumulator, counterbalance or pilot-operated check valve, suspended load on an actuator, intensifier, or a pressurized reservoir.

Step 4.   Install pressure gauge (1) in parallel with the connector at the external pilot pressure port located on the side of the directional control valve body.

Step 5.   Start the prime mover. Inspect the diagnostic equipment connector(s) for leaks.

Step 6.   Allow the system to warm up to approximately 130º F. (54º C.). Observe pyrometer (2).

Step 7.   Record the pressure indicated on pressure gauge (1).

Step 8.   At the conclusion of this test procedure, remove the diagnostic equipment. Reconnect the transmission lines and tighten the connectors securely.

Step 9.   Start the prime mover and inspect the connectors for leaks.

1. Pressure gauge
2. Pyrometer

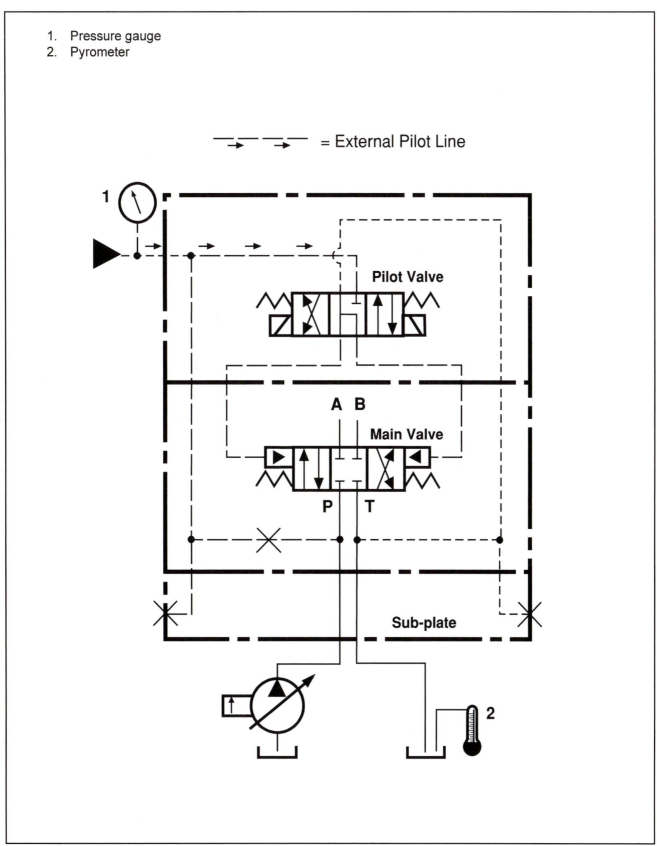

Figure 5-9   Placement of diagnostic equipment

ANALYZING THE TEST RESULTS

1. **Diagnostic Observation:** The pressure in the in the external pilot pressure port is less than 70 PSI (4.8 bar).

   Diagnosis: Low pilot pressure can be caused by a number of problems:

   a. Low external pilot pressure source-- Check the pressure source from which the external pilot pressure is being generated.

   b. Pressure reducing valve incorrectly adjusted, or malfunction -- The pilot pressure may be supplied from the secondary port of a pressure reducing valve. Refer to chapter four for pressure reducing valve test procedures.

   c. Directional control valve is not configured for external pilot operation-- Refer to chapter six for directional control valve conversion procedures.

   d. Incorrect pilot valve-- The pilot valve which controls the main valve spool has the incorrect center configuration. The main spool may operate. However, pilot pressure will be low when the pilot valve is in the neutral position.

   The most common spool configuration for a pilot valve is "float center."

   In the float center position, the "P" port is plugged while the "A," "B," and "T" ports are connected in parallel.

   This configuration is necessary to allow the spring-centered main valve to find its neutral (center) position.

Internal Drain Back-Pressure Test Procedure

What will this test procedure accomplish?

A solenoid-controlled, pilot-operated, directional control valve is made up of a smaller, direct-operated valve (pilot valve) mounted on top of a larger valve (main valve). The spool of the pilot valve is shifted with a solenoid, while the spool of the main valve is shifted with pilot pressure.

There are two internal oil passages in the main valve; the pilot pressure supply passage, and the pilot drain passage. The pilot drain passage drains the opposite spool end-cap when the main spool is shifted.

The pilot can be internally or externally drained, When internally drained, the drain passage is connected from the "T" port of the pilot valve, through an internal passage, to the "T" port of the main valve (Figure 5-10). It connects, in parallel, with the oil returning from the actuator.

Generally, a minimum pressure difference of 70 PSI (4.8 bar) is required across the "P" and "T" ports of the main valve to allow the main spool to shift. Any surges in the reservoir return-line could cause the pressure on both sides of the main spool to equalize. This could cause the spring-centered main spool to momentarily move to the neutral position resulting in erratic actuator operation.

The most common symptoms of problems associated with excessive drain-line back-pressure are:
    a. Main directional control valve does not respond when the pilot valve is actuated.
    b. Erratic actuator operation.

This test procedure will determine:
    a. Internal drain pressure on new system start-up.
    b. If erratic actuator operation is caused by return-line pressure surges.
    c. If the directional control valve is correctly configured for internal drain operation.

 **-WARNING-**   Do not work on or around hydraulic systems without wearing safety glasses which conform to ANSI Z87.1-1989 standard.

YOU WILL NEED THE FOLLOWING DIAGNOSTIC EQUIPMENT TO CONDUCT THIS TEST PROCEDURE

CAUTION! *The pressure and flow ratings of the diagnostic equipment which will be used to conduct this test procedure must be equal to, or greater than, the pressure and flow ratings of the system being tested. Refer to the system schematic for recommended pressure and flow data.*

---

1. Pressure gauge
2. Pressure gauge
3. Pyrometer

PREPARATION

To conduct this test safely and accurately, refer to Figure 5-10 for correct placement of diagnostic equipment.

To record the test data, make a copy of the test worksheet on page 5-56 (Figure 5-11).

TEST PROCEDURE

Step 1.   Shut the prime mover off.

Step 2.   Lock the electrical system out or tag the keylock switch.

Step 3.   Observe the system pressure gauge. Release any residual pressure trapped in the system by an accumulator, counterbalance or pilot-operated check valve, suspended load on an actuator, intensifier, or a pressurized reservoir.

Step 4.   Remove the plug from the pilot pressure test port located on the side of the main valve body. The test port is usually marked with an "X" or "PP" signature. If necessary, refer to the manufacturer's service literature for guidance.

Step 5.   Install pressure gauge (1) in series with the pilot pressure test port.

Step 6.   Remove the plug from the pilot drain test port located on the side of the main valve body. The test port is usually marked with a "Y" or "DR" signature. If necessary, refer to the manufacturer's service literature for guidance.

Step 7.   Install pressure gauge (2) in series with the pilot drain test port.

Step 8.   Start the prime mover. Inspect the diagnostic equipment connectors for leaks.

Step 9.   Allow the system to warm up to approximately 130º F. (54º C.). Observe pyrometer (3).

1. Pressure gauge
2. Pressure gauge
3. Pyrometer

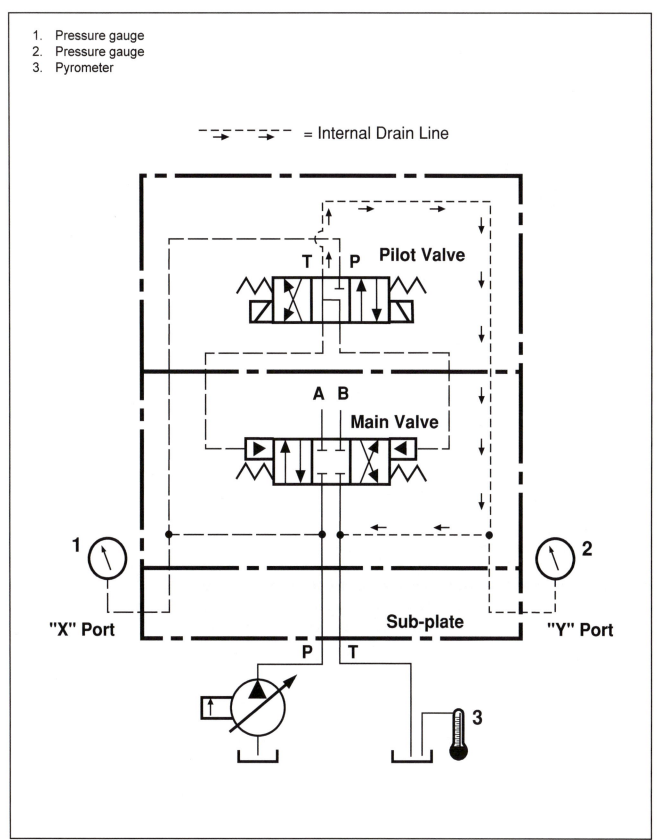

Figure 5-10   Placement of diagnostic equipment

Step 10. With the directional control valve in the neutral position record on the test worksheet:
      a. The pressure indicated on "X" port pressure gauge (1).
      b. The pressure indicated on "Y" port pressure gauge (2).

Step 11. Activate the directional control valve and operate the machine through a normal load cycle.

NOTE: *In a system using a linear actuator (cylinder), it is imperative that the highest pressure be recorded with the cylinder rod travelling in each direction (extend and retract).*

High drain-line resistance is sometimes caused by flow intensification - a characteristic of single-rod linear actuators. Flow intensification is caused by the area difference across the piston (rod-to-bore ratio).

The larger the area difference, the higher the flow rate relative to pump flow, in the rod retract position.

Flow intensification can also be caused by the tendency of a heavy load to accelerate the output shaft of an actuator. This will cause a proportionate increase in flow and higher return-line resistance.

Step 12. Record on the test worksheet the highest pressure indicated on:
      a. "X" port pressure gauge (1).
      b. "Y" port pressure gauge (2).

Step 13. Shut the prime mover off and analyze the test results.

Step 14. At the conclusion of this test procedure, remove the diagnostic equipment. Reconnect the transmission lines and tighten the connectors securely.

Step 15. Start the prime mover and inspect the connectors for leaks.

ANALYZING THE TEST RESULTS

1. **Diagnostic Observation:** The pressure in the internal drain passage indicated on pressure gauge (2), is equal to the pressure in the internal pilot passage indicated on pressure gauge (1).

Diagnosis: There appears to be excessive resistance in the directional control valve's main return-line.

Generally, there must be a pressure difference of at least 70 PSI (4.8 bar) across the "P" and "T" ports of the directional control valve to generate sufficient pilot pressure to shift the main valve spool.

Install pressure gauge (2) in the parallel with the next downstream connector in the return-line and repeat the test procedure.

Continue moving the pressure gauge through the main return-line until the root-cause has been determined.

If it is not possible to reduce the return-line resistance to a level which will allow the main valve spool to operate, convert the valve to an "external drain" configuration.

Refer to chapter six for directional control valve conversion procedures.

2. **Diagnostic Observation:** Pressure in the internal drain passage indicated on pressure gauge (2) momentarily equals the pressure in the internal pilot passage indicated on pressure gauge (1).

    Diagnosis: If the pressure in the internal drain passage momentarily equals the pressure in the pilot pressure passage, it will cause the spring centered main spool to momentarily shift to the center (neutral) position. This will cause erratic actuator operation.

    Since the internal drain passage of the pilot valve is in parallel with the directional control valve's tank return-line, pressure surges in the tank return-line will directly affect the pressure in the internal drain passage.

    Return-line pressure surges can be caused by:
    a. Flow intensification in linear actuator applications.
    b. Load induced actuator overspeed conditions.
    c. Return-line filters and/or heat exchangers
    d. Return-line manifolds.

    If return-line pressure surges, which appear to be an operational characteristic of the system, cause the main spool to operate erratically, it may be necessary to convert the directional control valve to external drain configuration.

    Refer to chapter six for external drain conversion procedures.

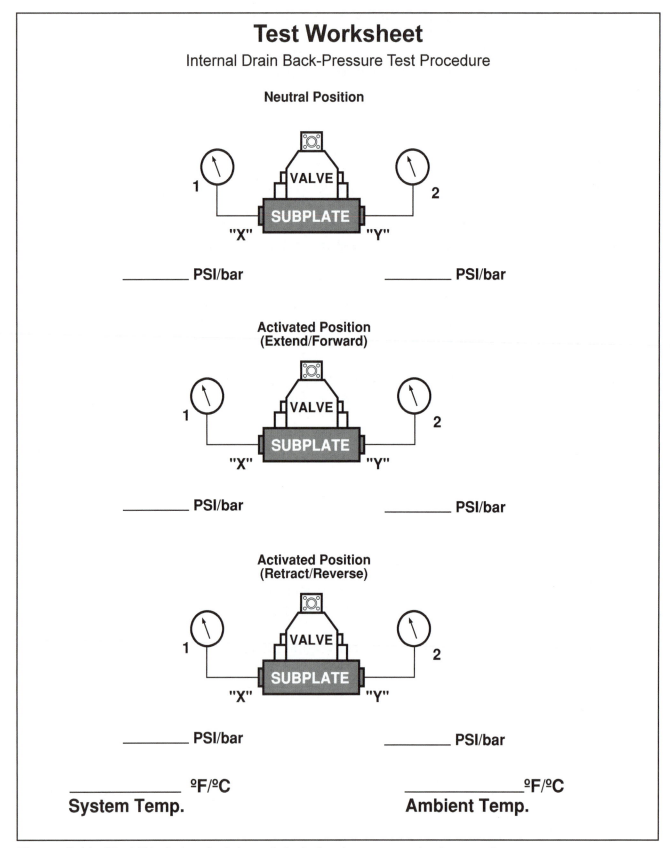

Figure 5-11   Test Worksheet - Internal drain back-pressure test procedure

External Drain Back-Pressure Test Procedure

What will this test procedure accomplish?

A solenoid-controlled, pilot-operated, directional control valve is made up of a smaller, direct-operated valve (pilot valve) mounted on top of a larger valve (main valve). The spool of the pilot valve is shifted with a solenoid, while the spool of the main valve is shifted with pilot pressure.

There are two internal oil passages in the main valve; the pilot pressure supply passage, and the pilot drain passage. The pilot drain passage drains the opposite spool end-cap when the main spool is shifted.

The pilot valve can be internally or externally drained. When externally drained, the internal drain passage is blocked. A plug, located in the main valve housing (port "Y"), is removed providing external access to the pilot valve drain (Figure 5-12). An independent transmission line, dedicated to draining the pilot valve, is connected from the valve sub-plate directly to the reservoir.

This configuration provides the advantage of having the pilot valve drain totally immune to main return-line back-pressure or pressure surges.

The most common symptoms of problems associated with excessive drain-line back-pressure are:
      a. Main directional control valve does not respond when the pilot valve is activated.
      b. Erratic actuator operation.

This test procedure will determine:
      a. External drain pressure on a new system start-up.
      b. If the directional control valve is correctly configured for external drain operation.
      c. If there are pressure surges in the external drain-line.
      d. If there is a restriction in the external drain-line.

 **-WARNING-** | Do not work on or around hydraulic systems without wearing safety glasses which conform to ANSI Z87.1-1989 standard.

YOU WILL NEED THE FOLLOWING DIAGNOSTIC EQUIPMENT TO CONDUCT THIS TEST PROCEDURE

CAUTION! *The pressure and flow ratings of the diagnostic equipment which will be used to conduct this test procedure must be equal to, or greater than, the pressure and flow ratings of the system being tested. Refer to the system schematic for recommended pressure and flow data.*

---

1. Pressure gauge
2. Pressure gauge
3. Pyrometer

PREPARATION

To conduct this test safely and accurately, refer to Figure 5-12 for correct placement of diagnostic equipment.

To record the test data, make a copy of the test worksheet on page 5-62 (Figure 5-13).

TEST PROCEDURE

Step 1. Shut the prime mover off.

Step 2. Lock the electrical system out or tag the keylock switch.

Step 3. Observe the system pressure gauge. Release any residual pressure trapped in the system by an accumulator, counterbalance or pilot-operated check valve, suspended load on an actuator, intensifier, or a pressurized reservoir.

Step 4. Install pressure gauge (2) in parallel with the connector at the external drain port in the main valve housing.

Step 5. Remove the plug from the pilot pressure test port located on the side of the main valve body. The test port is usually marked with an "X" or "PP" signature. If necessary, refer to the manufacturer's service literature for guidance.

Step 6. Install pressure gauge (1) in series with the pilot pressure test port in the main valve housing.

NOTE: *If the valve is externally piloted, install pressure gauge (1) in parallel with the connector at the inlet of the pilot pressure test port in the main valve housing.*

Step 7. Start the prime mover. Inspect the diagnostic equipment connectors for leaks.

Step 8. Allow the system to warm up to approximately 130° F. (54° C.). Observe pyrometer (3).

1. Pressure gauge
2. Pressure gauge
3. Pyrometer

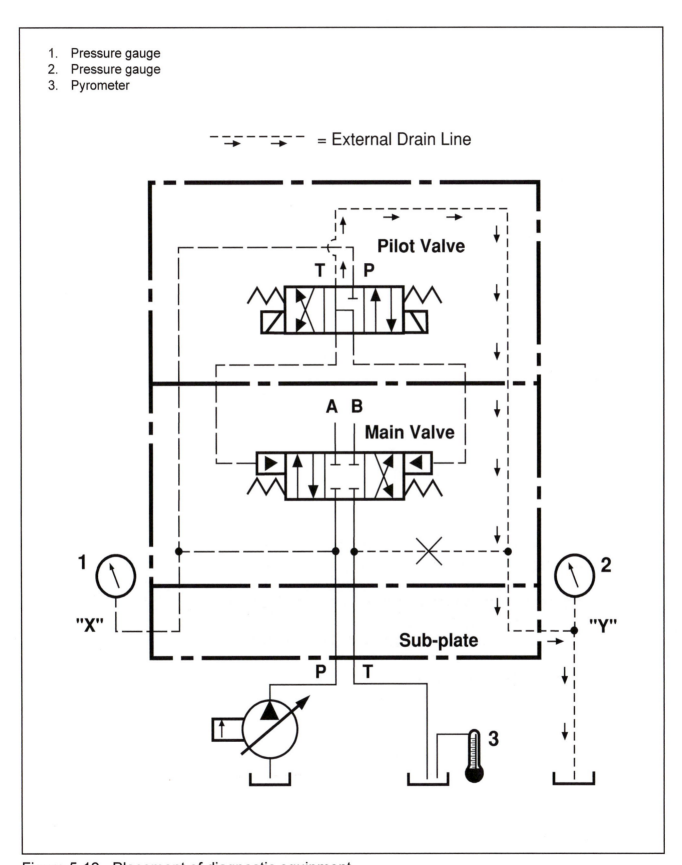

Figure 5-12   Placement of diagnostic equipment

Step 9.   With the directional control valve in the neutral position, record on the test worksheet:
   a. The pressure indicated on "X" port pressure gauge (1).
   b. The pressure indicated on "Y" port pressure gauge (2).

Step 10. Activate the directional control valve and operate the machine through a normal load cycle.

NOTE: *In a system using a linear actuator (cylinder), it is imperative that the highest pressure be recorded with the cylinder rod travelling in each direction (extend and retract).*

High drain-line resistance is sometimes caused by flow intensification-- a characteristic of single-rod linear actuators. Flow intensification is caused by the area difference across the piston (rod-to-bore ratio).

The larger the area difference, the higher the flow rate relative to pump flow, in the rod retract position.

Flow intensification can also be caused by the tendency of a heavy load to accelerate the output shaft of an actuator. This will cause a proportionate increase in flow and higher return-line resistance.

Step 11. Record on the test worksheet the highest pressure indicated on:
   a. "X" port pressure gauge (1).
   b. "Y" port pressure gauge (2).

Step 12. Shut the prime mover off and analyze the test results.

Step 13. At the conclusion of this test procedure, remove the diagnostic equipment. Reconnect the transmission lines and tighten the connectors securely.

Step 14. Start the prime mover and inspect the connectors for leaks.

ANALYZING THE TEST RESULTS

1.   **Diagnostic Observation:** The pressure in the internal drain passage indicated on pressure gauge (2), is equal to the pressure in the internal pilot passage indicated on pressure gauge (1).

Diagnosis: There appears to be excessive resistance in the directional control valve's external drain-line.

Generally, there must be a pressure difference of at least 70 PSI (4.8 bar) across the "P" and "T" ports of the directional control valve to generate sufficient pilot pressure to shift the main valve spool.

Install pressure gauge (2) in parallel with the next downstream connector in the return-line and repeat the test procedure.

Continue moving the pressure gauge through the main return-line until the root-cause has been determined.

2.   **Diagnostic Observation:** Pressure in the internal drain passage indicated on pressure gauge (2) momentarily equals the pressure in the internal pilot passage indicated on pressure gauge (1).

Diagnosis: If the pressure in the internal drain passage momentarily equals the pressure in the pilot pressure passage it will cause the spring-centered main spool to momentarily shift to the center (neutral) position. This will cause erratic actuator operation.

Since the internal drain passage of the pilot valve is in parallel with the directional control valve's main return-line, pressure surges in the return-line will directly affect the pressure in the internal drain passage.

Return-line pressure surges can be caused by:
   a.   Flow intensification in linear actuator applications.
   b.   Load induced actuator overspeed conditions.
   c.   Return-line filters and/or heat exchangers.
   d.   Return-line manifolds.

If return-line pressure surges, which appear to be an operational characteristic of the system, cause the main spool to operate erratically, it may be necessary to connect the external drain-line directly to the reservoir in parallel with the main return-line. This will prevent return-line components from influencing the pressure in the pilot valve's external drain-line.

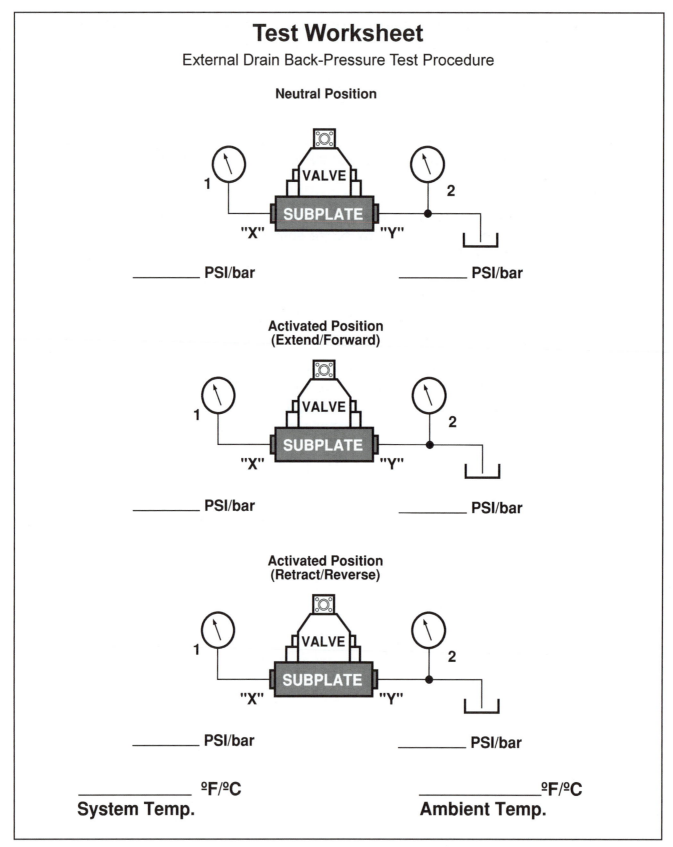

Figure 5-13   Test Worksheet-- External drain back-pressure test procedure

Back-Pressure Check Valve (Internal) Test Procedure

What will this test procedure accomplish?

When a pilot-operated directional control valve, with open-center or tandem-center spool configuration, is specified for a particular application it is important to consider pilot pressure source.

Generally, a minimum pilot pressure of 70 PSI (4.8 bar) is required to shift the main spool. It is usually not possible to rely on the pressure difference across the main spool ("P" and "T" ports) to provide sufficient pilot pressure.

A back-pressure check valve "option" is available to provide the minimum pilot pressure necessary to shift the main spool.

A back-pressure check valve is basically a special cartridge-type check valve. It consists of a cylindrical body with an inlet and outlet port. Inside the body is an active poppet which is biased toward the inlet port by a 70 PSI (4.8 bar) mechanical spring force.

It is usually installed inside the main valve body in series with the "P" port. When oil flows through the main valve body, the back-pressure check valve generates a resistance equal to the value of the bias spring-- 70 PSI (4.8 bar).

Since the internal pilot passage is located in parallel with the inlet port of the back-pressure check valve there is always a minimum of 70 PSI (4.8 bar) at the "P" port of the pilot valve (Figure 5-14).

The most common symptoms of problems associated with back-pressure check valves are:
    a. Actuator fails to respond when the directional control valve is activated.
    b. Erratic actuator operation.
    c. Actuator responds very slowly when the directional control valve is activated.

The most common causes of back-pressure check valve related problems are:
    a. Not installed during valve installation.
    b. Bias-spring fatigue.

This test procedure will determine:
    a. If the back-pressure check valve is operating at the correct pressure.
    b. If the directional control valve is correctly configured for external back-pressure application.
    c. If there are surges in the external drain line affecting the operation of the back-pressure check valve.

 **-WARNING-** Do not work on or around hydraulic systems without wearing safety glasses which conform to ANSI Z87.1-1989 standard.

YOU WILL NEED THE FOLLOWING DIAGNOSTIC EQUIPMENT TO CONDUCT THIS TEST PROCEDURE

CAUTION! *The pressure and flow ratings of the diagnostic equipment which will be used to conduct this test procedure must be equal to, or greater than, the pressure and flow ratings of the system being tested. Refer to the system schematic for recommended pressure and flow data.*

1. Pressure gauge   2. Pressure gauge   3. Pyrometer

PREPARATION

To conduct this test safely and accurately, refer to Figure 5-14 for correct placement of diagnostic equipment.

To record the test data, make a copy of the test worksheet on page 5-72 (Figure 5-15).

TEST PROCEDURE

Step 1.   Shut the prime mover off.

Step 2.   Lock the electrical system out or tag the keylock switch.

Step 3.   Observe the system pressure gauge. Release any residual pressure trapped in the system by an accumulator, counterbalance or pilot-operated check valve, suspended load on an actuator, intensifier, or a pressurized reservoir.

Step 4.   Remove the plug from the pilot pressure test port located on the side of the main valve body. The test port is usually marked with an "X" or "PP" signature. If necessary, refer to the manufacturer's service literature for guidance.

Step 5.   Install pressure gauge (1) in series with the pilot pressure test port.

Step 6.   Remove the plug from the pilot drain test port located on the side of the main valve body. The test port is usually marked with a "Y" or "DR" signature. If necessary, refer to the manufacturer's service literature for guidance.

Step 7.   Install pressure gauge (2) in series with the pilot drain test port.

Step 8.   Start the prime mover. Inspect the diagnostic equipment connectors for leaks.

1. Pressure gauge
2. Pressure gauge
3. Pyrometer

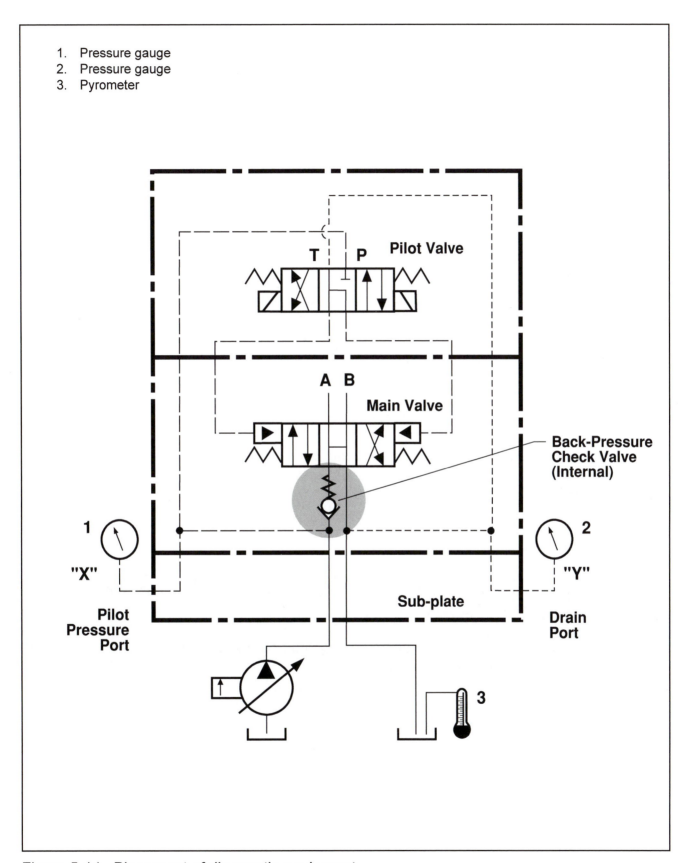

Figure 5-14   Placement of diagnostic equipment

Step 9.   Allow the system to warm up to approximately 130º F. (54º C.). Observe pyrometer (3).

Step 10.   With the directional control valve in the neutral position, record on the test worksheet:
    a. The pressure indicated on "X" port pressure gauge (1).
    b. The pressure indicated on "Y" port pressure gauge (2).

Step 11.   Shut the prime mover off and analyze the test results.

Step 12.   At the conclusion of this test procedure, remove the diagnostic equipment. Reconnect the transmission lines and tighten the connectors securely.

Step 13.   Start the prime mover and inspect the connectors for leaks.

ANALYZING THE TEST RESULTS

1.   **Diagnostic Observation:** The pressure indicated on pressure gauge (1) is approximately 70 PSI (4.8 bar) higher than the pressure indicated on pressure gauge (2).

       Diagnosis: The pilot pressure appears to be within design specification.

2.   **Diagnostic Obervation:** The pressure indicated on pressure gauge (1) is equal to or slightly greater than the pressure indicated on pressure gauge (2). Both pressures are nominal.

       Diagnosis: There are indications that the back-pressure check valve is either not installed, or has malfunctioned.

       There must be at pressure difference of at least 70 PSI (4.8 bar) across the "X" and "Y" test ports for the valve to function.

3.   **Diagnostic Observation:** The pressure indicated on pressure gauge (1) is higher than nominal but somewhat less than 70 PSI (4.8 bar).

       Diagnosis: There are indications that the pressure setting of the back-pressure check valve is incorrect. This could also be caused by back-pressure check valve spring fatigue or damage.

Back-Pressure Check Valve (External)
Test Procedure

What will this test procedure accomplish?

When a pilot-operated directional control valve with open-center or tandem-center spool configuration is specified for a particular application, it is important to consider the pilot pressure source.

Generally, a minimum pilot pressure of 70 PSI (4.8 bar) is required to shift the main spool. It is usually not possible to rely on the pressure difference across the main spool ("P" and "T" ports) to provide sufficient pilot pressure.

Back-pressure can be generated by installing a check valve with a 70 PSI (4.8 bar) bias spring in series with the main return-line at the outlet port of the directional control valve.

This method of generating pilot pressure is frequently used. However, since the internal pilot pressure passage and the external drain passage are in parallel with the inlet port of the check valve, the directional control valve must be converted to an external drain configuration.

Failure to convert the valve will cause pilot pressure to equalize on both sides of the spring-centered main spool. This will cause the spool to remain in the spring-centered, neutral position.

Refer to chapter six for external drain conversion procedures.

The most common symptoms of problems associated with back-pressure check valves are:
    a. Actuator fails to respond when the directional control valve is activated.
    b. Erratic actuator operation.
    c. Actuator responds very slowly when the directional control valve is activated.

This test procedure will determine:
    a. If the back-pressure check valve is operating at the correct pressure.
    b. If the directional control valve is correctly configured for external back-pressure application.
    c. If there are surges in the external drain-line affecting the operation of the valve.

-WARNING- Do not work on or around hydraulic systems without wearing safety glasses which conform to ANSI Z87.1-1989 standard.

YOU WILL NEED THE FOLLOWING DIAGNOSTIC EQUIPMENT TO CONDUCT THIS TEST PROCEDURE

CAUTION! *The pressure and flow ratings of the diagnostic equipment which will be used to conduct this test procedure must be equal to, or greater than, the pressure and flow ratings of the system being tested. Refer to the system schematic for recommended pressure and flow data.*

1. Pressure gauge
2. Pressure gauge

3. Pyrometer

PREPARATION

To conduct this test safely and accurately, refer to Figure 5-16 for correct placement of diagnostic equipment.
To record the test data, make a copy of the test worksheet on page 5-71 (Figure 5-17).

TEST PROCEDURE

Step 1. Shut the prime mover off.

Step 2. Lock the electrical system out or tag the keylock switch.

Step 3. Observe the system pressure gauge. Release any residual pressure trapped in the system by an accumulator, counterbalance or pilot-operated check valve, suspended load on an actuator, intensifier, or a pressurized reservoir.

Step 4. Remove the plug from the pilot pressure test port located on the side of the main valve body. The test port is usually marked with an "X" or "PP" signature. If necessary, refer to the manufacturer's service literature for guidance.

Step 5. Install pressure gauge (1) in series with the pilot pressure test port.

Step 6. Remove the plug from the pilot drain test port located on the side of the main valve body. The test port is usually marked with a "Y" or "DR" signature. If necessary, refer to the manufacturer's service literature for guidance.

Step 7. Install pressure gauge (2) in series with the pilot drain test port.

Step 8. Start the prime mover. Inspect the diagnostic equipment connectors for leaks.

1. Pressure gauge
2. Pressure gauge
3. Pyrometer

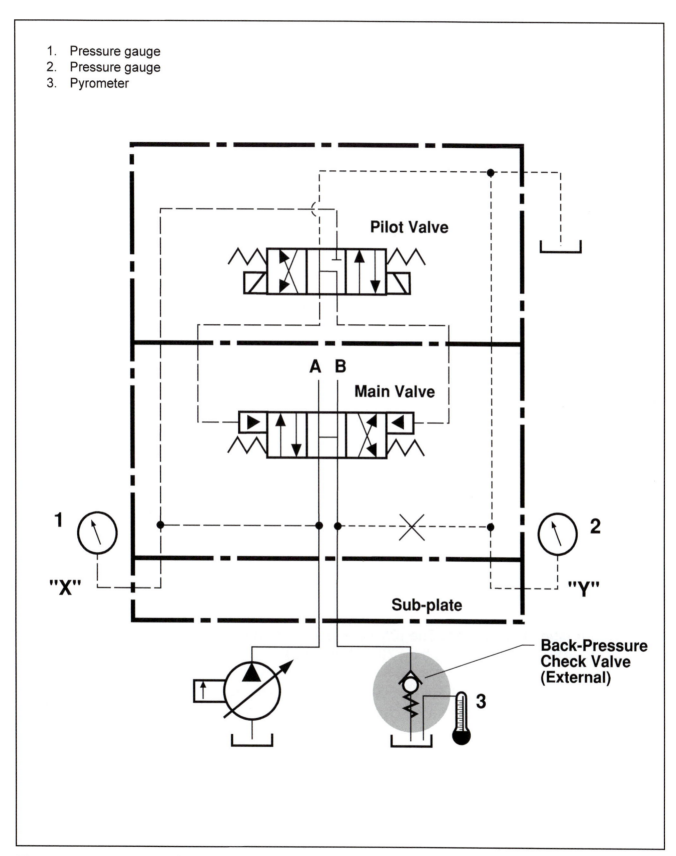

Figure 5-16   Placement of diagnostic equipment

Step 9.   Allow the system to warm up to approximately 130° F. (54° C.). Observe pyrometer (3).

Step 10. With the directional control valve in the neutral position, record on the test worksheet:
        a. The pressure indicated on "X" port pressure gauge (1).
        b. The pressure indicated on "Y" port pressure gauge (2).

Step 11. Shut the prime mover off and analyze the test results.

Step 12. At the conclusion of this test procedure, remove the diagnostic equipment. Reconnect the transmission lines and tighten the connectors securely.

Step 13. Start the prime mover and inspect the connectors for leaks.

ANALYZING THE TEST RESULTS

1.   **Diagnostic Observation:** The pressure indicated on pressure gauge (1) is approximately 70 PSI (4.8 bar) higher than the pressure indicated on pressure gauge (2).

    Diagnosis: The pilot pressure appears to be within design specification.

2.   **Diagnostic Obervation:** The pressure indicated on pressure gauge (1) is equal to or slightly greater than the pressure indicated on pressure gauge (2). Both pressures are nominal.

    Diagnosis: There are indications that the back-pressure check valve is either not installed or has malfunctioned.

    There must be at pressure difference of at least 70 PSI (bar) across the "X" and "Y" test ports for the valve to function.

3.   **Diagnostic Observation:** The pressure indicated on pressure gauge (1) is higher than nominal but somewhat less than 70 PSI (4.8 bar).

    Diagnosis: There are indications that the pressure setting of the back-pressure check valve is incorrect. This could also be caused by back-pressure check valve spring fatigue or damage.

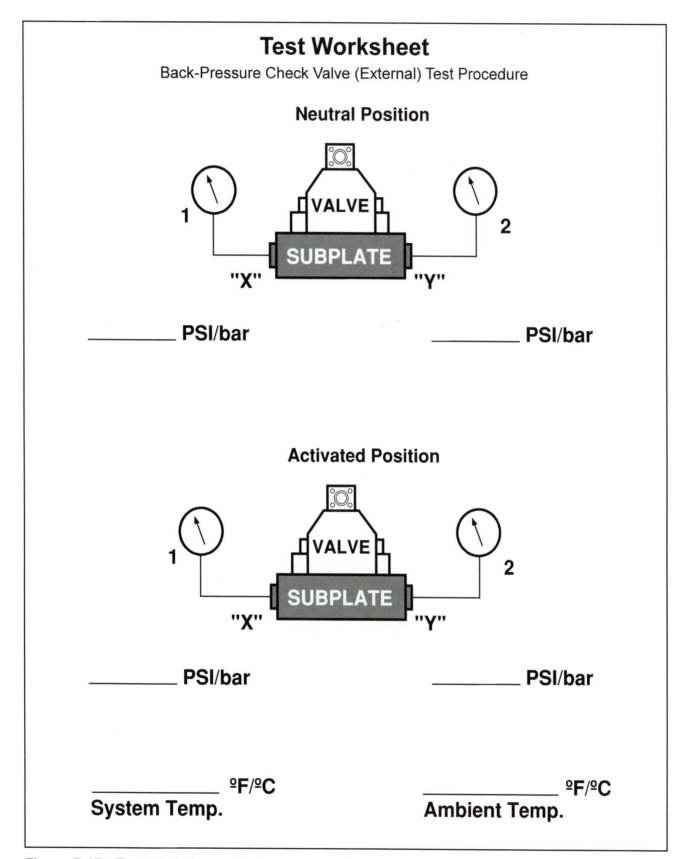

Figure 5-17   Test Worksheet - Back-pressure check valve (external) test procedure

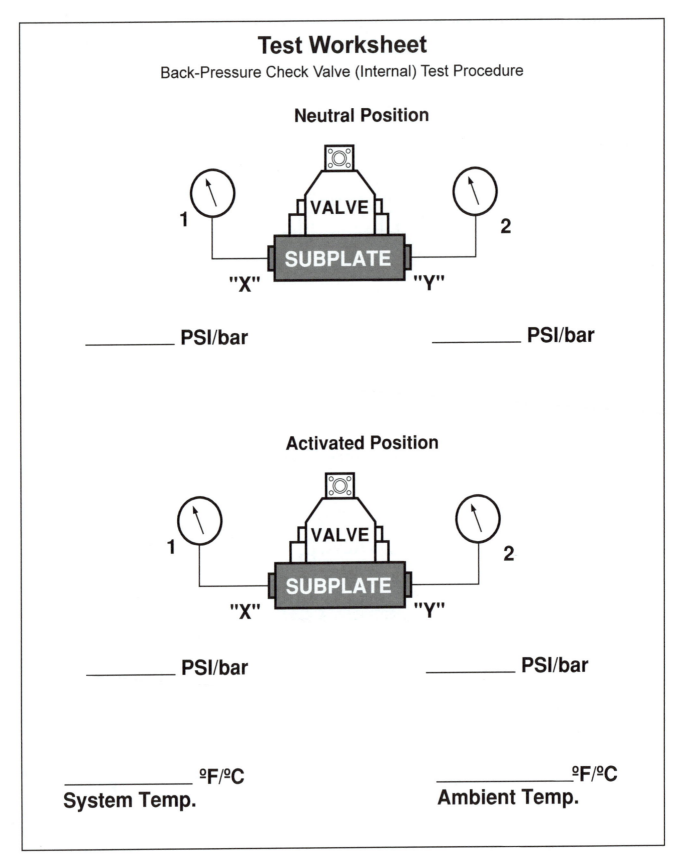

Figure 5-15   Test Worksheet - Back-pressure check valve (internal) test procedure

chapter six

Directional Control Valve Conversion Procedures

Directional Control Valve Internal Pilot Conversion

What will this conversion accomplish?

The pilot valve of a solenoid-controlled, pilot-operated, directional control valve can be internally or externally piloted.

When the valve is internally piloted, oil flows from the inlet port "P" in the main valve, through an internal passage to the "P" port of the pilot valve. The composite drawing (Figure 6-2), illustrates the oil passage. Valve application will determine internal or external pilot configuration.

This conversion will modify a solenoid-controlled, pilot-operated directional control valve to an internal pilot configuration.

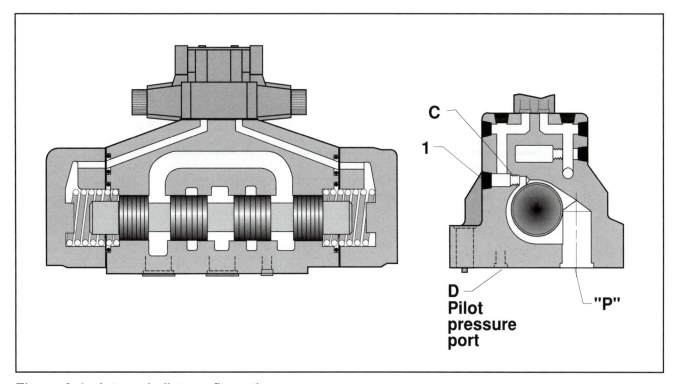

Figure 6-1   Internal pilot configuration

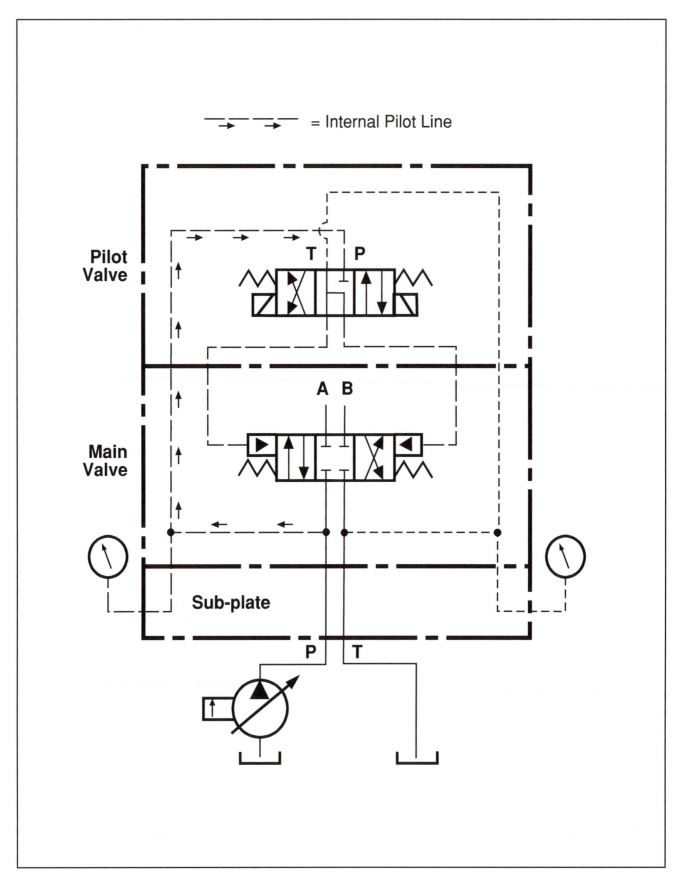

Figure 6-2   Internal pilot oil flow path

 **-WARNING-** | Do not work on or around hydraulic systems without wearing safety glasses which conform to ANSI Z87.1-1989 standard.

CAUTION! If the valve is mounted on a sub-plate in an active system, be sure to follow all recommended safety procedures while performing the conversion procedure.

PREPARATION

Step 1.   Shut the prime mover off.

Step 2.   Lock the electrical system out or tag the keylock switch.

Step 3.   Observe the system pressure gauge. Release any residual pressure trapped in the system by an accumulator, counterbalance or pilot-operated check valve, suspended load on an actuator, intensifier, or a pressurized reservoir.

CONVERSION PROCEDURE

Step 1.   Remove the valve from the sub-plate (if necessary).

Step 2.   Remove plug (1) located in the side of the the main valve body (Figure 6-1).

NOTE: *It may be necessary to review the specific manufacturer's product information to locate the correct plug.*

Step 3.   Remove the 1/8" NPT pipe plug from port "C" (Figure 6-1).

NOTE: *Threaded port "C" must be open (unplugged) for correct internal pilot configuration.*

Step 4.   Replace plug (1) in the main valve body. Tighten it securely.

Step 5.   Lay the valve on its side.

Step 6.   Install the 1/8" NPT pipe plug in pilot pressure port "D" (Figure 6-1).

NOTE: *Threaded pilot pressure port "D" must be closed (plugged) for correct internal pilot configuration.*

Step 7.   Install the valve on the subplate. Tighten the retaining bolts to the recommended torque values.

notes

Directional Control Valve
External Pilot Conversion

What will this conversion accomplish?

The pilot valve of a solenoid-controlled, pilot-operated directional control valve can be internally or externally piloted.

When the valve is externally piloted, the internal pilot passage between the main valve inlet port "P" and the pilot valve inlet port "P" is plugged with a 1/8" NPT pipe plug (C). The pilot valve receives pressure from an external source through a port in the main valve body.

The composite drawing (Figure 6-4) illustrates the external pilot source connected to the main valve body. It also indicates the location of the 1/8" NPT pipe plug blocking the internal pilot passage.

This conversion will modify a solenoid-controlled, pilot-operated directional control valve to an external pilot configuration.

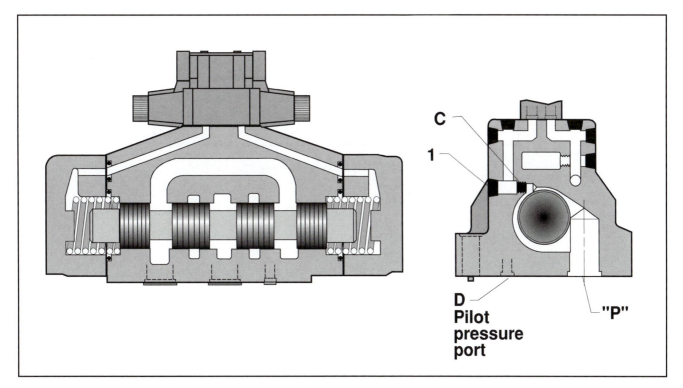

Figure 6-3   External pilot configuration

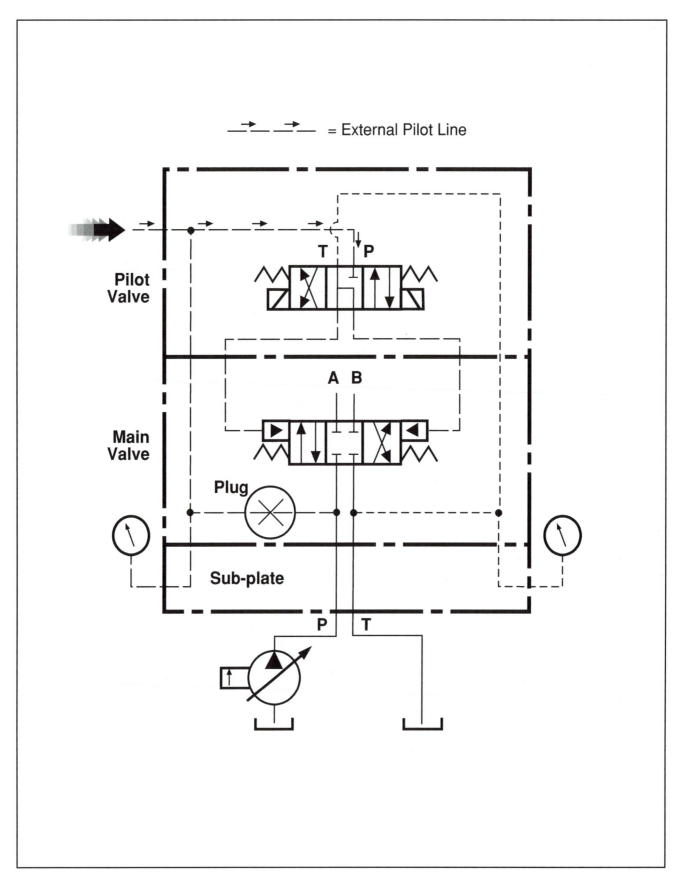

Figure 6-4   External pilot oil flow path

-WARNING- | Do not work on or around hydraulic systems without wearing safety glasses which conform to ANSI Z87.1-1989 standard.

CAUTION! If the valve is mounted on a sub-plate in an active system, be sure to follow all recommended safety procedures while performing the conversion procedure.

PREPARATION

Step 1.   Shut the prime mover off.

Step 2.   Lock the electrical system out or tag the keylock switch.

Step 3 .   Observe the system pressure gauge. Release any residual pressure trapped in the system by an accumulator, counterbalance or pilot-operated check valve, suspended load on an actuator, intensifier, or a pressurized reservoir.

CONVERSION PROCEDURE

Step 1.   Remove the valve from the sub-plate (if necessary).

Step 2.   Remove plug (1) located in the side of the main valve body (Figure 6-3).

NOTE: *It may be necessary to review the specific manufacturer's product information to locate the correct plug.*

Step 3.   Lay the valve on its side.

Step 4.   Remove the 1/8" NPT pipe plug from pilot pressure port "D" (Figure 6-3).

NOTE: *Threaded pilot pressure port "D" must be open (unplugged) for correct external pilot configuration.*

Step 5.   Install the 1/8" NPT pipe plug in the internal threaded port "C" (Figure 6-3). Tighten it securely.

NOTE: *Internal threaded port "C" must be closed (plugged) for correct external pilot configuration.*

Step 6.   Replace plug (1) in the valve body. Tighten it securely.

Step 7.   Install the valve on the subplate. Tighten the retaining bolts to the recommended torque values.

notes

Directional Control Valve
Internal Drain Conversion

What will this conversion accomplish?

The pilot valve of a solenoid-controlled, pilot-operated directional control valve can be internally or externally drained.

When a directional control valve is configured for external drain, the drain port of the pilot valve is connected through an internal passage, to the drain port of the main valve. The composite drawing (Figure 6-6) illustrates the oil flow path between the pilot valve and the main valve. Note that the drain-line from the pilot valve connects in parallel with the main return-line. Thus, the pilot valve will be exposed to any pressure fluctuations or resistance in the main return-line. This should be considered preceeding valve application.

Valve application will determine the drain configuration of a directional control valve.

This conversion will modify a solenoid-controlled, pilot-operated, directional control valve to an internal drain configuration.

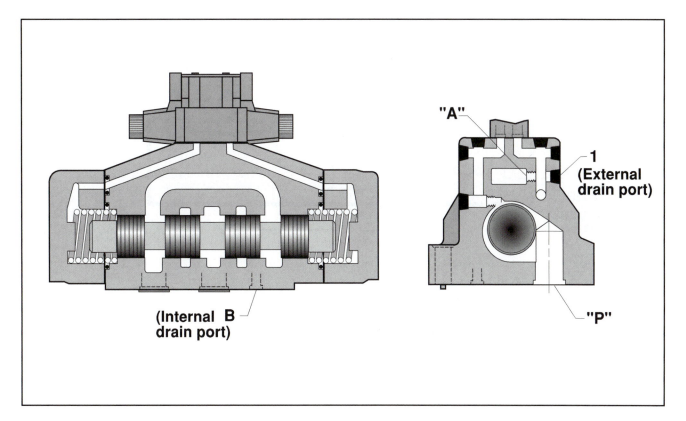

Figure 6-5   Internal drain configuration

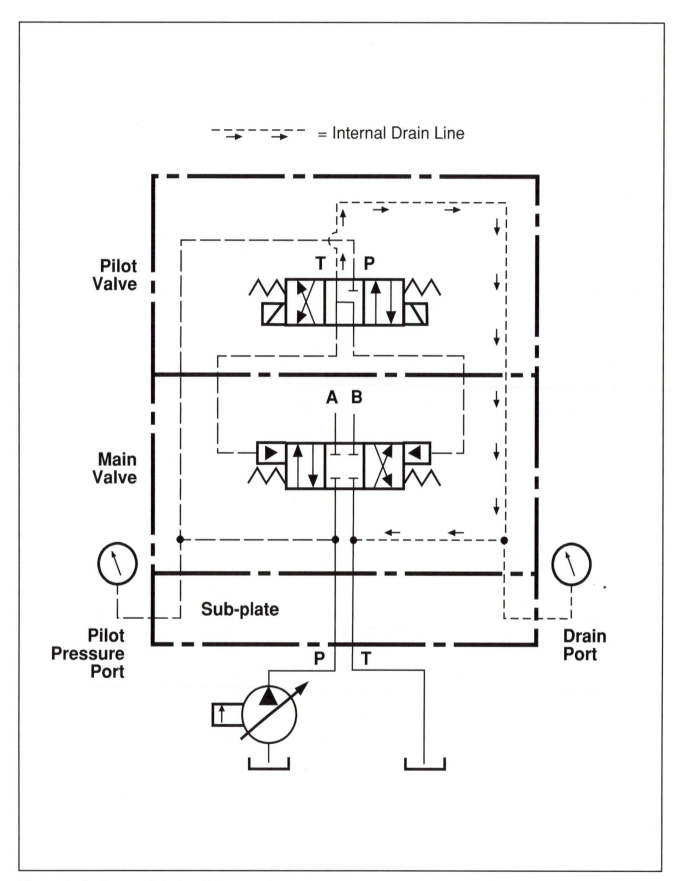

Figure 6-6   Internal drain oil flow path

-WARNING- | Do not work on or around hydraulic systems without wearing safety glasses which conform to ANSI Z87.1-1989 standard.

CAUTION! If the valve is mounted on a sub-plate in an active system, be sure to follow all recommended safety procedures while performing the conversion procedure.

PREPARATION

Step 1.   Shut the prime mover off.

Step 2.   Lock the electrical system out or tag the keylock switch.

Step 3.   Observe the system pressure gauge. Release any residual pressure trapped in the system by an accumulator, counterbalance or pilot-operated check valve, suspended load on an actuator, intensifier, or a pressurized reservoir.

CONVERSION PROCEDURE

Step 1.   Remove the valve from the sub-plate.

Step 2.   Remove plug (1) located in the side of the main valve body (Figure 6-5).

NOTE: *It may be necessary to review the specific manufacturer's product information to locate the correct plug.*

Step 3.   Remove the 1/8" NPT pipe plug from the internal, threaded port "A" (Figure 6-5).

NOTE: *Threaded port "A" must be open (unplugged) for correct internal drain configuration.*

Step 4.   Replace plug (1) in the main valve body. Tighten it securely.

Step 5.   Lay the valve on its side.

Step 6.   Install the 1/8" NPT pipe plug in threaded drain port "B" (Figure 6-5). Tighten it securely.

NOTE: *Threaded drain port "B" must be closed (plugged) for correct internal drain configuration.*

Step 7.   Install the valve on the sub-plate. Tighten the retaining bolts to the recommended torque values.

notes

Directional Control Valve
External Drain Conversion

What will this conversion accomplish?

The pilot valve of a solenoid-controlled, pilot-operated directional control valve can be internally or externally drained. Generally, the valve body is designed to be converted to either configuration by following a few simple steps.

Basically, to convert an internally drained valve to external drain, the internal drain passage is plugged with a 1/8" NPT pipe plug. This prevents the internal drain passage from connecting internally with the main tank return-line.

Access to the internal drain passage is made by removing a threaded plug from a pre-drilled port in the side of the main valve housing. A suitable drain-line can be connected from this port directly to the reservoir.

This conversion will make the pilot drain immune to pressure fluctuations and/or high resistance in the main reservoir return-line which could cause erratic main spool operation.

The composite drawing (Figure 6-8) illustrates the position of the internal plug and the pilot drain oil flow path.

This conversion will modify a solenoid-controlled, pilot-operated directional control valve to an external drain configuration.

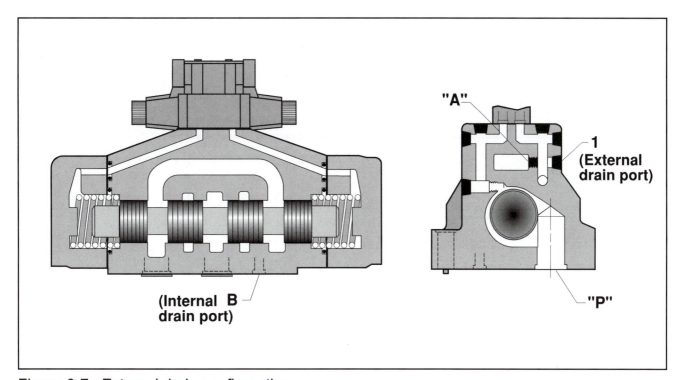

Figure 6-7   External drain configuration

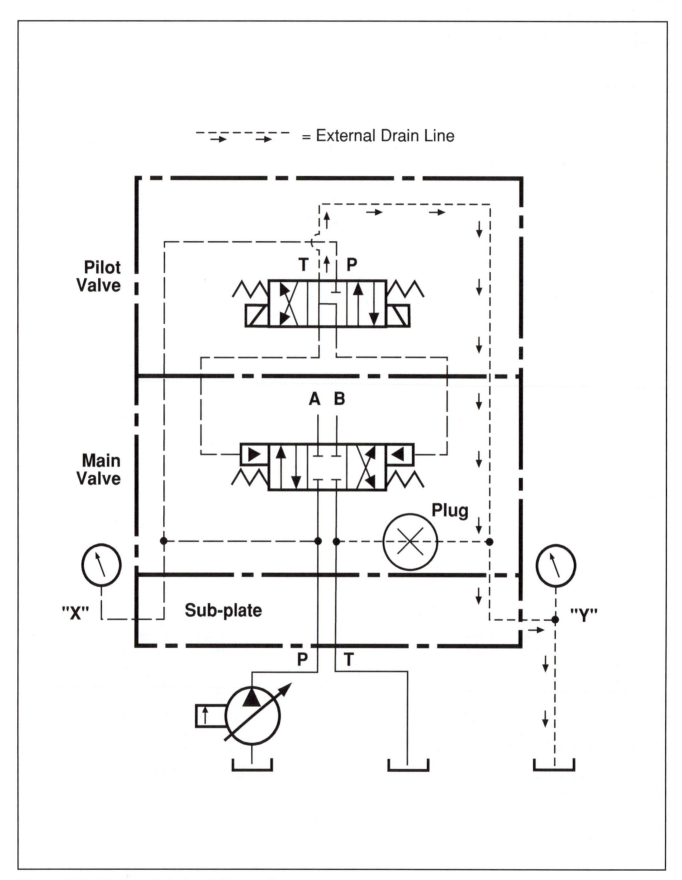

Figure 6-8   External drain oil flow path

-WARNING- | Do not work on or around hydraulic systems without wearing safety glasses which conform to ANSI Z87.1-1989 standard.

CAUTION! If the valve is mounted on a sub-plate in an active system, be sure to follow all recommended safety procedures while performing the conversion procedure.

PREPARATION

Step 1. Shut the prime mover off.

Step 2. Lock the electrical system out or tag the keylock switch.

Step 3. Observe the system pressure gauge. Release any residual pressure trapped in the system by an accumulator, counterbalance or pilot-operated check valve, suspended load on an actuator, intensifier, or a pressurized reservoir.

CONVERSION PROCEDURE

Step 1. Remove the valve from the sub-plate (if necessary).

Step 2. Lay the valve on its side.

Step 3. Remove the 1/8" NPT pipe plug from drain port "B" (Figure 6-7).

NOTE: *Threaded drain port "B" must be open (unplugged) for correct external drain conversion.*

Step 4. Remove plug (1) located in the side of the main valve body (Figure 6-7).

NOTE: *It may be necessary to review the specific manufacturer's product information to locate the correct plug.*

Step 5. Install the 1/8" NPT pipe plug in the internal threaded port "A" (Figure 6-7). Tighten it securely.

NOTE: *Threaded port "A" must be closed (plugged) for correct external drain configuration.*

Step 6. Replace threaded plug (1) in the main valve body. Tighten it securely.

Step 7. Install the valve on the subplate. Tighten the retaining bolts to the recommended torque values.

notes

chapter seven

Procedures for Testing
Mobile Directional Control Valves

"A" Port Passage
Direct-Access Test Procedure

What will this test procedure accomplish?

Generally, directional control valves have radial clearances between the valve bore and spool of between 5 and 15 micrometers.

The production of perfectly round and straight bores is exceptionally difficult to achieve, so it is highly unlikely that any spool will lie precisely concentric within its bore.

Valve manufacturers strive to develop casting and machining methods that will improve bore and spool production thereby reducing cross-port leakage. However, since the spool moves in relation to the bore, some leakage is expected.

If the fluid is heavily contaminated with particles which are equal in size to the clearances between the spool and bore, it will cause abrasive wear. Abrasive wear will gradually increase the radial and diametrical clearances, resulting in increased cross-port leakage.

High leakage will cause system overheating and/or a decline in actuator speed. It can also cause increased cylinder rod drifting in pressure manifold applications.

Clearance-related problems are generally viscosity sensitive, i.e. increased leakage will be less obvious when the fluid temperature is low and will increase as the fluid temperature increases.

The most common symptoms of problems associated with excessive cross-port leakage in directional control valves are:

    a. Linear actuator drifting in the rod extend position while the directional control valve is in the neutral (hold) position.

    b. Linear actuator vertical load "down-drift" while the directional control valve is in the neutral (hold) position.

    c. Increased rotary actuator drifting while the directional control valve is in the neutral (hold) position.

    d. Moderate to high increase in the operating temperature of the fluid.

    e. Loss of precise position control.

This test procedure will determine the amount of leakage across the "A" port passage of a directional control valve.

-WARNING- | Do not work on or around hydraulic systems without wearing safety glasses which conform to ANSI Z87.1-1989 standard.

YOU WILL NEED THE FOLLOWING DIAGNOSTIC EQUIPMENT TO CONDUCT THIS TEST PROCEDURE

CAUTION! The pressure and flow ratings of the diagnostic equipment which will be used to conduct this test procedure must be equal to, or greater than, the pressure and flow ratings of the system being tested. Refer to the system schematic for recommended pressure and flow data.

1. Pressure gauge
2. MicroLeak analyzer
3. Stopwatch (not shown)

PREPARATION

To conduct this test safely and accurately, refer to Figure 7-1 for correct placement of diagnostic equipment.

TEST PROCEDURE

Step 1. Shut the prime mover off.

Step 2. Lock the electrical system out or tag the keylock switch.

Step 3. Observe the system pressure gauge. Release any residual pressure trapped in the system by an accumulator, counterbalance or pilot-operated check valve, suspended load on an actuator, intensifier, or a pressurized reservoir.

NOTE: *It is not necessary to remove the "P" or "T" port transmission lines from the valve ports to conduct this test procedure.*

Step 4. Remove the transmission line from the "A" port of the directional control valve.

Step 5. Install MicroLeak analyzer (2) in series with the "A" port of the directional control valve.

Step 6. Install pressure gauge (1) in parallel with the connector at the outlet port of MicroLeak analyzer (2).

Step 7. Remove the transmission line from the "B" port of the directional control valve.

Step 8. Gradually pressurize the "A" port passage with MicroLeak analyzer (2). Stop when the value of the system's main pressure relief valve or compensator setting is reached.

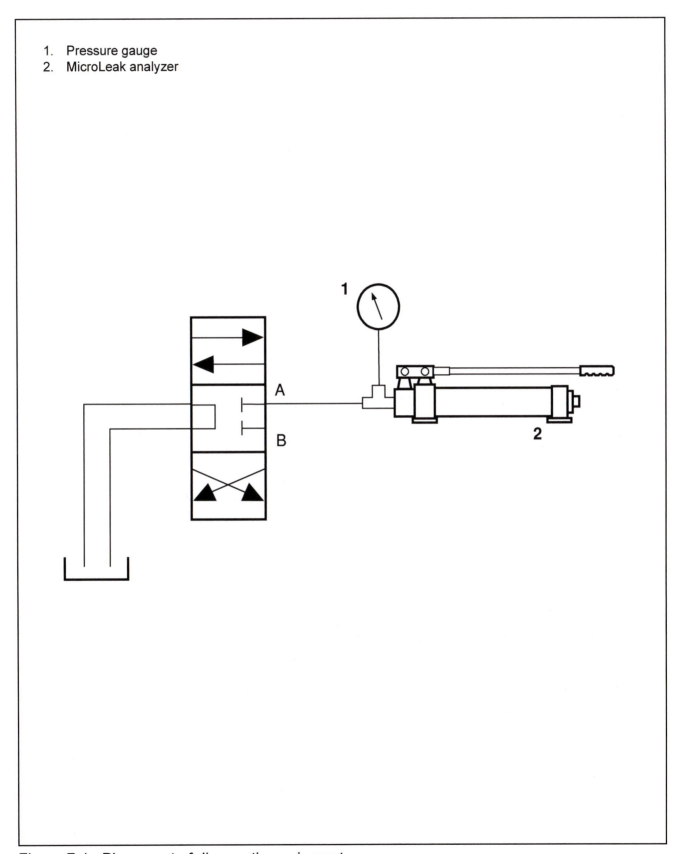

1. Pressure gauge
2. MicroLeak analyzer

Figure 7-1   Placement of diagnostic equipment

NOTE: *Oil is now trapped between the MicroLeak analyzer and the spool in the "A" port passage of the directional control valve.*

Step 9.   Observe the needle on pressure gauge (1). Using the stopwatch, time how long it takes for the pressure to drop to 200 PSI (13.8 bar).

NOTE: *The condition of the "A" port passage is determined by timing how long it takes for the pressure to drop from the value of the system's main pressure relief valve setting down to approximately 200 PSI (13.8 bar).*

Step 10.   Repeat the test procedure in the "B" port passage of the valve and compare the leakage rates.

ANALYZING THE TEST RESULTS

1.   **Diagnostic Observation:** The pressure indicated on pressure gauge (1) immediately begins to decrease when the pumping action of the MicroLeak analyzer stops.

Initially, the pressure drops fairly rapidly due to the high pressure differential across the ports. The pressure gauge indicates that the leakage across the ports decreases progressively as the pressure differential narrows.

It takes a "reasonable" length of time for the pressure to drop from the value of the system's main pressure relief valve setting down to approximately 200 PSI (13.8 bar).

NOTE: *There is no general rule-of-thumb regarding the volumetric efficiency of directional control valves. If it is necessary to determine accurate leakage rates, refer to the valve manufacturer's specifications.*

Diagnosis: Leakage across the "A" port passage appears to be within design specification.

2.   **Diagnostic Observation:** The pressure indicated on pressure gauge (1) attempts to increase when the MicroLeak analyzer is stroked. However, between pumping strokes the pressure drops rapidly.

If the "P," "T," and "B" ports are open, a steady stream of oil can be seen pouring from the valve.

The MicroLeak analyzer fails to pressurize the "A" port passage to the value of the system's main pressure relief valve setting.

Diagnosis: Leakage across the "A" port passage appears to be excessive. Although this amount of leakage may appear to be insignificant, it could cause actuator drifting and/or a marked increase in the operating temperature of the fluid.

If it is not possible to determine if cross-port leakage is excessive, conduct an actuator leakage path analysis test procedure. If actuator leakage is within specification, replace the directional control valve.

If necessary, refer to the manufacturer's specifications for recommended leakage data.

3. **Diagnostic Observation:** The MicroLeak analyzer fails to pressurize the "A" port passage. Oil leaks profusely from the open ports in the valve.

    Diagnosis: Leakage across the "A" port passage is excessive. Replace the directional control valve.

NOTE: *Before replacing the directional control valve check if excessive leakage is caused by the spool either not centering or binding.*

The spool may not be centering because there is an accumulation of contamination between either the spool and bore, or the end-caps.

Spool binding could also be caused by valve body mis-alignment due to sub-standard mounting procedures, over-tightened mounting bolts, or tapered pipe connectors.

CAUTION! Component repair procedures must be conducted by trained, authorized personnel. Incorrect parts, or parts which are improperly installed, can cause catastrophic component malfunction.

---

notes

"B" Port Passage
Direct-Access Test Procedure

What will this test procedure accomplish?

Generally, directional control valves have radial clearances between the valve bore and spool of between 5 and 15 micrometers.

The production of perfectly round and straight bores is exceptionally difficult to achieve, so it is highly unlikely that any spool will lie precisely concentric within its bore.

Valve manufacturers strive to develop casting and machining methods that will improve bore and spool production thereby reducing cross-port leakage. However, since the spool moves in relation to the bore, some leakage is expected.

If the fluid is heavily contaminated with particles which are equal in size to the clearances between the spool and bore, it will cause abrasive wear. Abrasive wear will gradually increase the radial and diametrical clearances, resulting in increased cross-port leakage.

High leakage will cause system overheating and a decline in actuator speed. It can also cause increased cylinder rod drifting in pressure manifold applications.

Clearance-related problems are generally viscosity sensitive, i.e. increased leakage will be less obvious when the fluid temperature is low and will increase as the fluid temperature increases.

The most common symptoms of problems associated with excessive cross-port leakage in directional control valves are:

    a. Linear actuator drifting in the rod extend position while the directional control valve is in the neutral (hold) position.

    b. Linear actuator vertical load "down-drift" while the directional control valve is in the neutral (hold) position.

    c. Increased rotary actuator drifting while the directional control valve is in the neutral (hold) position.

    d. Moderate to high increase in the operating temperature of the fluid.

    e. Loss of precise position control.

This test procedure will determine the amount of leakage across the "B" port passage of a directional control valve.

-WARNING- | **Do not work on or around hydraulic systems without wearing safety glasses which conform to ANSI Z87.1-1989 standard.**

YOU WILL NEED THE FOLLOWING DIAGNOSTIC EQUIPMENT
TO CONDUCT THIS TEST PROCEDURE

CAUTION! The pressure and flow ratings of the diagnostic equipment which will be used to conduct this test procedure must be equal to, or greater than, the pressure and flow ratings of the system being tested. Refer to the system schematic for recommended pressure and flow data.

---

1. MicroLeak analyzer
2. Pressure gauge
3. Stopwatch (not shown)

PREPARATION

To conduct this test safely and accurately, refer to Figure 7-2 for correct placement of diagnostic equipment.

TEST PROCEDURE

Step 1.   Shut the prime mover off.

Step 2.   Lock the electrical system out or tag the keylock switch.

Step 3.   Observe the system pressure gauge. Release any residual pressure trapped in the system by an accumulator, counterbalance or pilot-operated check valve, suspended load on an actuator, intensifier, or a pressurized reservoir.

NOTE: *It is not necessary to remove the "P" or "T" port transmission lines from the valve ports to conduct this test procedure.*

Step 4.   Remove the transmission line from the "B" port of the directional control valve.

Step 5.   Install MicroLeak analyzer (1) in series with the "B" port of the directional control valve.

Step 6.   Install pressure gauge (2) in parallel with the connector at the outlet port of MicroLeak analyzer (1).

Step 7.   Remove the transmission line from the "A" port of the directional control valve.

Step 8.   Gradually pressurize the "B" port passage with MicroLeak analyzer (1). Stop when the value of the system's main pressure relief valve or compensator setting is reached.

NOTE: *Oil is now trapped between MicroLeak analyzer (1) and the spool in the "B" port passage of the directional control valve.*

1. MicroLeak analyzer
2. Pressure gauge

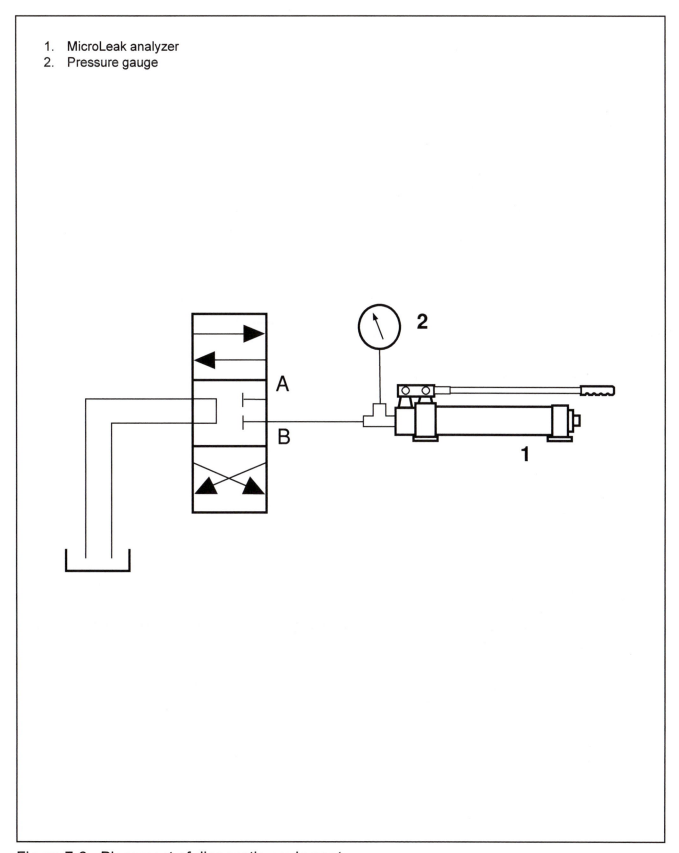

Figure 7-2   Placement of diagnostic equipment

Step 9.   Observe the needle on pressure gauge (2). Using the stopwatch, time how long it takes for the pressure to drop to 200 PSI (13.8 bar).

NOTE: *The condition of the "B" port passage is determined by timing how long it takes for the pressure to drop from the value of the system's main pressure relief valve setting down to approximately 200 PSI (13.8 bar).*

Step 10.   Repeat the test procedure in the "A" port passage of the valve and compare the leakage rates.

ANALYZING THE TEST RESULTS

1.   **Diagnostic Observation:** The pressure indicated on pressure gauge (2) immediately begins to decrease when the pumping action of the portable Mini-Tester stops.

       Initially, the pressure drops fairly rapidly due to the high pressure differential across the ports. The pressure gauge indicates that the leakage across the ports decreases progressively as the pressure differential narrows.

       It takes a "reasonable" length of time for the pressure to drop from the value of the system's main pressure relief valve setting down to approximately 200 PSI (13.8 bar)

NOTE: *There is no general rule-of-thumb regarding the volumetric efficiency of directional control valves. If it is necessary to determine accurate leakage rates, refer to the valve manufacturer's specifications.*

       Diagnosis: Leakage across the "B" port passage appears to be within design specification.

2.   **Diagnostic Observation:** The pressure indicated on pressure gauge (2) attempts to increase when the MicroLeak analyzer is stroked. However, between pumping strokes the pressure drops rapidly.

       If the "P," "T," and "A" ports are open, a steady stream of oil can be seen pouring from the valve

       The MicroLeak analyzer fails to pressurize the "B" port passage to the value of the system's main pressure relief valve setting.

       Diagnosis: Leakage across the "B" port passage appears to be excessive. Although this amount of leakage may appear to be insignificant it could cause actuator drifting and/or a marked increase in the operating temperature of the fluid.

       If it is not possible to determine if the cross-port leakage is excessive conduct an actuator leakage path analysis test procedure. If actuator leakage is within specification replace the directional control valve.

If necessary, refer to the manufacturer's specifications for recommended leakage data.

3.   **Diagnostic Observation:** The MicroLeak analyzer fails to pressurize the "B" port passage. Oil leaks profusely from the open ports in the valve.

Diagnosis: Leakage across the "B" port passage is excessive. Replace the directional control valve.

NOTE: *Before replacing the directional control valve check if excessive leakage is caused by the spool either not centering or binding.*

The spool may not be centering because there is an accumulation of contamination between either the spool and bore, or the end-caps.

Spool binding could also be caused by valve body mis-alignment due to sub-standard mounting procedures, over-tightened mounting bolts, or tapered pipe connectors.

notes

"P" Port to "A" Port Passage
Direct-Access Test Procedure

What will this test procedure accomplish?

Generally, directional control valves have radial clearances between the valve bore and spool of between 5 and 15 micrometers.

The production of perfectly round and straight bores is exceptionally difficult to achieve, so it is highly unlikely that any spool will lie precisely concentric within its bore.

Valve manufacturers strive to develop casting and machining methods that will improve bore and spool production thereby reducing cross-port leakage. However, since the spool moves in relation to the bore, some leakage is expected.

If the fluid is heavily contaminated with particles which are equal in size to the clearances between the spool and bore, it will cause abrasive wear. Abrasive wear will gradually increase the radial and diametrical clearances, resulting in increased cross-port leakage.

High leakage will cause system overheating and a decline in actuator speed. It can also cause increased cylinder rod drifting in pressure manifold applications.

Clearance-related problems are generally viscosity sensitive, i.e. increased leakage will be less obvious when the fluid temperature is low and will increase as the fluid temperature increases.

The most common symptoms of problems associated with excessive cross-port leakage in directional control valves leakage are:

    a. Linear actuator drifting in the rod extend position while the directional control valve is in the neutral (hold) position.

    b. Linear actuator vertical load "down-drift" while the directional control valve is in the neutral (hold) position.

    c. Increased rotary actuator drifting while the directional control valve is in the neutral (hold) position.

    d. Moderate to high increase in the operating temperature of the fluid.

    e. Loss of precise position control.

This test procedure will determine the amount of leakage across the "P" port to "A" port passage of a directional control valve.

-WARNING-	Do not work on or around hydraulic systems without wearing safety glasses which conform to ANSI Z87.1-1989 standard.

YOU WILL NEED THE FOLLOWING DIAGNOSTIC EQUIPMENT
TO CONDUCT THIS TEST PROCEDURE

CAUTION! The pressure and flow ratings of the diagnostic equipment which will be used to conduct this test procedure must be equal to, or greater than, the pressure and flow ratings of the system being tested. Refer to the system schematic for recommended pressure and flow data.

1. MicroLeak analyzer
2. Pressure gauge
3. Stopwatch (not shown)

PREPARATION

To conduct this test safely and accurately, refer to Figure 7-3 for correct placement of diagnostic equipment.

TEST PROCEDURE

Step 1.   Shut the prime mover off.

Step 2.   Lock the electrical system out or tag the keylock switch.

Step 3.   Observe the system pressure gauge. Release any residual pressure trapped in the system by an accumulator, counterbalance or pilot-operated check valve, suspended load on an actuator, intensifier, or a pressurized reservoir.

Step 4.   Remove the transmission line from the "P" port of the directional control valve.

Step 5.   Install MicroLeak analyzer (1) in series with the "P" port of the directional control valve.

Step 6.   Install pressure gauge (2) in parallel with the connector at the outlet port of MicroLeak analyzer (1).

Step 7.   Remove the transmission line from the "A" port of the directional control valve.

Step 8.   Plug the "A" port of the directional control valve.

Step 9.   Activate the directional control valve spool. Select the position which aligns the "P" port to the "A" port.

1. MicroLeak analyzer
2. Pressure gauge

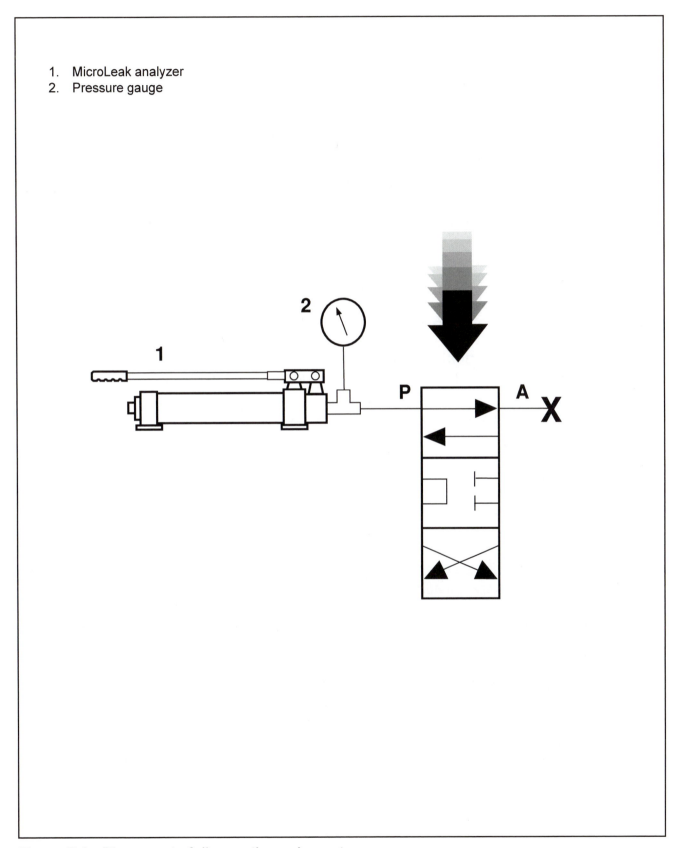

Figure 7-3   Placement of diagnostic equipment

Step 10. Gradually pressurize the "P" port to "A" port passage with MicroLeak analyzer (1). Stop when the value of the system's main pressure relief valve or compensator setting is reached.

NOTE: *Oil is now trapped between MicroLeak analyzer (1) and the "P" port to "A" port passage.*

Step 11. Observe the needle on pressure gauge (2). Using the stopwatch, time how long it takes for the pressure to drop to 200 PSI (13.8 bar).

NOTE: *The condition of the "P" port to "A" port passage is determined by timing how long it takes for the pressure to drop from the value of the system's main pressure relief valve setting down to approximately 200 PSI (13.8 bar).*

Step 12. Repeat the test procedure in the "P" port to "B" port passage of the valve and compare the leakage rates.

ANALYZING THE TEST RESULTS

1. **Diagnostic Observation:** The pressure indicated on pressure gauge (2) immediately begins to decrease when the pumping action of the MicroLeak analyzer stops.

    Initially, the pressure drops fairly rapidly due to the high pressure differential across the ports. The pressure gauge indicates that the leakage across the ports decreases progressively as the pressure differential narrows.

    It takes a "reasonable" length of time for the pressure to drop from the value of the system's main pressure relief valve setting down to approximately 200 PSI (13.8 bar)

NOTE: *There is no general rule-of-thumb regarding the volumetric efficiency of directional control valves. If it is necessary to determine accurate leakage rates, refer to the valve manufacturer's specifications.*

    Diagnosis: Leakage across the "P" port to "A" port passage appears to be within design specification.

2. **Diagnostic Observation:** The pressure indicated on pressure gauge (2) attempts to increase when the MicroLeak analyzer is stroked. However, between pumping strokes the pressure drops rapidly.

    If the "T" and "B" ports are open, a steady stream of oil can be seen pouring from the valve.

    The MicroLeak analyzer fails to pressurize the "P" port to "A" port passage to the value of the system's main pressure relief valve setting.

Diagnosis: Leakage across the "P" port to "A" port passage appears to be excessive. Although this amount of leakage may appear to be insignificant, it could cause actuator drifting and/or a marked increase in the operating temperature of the fluid.

If it is not possible to determine if the cross-port leakage is excessive, conduct an actuator leakage path analysis test procedure. If actuator leakage is within specification, replace the directional control valve.

If necessary, refer to the manufacturer's specifications for recommended leakage data.

3. **Diagnostic Observation:** The MicroLeak analyzer fails to pressurize the "P" port to "A" port passage. Oil leaks profusely from the open ports in the valve.

   Diagnosis: Leakage across the "P" port to "A" port passage is excessive. Replace the directional control valve.

NOTE: *Before replacing the directional control valve check if excessive leakage is caused by the spool either not centering or binding.*

The spool may not be centering because there is an accumulation of contamination between either the spool and bore, or the end-caps.

Spool binding could also be caused by valve body mis-alignment due to sub-standard mounting procedures, over-tightened mounting bolts, or tapered pipe connectors.

notes

"P" Port to "B" Port Passage
Direct Access Test Procedure

What will this test procedure accomplish?

Generally, directional control valves have radial clearances between the valve bore and spool of between 5 and 15 micrometers.

The production of perfectly round and straight bores is exceptionally difficult to achieve, so it is highly unlikely that any spool will lie precisely concentric within its bore.

Valve manufacturers strive to develop casting and machining methods that will improve bore and spool production thereby reducing cross-port leakage. However, since the spool moves in relation to the bore, some leakage is expected.

If the fluid is heavily contaminated with particles which are equal in size to the clearances between the spool and bore, it will cause abrasive wear. Abrasive wear will gradually increase the radial and diametrical clearances, resulting in increased cross-port leakage.

High leakage will cause system overheating and a decline in actuator speed. It can also cause increased cylinder rod drifting in pressure manifold applications.

Clearance-related problems are generally viscosity sensitive, i.e. increased leakage will be less obvious when the fluid temperature is low and will increase as the fluid temperature increases.

The most common symptoms of problems associated with excessive cross-port leakage in directional control valves are:
   a. Linear actuator drifting in the rod extend position while the directional control valve is in the neutral (hold) position.
   b. Linear actuator vertical load "down-drift" while the directional control valve is in the neutral (hold) position.
   c. Increased rotary actuator drifting while the directional control valve is in the neutral (hold) position.
   d. Moderate to high increase in the operating temperature of the fluid.
   e. Loss of precise position control.

This test procedure will determine the amount of leakage across the "P" port to "B" port passage of a directional control valve.

-WARNING- | **Do not work on or around hydraulic systems without wearing safety glasses which conform to ANSI Z87.1-1989 standard.**

YOU WILL NEED THE FOLLOWING DIAGNOSTIC EQUIPMENT
TO CONDUCT THIS TEST PROCEDURE

CAUTION! The pressure and flow ratings of the diagnostic equipment which will be used to conduct this test procedure must be equal to, or greater than, the pressure and flow ratings of the system being tested. Refer to the system schematic for recommended pressure and flow data.

---

1. MicroLeak analyzer
2. Pressure gauge
3. Stopwatch (not shown)

PREPARATION

To conduct this test safely and accurately, refer to Figure 7-4 for correct placement of diagnostic equipment.

TEST PROCEDURE

Step 1. Shut the prime mover off.

Step 2. Lock the electrical system out or tag the keylock switch.

Step 3. Observe the system pressure gauge. Release any residual pressure trapped in the system by an accumulator, counterbalance or pilot-operated check valve, suspended load on an actuator, intensifier, or a pressurized reservoir.

Step 4. Remove the transmission line from the "P" port of the directional control valve.

Step 5. Install MicroLeak analyzer (1) in series with the "P" port of the directional control valve.

Step 6. Install pressure gauge (2) in parallel with the connector at the outlet port of MicroLeak analyzer (1).

Step 7. Remove the transmission line from the "B" port of the directional control valve.

Step 8. Plug the "B" port of the directional control valve.

Step 9. Activate the directional control valve spool. Select the position which aligns the "P" port to the "B" port.

Step 10. Gradually pressurize the "P" port to "B" port passage with MicroLeak analyzer (1). Stop when the value of the system's main pressure relief valve or compensator setting is reached.

1. MicroLeak analyzer
2. Pressure gauge

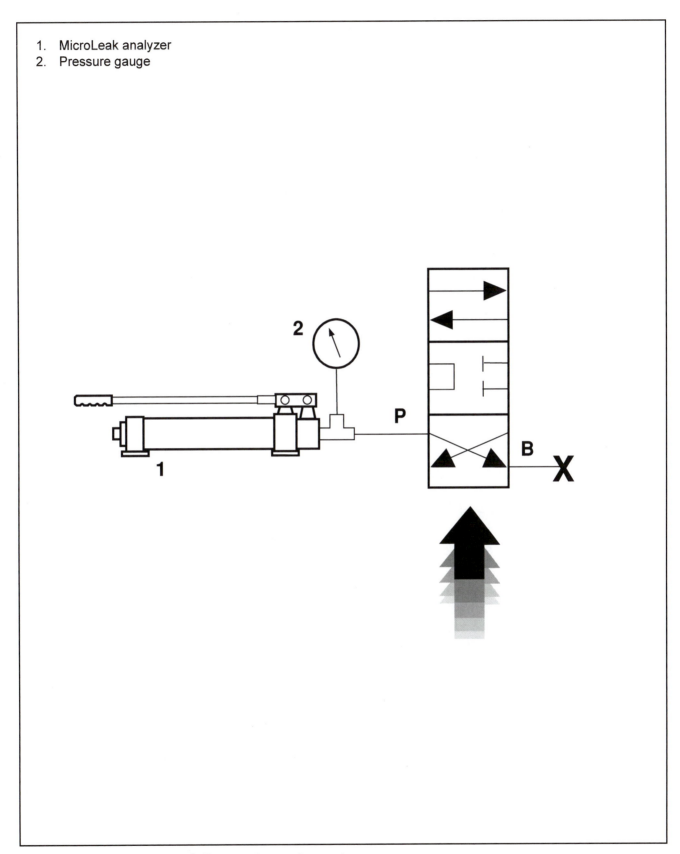

Figure 7-4   Placement of diagnostic equipment

NOTE: *Oil is now trapped between MicroLeak analyzer (1) and the "P" port to "B" port passage.*

Step 11. Observe the needle on pressure gauge (2). Using the stopwatch, time how long it takes for the pressure to drop to 200 PSI (13.8 bar).

NOTE: *The condition of the "P" port to "B" port passage is determined by timing how long it takes for the pressure to drop from the value of the system's main pressure relief valve setting down to approximately 200 PSI (13.8 bar).*

Step 12. Repeat the test procedure in the "P" port to "A" port passage of the valve and compare the leakage rates.

ANALYZING THE TEST RESULTS

1. **Diagnostic Observation:** The pressure indicated on pressure gauge (2) immediately begins to decrease when the pumping action of the MicroLeak analyzer stops.

   Initially, the pressure drops fairly rapidly due to the high pressure differential across the ports. The pressure gauge indicates that the leakage across the ports decreases progressively as the pressure differential narrows.

   It takes a "reasonable" length of time for the pressure to drop from the value of the system's main pressure relief valve setting down to approximately 200 PSI (13.8 bar).

   NOTE: *There is no general rule-of-thumb regarding the volumetric efficiency of directional control valves. If it is necessary to determine accurate leakage rates, refer to the valve manufacturer's specifications.*

   Diagnosis: Leakage across the "P" port to "B" port passage appears to be within design specification.

2. **Diagnostic Observation:** The pressure indicated on pressure gauge (2) attempts to increase when the MicroLeak analyzer is stroked. However, between pumping strokes the pressure drops rapidly.

   If the "T" and "A" ports are open, a steady stream of oil can be seen pouring from the valve.

   The MicroLeak analyzer fails to pressurize the "P" port to "B" port passage to the value of the system's main pressure relief valve setting.

   Diagnosis: Leakage across the "P" port to "B" port passage appears to be excessive. Although this amount of leakage may appear to be insignificant, it could cause actuator drifting and/or a marked increase in the operating temperature of the fluid.

If it is not possible to determine if the cross-port leakage is excessive, conduct an actuator leakage path analysis test procedure. If actuator leakage is within specification, replace the directional control valve.

If necessary, refer to the manufacturer's specifications for recommended leakage data.

3. **Diagnostic Observation:** The MicroLeak analyzer fails to pressurize the "P" port to "B" port passage. Oil leaks profusely from the open ports in the valve.

    Diagnosis: Leakage across the "P" port to "B" port passage is excessive. Replace the directional control valve.

NOTE: *Before replacing the directional control valve check if excessive leakage is caused by the spool either not centering or binding.*

   The spool may not be centering because there is an accumulation of contamination between either the spool and bore, or the end-caps.

   Spool binding could also be caused by valve body mis-alignment due to sub-standard mounting procedures, over-tightened mounting bolts, or tapered pipe connectors.

notes

Cylinder Port Relief Valve
Direct-Access Test Procedure

What will this test procedure accomplish?

A cylinder port relief valve is basically a cartridge type adjustable pressure relief valve which is designed into a hydraulic system to offer overload protection on the work-port or actuator side of a directional control valve. It is normallly set at approximately 300 PSI (20.7 bar) to 500 PSI (34.5 bar) higher than the main pressure relief valve setting.

Generally, manufacturers of mobile directional control valves design valve bodies for maximum application versatility. They are usually designed with cartridge cavities in the actuator work-ports. Cartridge-type valve integration means fewer hoses and connectors. The system is "stiffer" and is more efficient.

The most common concerns in mobile valve application are cylinder port (work-port) overload protection, and cavitation prevention. For cylinder port overload protection cylinder port relief valves can be installed. To prevent cavitation, anti-cavitation valves can be installed. Generally, the cylinder port relief valves and anti-cavitation valves are combined in a single cartridge.

The primary function of a cylinder port relief valve is to offer work-port overload protection should an actuator be subjected to a mechanical force while the directional control valve is in the neutral position: "A" and "B" work ports closed by the position of the spool.

They also function in-concert with load check valves. If an actuator should receive a shock load while the directional control valve is in the activated position, the shock would travel from the actuator, through the directional control valve, to the pump outlet port. The pump would be vulnerable to catastrophic failure.

However, since the load check valve is located within the directional control valve, in series with the pump flow, the potential shock cannot reach the pump. It is stopped by the load check valve and is dissipated by the cylinder port relief valve. When a shock-load closes the load check valve, the pump flow momentarily passes over the main pressure relief valve.

The most common symptoms of problems associated with cylinder port relief valves are:
    a. Actuator travels in one direction only.
    b. Reduced actuator speed in one direction only.
    c. System overheating.
    d. Actuator speed reduction as load increases.
    e. Actuator drifting with the directional control valve in the neutral (hold) position.
    f. Cylinder rod bending.
    g. Mounting clevis failure.
    h. Structural fatigue.

This test procedure will determine:

    a. How much fluid leaks across a cylinder port relief valve.

    b. The root-cause of a cross-port leakage path.

 **-WARNING-**   **Do not work on or around hydraulic systems without wearing safety glasses which conform to ANSI Z87.1-1989 standard.**

YOU WILL NEED THE FOLLOWING DIAGNOSTIC EQUIPMENT TO CONDUCT THIS TEST PROCEDURE

CAUTION! The pressure and flow ratings of the diagnostic equipment which will be used to conduct this test procedure must be equal to, or greater than, the pressure and flow ratings of the system being tested. Refer to the system schematic for recommended pressure and flow data.

1. MicroLeak analyzer   3. Pyrometer

2. Pressure gauge   4. Stopwatch (not shown)

PREPARATION

To conduct this test safely and accurately, refer to Figure 7-5 for correct placement of diagnostic equipment.

TEST PROCEDURE

Step 1.   Shut the prime mover off.

Step 2.   Lock the electrical system out or tag the keylock switch.

Step 3.   Observe the system pressure gauge. Release any residual pressure trapped in the system by an accumulator, counterbalance or pilot-operated check valve, suspended load on an actuator, intensifier, or a pressurized reservoir.

Step 4.   Study the system schematic and determine in which port the cylinder port relief valve is installed.

NOTE: *It is possible for cylinder port relief valves to be installed in both work-ports. This test procedure will show the cylinder port relief valve installed in the "B" port of the directional control valve.*

Step 5.   Remove the transmission line from the "B" port of the directional control valve.

Step 6.   Connect pressure gauge (2) in parallel with the connector at the "B" port of the directional control valve.

1. MicroLeak analyzer
2. Pressure gauge
3. Pyrometer

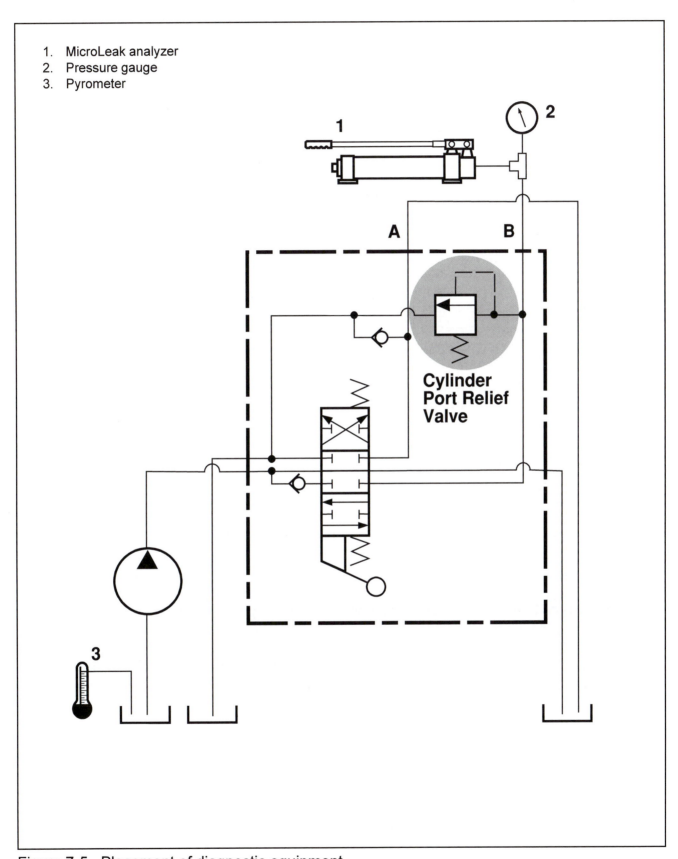

Figure 7-5   Placement of diagnostic equipment

Step 7.   Connect MicroLeak analyzer (1) in series with the "B" port of the directional control valve.

Step 8.   Gradually pressurize the "B" port passage with MicroLeak analyzer (1). Stop when the cylinder port relief valve setting is reached.

NOTE: *Oil is now trapped between the MicroLeak analyzer (1), blocked valve spool, and the cylinder port relief valve.*

Step 9.   Observe the needle on pressure gauge (2). Using the stopwatch, time how long it takes for the pressure to drop to 200 PSI (13.8 bar).

Step 10.   Repeat the test procedure in the "A" port passage of the directional control valve and compare the leakage rates.

ANALYZING THE TEST RESULTS

1.   **Diagnostic Observation:** The pressure indicated on pressure gauge (2) immediately begins to decrease when the pumping action of the MicroLeak analyzer stops.

Initially, the pressure drops fairly rapidly due to the high pressure differential across the clearances in the directional control valve and/or the cylinder port relief valve. The pressure gauge indicates that the leakage across the clearances decreases progressively as the pressure differential narrows.

It takes a "reasonable" length of time for the pressure to drop from the value of the system's main pressure relief valve setting down to approximately 200 PSI (13.8 bar)

NOTE: *There is no general rule-of-thumb regarding the volumetric efficiency of directional control valves. If it is necessary to determine accurate leakage rates, refer to the valve manufacturer's specifications.*

Diagnosis: Leakage across the spool and the cylinder port relief valve appears to be within design specification.

2.   **Diagnostic Observation:** The pressure indicated on pressure gauge (2) attempts to increase when the MicroLeak analyzer is stroked. However, between pumping strokes the pressure drops rapidly.

If the "T" port is open, a steady stream of oil can be seen pouring from the valve.

The MicroLeak analyzer fails to pressurize the "B" port passage of the directional control valve to the value of the cylinder port relief valve setting.

Diagnosis: Leakage across the spool and/or the cylinder port relief valve appears to be excessive. Although this amount of leakage may appear to be insignificant, it could cause actuator drifting and/or a marked increase in the operating temperature of the fluid.

Since the cylinder port relief valve is in parallel with the valve spool and bore, it will be necessary to conduct further testing to determine the root-cause:

Step 1.   Remove the cylinder port relief valve from the suspect port and substitute it for one from the opposite port or from an adjoining valve section. If necessary, a new cylinder port cartridge assembly can be used to determine the root-cause.

Step 2.   Repeat the test procedure. If it is now possible to pressurize the valve port and hold pressure for a reasonable length of time, the cylinder port relief valve is suspect.

If the leakage remains unchanged after testing the valve with a good cylinder port relief valve, the spool/bore clearance is suspect.

If necessary, refer to the manufacturer's specifications for recommended leakage data.

3.   Diagnostic Observation: The MicroLeak analyzer fails to pressurize the "B" port passage of the valve. Oil leaks profusely from the open ports in the valve.

Diagnosis: Leakage across the spool and/or the cylinder port relief valve is excessive.

Since the cylinder port relief valve and valve spool are in parallel, it will be necessary to conduct further testing to determine the root-cause.

Step 1.   Remove the cylinder port relief valve from the suspect port and substitute it for one from the opposite port or from an adjoining valve section. If necessary, a new cylinder port cartridge assembly can be used to determine the root-cause.

Step 2.   Repeat the test procedure. If it is now possible to pressurize the valve port and hold pressure for a reasonable length of time, the cylinder port relief valve is suspect.

If the leakage remains unchanged after testing the valve with a good cylinder port relief valve, the spool is suspect.

If necessary, refer to the manufacturer's specifications for recommended leakage data.

NOTE: *Before replacing the directional control valve check if excessive leakage is caused by the spool either not centering or binding.*

The spool may not be centering because there is an accumulation of contamination between either the spool and bore, or the end-caps.

Spool binding could also be caused by valve body mis-alignment due to sub-standard mounting procedures, over-tightened mounting bolts, or tapered pipe connectors.

notes

Anti-Cavitation Valve
Direct-Access Test Procedure

What will this test procedure accomplish?

In variable speed prime-mover applications (primarily mobile), it is possible for the flow required by an actuator to exceed pump flow. This condition occurs primarily when the prime mover is operating at low speed.

In certain applications, it is possible to operate a loaded actuator when the prime mover is shut-down. Both these conditions could result in serious cavitation.

Generally, provision is made in cast directional control valve bodies for the installation, if necessary, of cartridge-type anti-cavitation valves. Actuator application will determine the need for an anti-cavitation valve.

An anti-cavitation valve is basically a check valve. It is usually located in parallel with a cylinder port and the reservoir return-line. Depending on the application, one or two anti-cavitation valves can be used-- one per cylinder port. They are usually positioned in the actuator port which is most likely to suffer cavitation.

An anti-cavitation valve allows the actuator to function much like the operation at the suction side of a pump.

If an actuator, influenced by an external load, moves at a speed which causes it to require more oil than the pump produces at maximum speed, a partial vacuum would be created on the pump supply side of the actuator.

Oil returning through the directional control valve, from the opposite side of the actuator, would sense the vacuum through the anti-cavitation valve. A condition similar to the suction side of a pump has now been created.

Instead of the return oil flowing back to the reservoir, it circulates through the anti-cavitation valve into the opposite work-port and combines with the pump flow. The combined flows are usually sufficient to supply the actuator, thus preventing cavitation.

The most common symptoms of problems associated with anti-cavitation valves are:
     a. Actuator drifting with the directional control valve in the neutral (hold) position.
     b. A reduction in actuator speed in one direction of travel.
     c. Total loss of actuator travel in one direction only.
     d. System overheating.
     e. Actuator speed loss with an increase in load.

The most common causes of anti-cavitation valve malfunction are:
    a. Contamination.
    b. Seat and poppet wear or damage.
    c. Cartridge valve seal failure.

This test procedure will determine:
    a. How much fluid leaks across an anti-cavitation valve.
    b. The root-cause of a cross-port leakage path.

-WARNING-   **Do not work on or around hydraulic systems without wearing safety glasses which conform to ANSI Z87.1-1989 standard.**

YOU WILL NEED THE FOLLOWING DIAGNOSTIC EQUIPMENT TO CONDUCT THIS TEST PROCEDURE

CAUTION! The pressure and flow ratings of the diagnostic equipment which will be used to conduct this test procedure must be equal to, or greater than, the pressure and flow ratings of the system being tested. Refer to the system schematic for recommended pressure and flow data.

    1.   MicroLeak analyzer   3.   Pyrometer
    2.   Pressure gauge   4.   Stopwatch (not shown)

PREPARATION

To conduct this test safely and accurately, refer to Figure 7-6 for correct placement of diagnostic equipment.

TEST PROCEDURE

Step 1.   Shut the prime mover off.

Step 2.   Lock the electrical system out or tag the keylock switch.

Step 3.   Observe the system pressure gauge. Release any residual pressure trapped in the system by an accumulator, counterbalance or pilot-operated check valve, suspended load on an actuator, intensifier, or a pressurized reservoir.

Step 4.   Study the system schematic and determine in which port the anti-cavitation valve is installed.

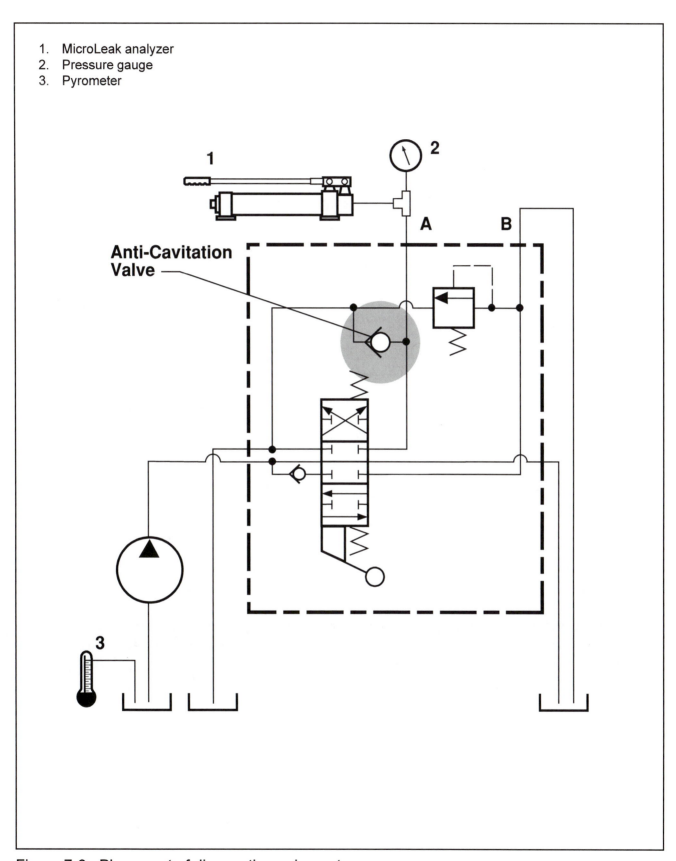

1. MicroLeak analyzer
2. Pressure gauge
3. Pyrometer

Anti-Cavitation Valve

Figure 7-6   Placement of diagnostic equipment

NOTE: *It is possible for anti-cavitation valves to be installed in both work ports. This test procedure will show the anti-cavitation valve installed in the "A" port of the directional control valve.*

Step 5.   Remove the transmission line from the "A" port of the directional control valve.

Step 6.   Connect pressure gauge (2) in parallel with the connector at the "A" port of the directional control valve.

Step 7.   Connect MicroLeak analyzer (1) in series with the "A" port of the directional control valve.

Step 8.   Gradually pressurize the "A" port passage with MicroLeak analyzer (1). Stop when the value of the main pressure relief valve or compensator setting is reached.

NOTE: *Oil is now trapped between the MicroLeak analyzer (1), the blocked valve spool, and the anti-cavitation valve.*

Step 9.   Observe the needle on pressure gauge (2). Using the stopwatch, time how long it takes for the pressure to drop to 200 PSI (13.8 bar).

Step 10.   Repeat the test procedure in the "B" port passage of the directional control valve and compare the leakage rates.

ANALYZING THE TEST RESULTS

1.   **Diagnostic Observation:** The pressure indicated on pressure gauge (2) immediately begins to decrease when the pumping action of the MicroLeak analyzer stops.

Initially, the pressure drops fairly rapidly due to the high pressure differential across the clearances in the directional control valve and/or the anti-cavitation valve. The pressure gauge indicates that the leakage across the clearances decreases progressively as the pressure differential narrows.

It takes a "reasonable" length of time for the pressure to drop from the value of the system's main pressure relief valve setting down to approximately 200 PSI (13.8 bar).

NOTE: *There is no general rule-of-thumb regarding the volumetric efficiency of directional control valves. If it is necessary to determine accurate leakage rates, refer to the valve manufacturer's specifications.*

Diagnosis: Leakage across the spool and/or the anti-cavitation valve appears to be within design specification.

2.   **Diagnostic Observation:** The pressure indicated on pressure gauge (2) attempts to increase when the MicroLeak analyzer is stroked. However, between pumping strokes the pressure drops rapidly.

If the "T" port of the directional control valve is open, a steady stream of oil can be seen pouring from the valve.

The MicroLeak analyzer fails to pressurize the "A" port passage to the value of the system's main pressure relief valve setting.

Diagnosis: Leakage across the spool and/or the anti-cavitation valve appears to be excessive. Although this amount of leakage may appear to be insignificant, it could cause actuator drifting and/or a marked increase in the operating temperature of the fluid.

Since the anti-cavitation valve and valve spool are in parallel it will be necessary to conduct further testing to determine the root-cause:

Step 1.   Remove the anti-cavitation valve from the suspect port and substitute it for one from the opposite port or from an adjoining valve section. If necessary, a new anti-cavitation valve cartridge assembly can be used to help determine the root-cause.

Step 2.   Repeat the test procedure. If it is now possible to pressurize the "A" port passage and hold pressure for a reasonable length of time, the anti-cavitation valve is suspect.

If the leakage remains unchanged after testing the valve with a good anti-cavitation valve, the spool is suspect.

If necessary, refer to the manufacturer's specification for recommended leakage data.

3.   **Diagnostic Observation:** The MicroLeak analyzer fails to pressurize the "A" port passage of the directional control valve. Oil leaks profusely from the open ports in the valve.

Diagnosis: Leakage across the spool and/or the anti-cavitation valve is excessive.

Since the anti-cavitation valve and valve spool are in parallel it will be necessary to conduct further testing to determine the root-cause:

Step 1.   Remove the anti-cavitation valve from the suspect port and substitute it for one from the opposite port or from an adjoining valve section. If necessary, a new anti-cavitation valve cartridge assembly can be used to help determine the root-cause.

Step 2.   Repeat the test procedure. If it is now possible to pressurize the valve port and hold pressure for a reasonable length of time, the anti-cavitation valve is suspect.

If the leakage remains unchanged after testing the valve with a good anti-cavitation valve, the spool is suspect.

If necessary, refer to the manufacturer's specification for recommended leakage data.

NOTE: *Before replacing the directional control valve check if excessive leakage is caused by the spool either not centering or binding.*

The spool may not be centering because there is an accumulation of contamination between either the spool and bore, or the end-caps.

Spool binding could also be caused by valve body mis-alignment due to sub-standard mounting procedures, over-tightened mounting bolts, or tapered pipe connectors.

Load Check Valve
Direct-Access Test Procedure

What will this test procedure accomplish?

A load check valve is basically a check valve which is installed in series with the inlet port of a directional control valve.

A load check valve performs four vital functions in hydraulic energy control.

1. **Load-check** - Holds a load in the event of unexpected prime mover shut-down with the directional control valve activated. Should this occur, it is possible for a vertically mounted actuator, which may be loaded, to push the fluid back through the system into the pump. If fluid is pushed into the pump it will tend to behave like a motor and rotate in reverse.

   If oil is pushed through a hydraulic system in reverse it could cause possible damage in three areas. The most common problem is pump shaft seal failure. If a high-pressure filter is installed in series with the pump outlet port transmission line, it will usually cause the filter element to split across the pleated media. In engine driven applications it may cause the engine to start in reverse.

2. **Load balance** - When two or more actuators, connected in parallel each with its own directional control valve, are activated simultaneously, pump flow will always go to the actuator with the lowest resistance.

   However, it is also possible for the actuator with the heaviest load to push fluid back through the directional control valve to the actuator with the lower resistance. This would cause a momentary loss of control.

   The load check valve will prevent the fluid in the loaded actuator from flowing to the actuator with the lower resistance.

3. **Shock Protection** - In certain applications an actuator could be forced to move by an external mechanical force.

   Should this occur while the directional control valve is activated, the shock would travel through the directional control valve and back to the pump. This would make the pump vulnerable to catastrophic failure.

In the event of a shock originating at the actuator, the load check valve, which is in series with the pump output transmission line, closes and prevents the shock from travelling back to the pump.

Pump flow momentarily passes over the main pressure relief valve while the shock is dissipated by a cylinder port relief valve.

4. **Pump Protection** - When a directional control valve is activated to stop a loaded actuator in an intermediate position it will trap load-generated pressure between the work-port of the directional control valve and the actuator port.

The pressure in the opposite work-port, the pump inlet port, and the tank port, may be significantly lower.

When the directional control valve is activated to raise the actuator, the load is inclined to momentarily drop or hesitate. This is caused by the pressure differential across the valve spool.

This situation could also produce undesireable pressure spikes in the system.

A load check valve protects the system by holding the load, when the directional control valve is activated, until the pressure between the pump and the load check valve is equal to the pressure between the load check valve and the loaded actuator.

The most common symptoms of problems associated with load check valves are:
a. The load drops or "hesitates" momentarily when the directional control valve is activated to lift a load from an intermediate position.
b. Pump shaft failure, shaft keyway shearing, or premature shaft spline wear.
c. Pump shaft seal failure and/or high pressure filter element distortion or catastrophic failure.
d. Momentary loss of actuator control with simultaneous operation of actuators in parallel directional control valve applications.

-WARNING- | Do not work on or around hydraulic systems without wearing safety glasses which conform to ANSI Z87.1-1989 standard.

YOU WILL NEED THE FOLLOWING DIAGNOSTIC EQUIPMENT TO CONDUCT THIS TEST PROCEDURE

CAUTION! The pressure and flow ratings of the diagnostic equipment which will be used to conduct this test procedure must be equal to, or greater than, the pressure and flow ratings of the system being tested. Refer to the system schematic for recommended pressure and flow data.

---

1. MicroLeak analyzer
2. Pressure gauge
3. Pyrometer
4. Stopwatch (not shown)

PREPARATION

To conduct this test safely and accurately, refer to Figure 7-7 for correct placement of diagnostic equipment.

TEST PROCEDURE

NOTE: *Before conducting this test procedure, determine from the circuit schematic if the directional control valve is equipped with a cylinder port relief valve or an anti-cavitation valve.*

If it is, conduct a cylinder port relief valve or an anti-cavitation valve test procedure before attempting to test the load check valve.

A cylinder port relief valve or an anti-cavitation valve is usually installed between the load check valve and the actuator, in parallel with the tank return-line.

If there is leakage across the cylinder port relief valve or the anti-cavitation valve when conducting a load check valve test procedure, the test result could be misleading.

Step 1. Shut the prime mover off.

Step 2. Lock the electrical system out or tag the keylock switch.

Step 3. Observe the system pressure gauge. Release any residual pressure trapped in the system by an accumulator, counterbalance or pilot-operated check valve, suspended load on an actuator, intensifier, or a pressurized reservoir.

Step 4. Study the system schematic and determine if the "A" and "B" ports of the directional control valve are serviced by a single load check valve or individual load check valves.

NOTE: *If the directional control valve is equipped with individual load check valves, conduct this test procedure on both the "A" and "B" ports.*

This test procedure will show the load check valve installed in the "A" port of the directional control valve.

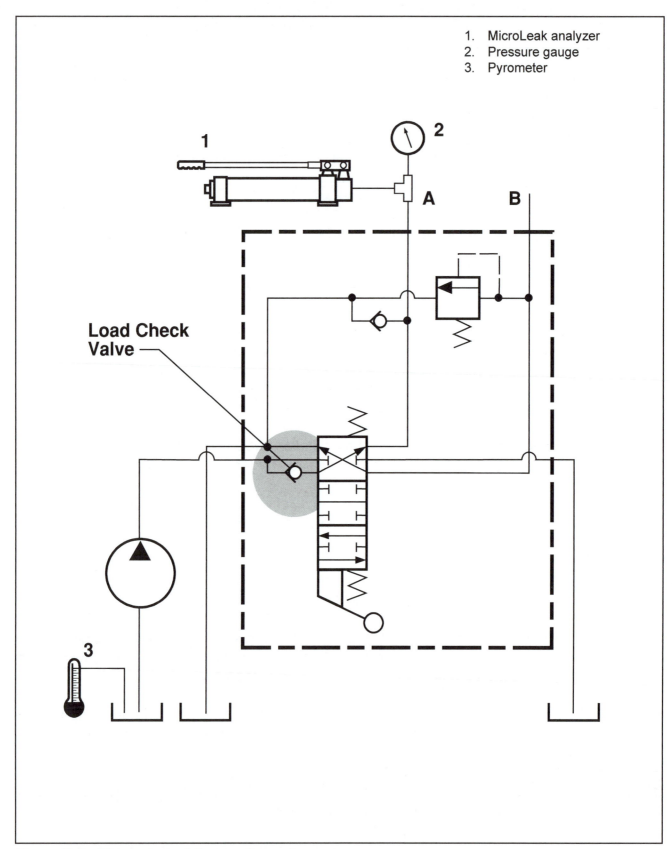

1. MicroLeak analyzer
2. Pressure gauge
3. Pyrometer

Load Check Valve

Figure 7-7   Placement of diagnostic equipment

Step 5. Remove the transmission line from the "A" port of the directional control valve.

Step 6. Connect pressure gauge (2) in parallel with the connector at the "A" port of the directional control valve.

Step 7. Connect MicroLeak analyzer (1) in series with the "A" port of the directional control valve.

Step 8. Activate the directional control valve (Valve position must be held while the test is being conducted) and align the "A" port with the "P" port.

Step 9. Gradually pressurize the "A" port passage of the directional control valve with MicroLeak analyzer (1). Stop when the value of the system's main pressure relief valve is reached.

NOTE: *Oil is now trapped between the MicroLeak analyzer (1), the valve spool, and the load check valve.*

Step 10. Observe the needle on pressure gauge (2). Using the stopwatch, time how long it takes for the pressure to drop to 200 PSI (13.8 bar).

ANALYZING THE TEST RESULTS

1. **Diagnostic Observation:** The pressure indicated on pressure gauge (2) immediately begins to decrease when the pumping action of the MicroLeak analyzer stops.

    Initially, the pressure drops fairly rapidly due to the high pressure differential across the clearances in the directional control valve and/or the load-check valve. The pressure gauge indicates that the leakage across the clearances decreases progressively as the pressure differential narrows.

    It takes a "reasonable" length of time for the pressure to drop from the value of the system's main pressure relief valve setting down to approximately 200 PSI (13.8 bar).

NOTE: *There is no general rule-of-thumb regarding the volumetric efficiency of directional control valves. If it is necessary to determine accurate leakage rates, refer to the valve manufacturer's specifications.*

    Diagnosis: Leakage across the anti-cavitation valve, the spool, and the load check valve appears to be within design specifiaction.

2. **Diagnostic Observation:** The pressure indicated on pressure gauge (2) attempts to increase when the MicroLeak analyzer is stroked. However, between pumping strokes the pressure drops rapidly.

If the "T" port of the directional control valve is open, a steady stream of oil can be seen pouring from the valve.

The MicroLeak analyzer fails to pressurize the "A" port passage to the value of the system's main pressure relief valve setting.

Diagnosis: Leakage across the spool and/or the load check valve appears to be within design specifiaction. Although this amount of leakage may appear to be insignificant, it could cause an actuator to drift when the directional control valve is activated.

Since the load-check valve and valve spool are in parallel it will be necessary to conduct further testing to determine the root-cause:

Step 1.   Remove the load-check valve from the suspect port and substitute it for one from the opposite port or from an adjoining valve section. If necessary, a new load-check valve assembly can be used to help determine the root-cause.

Step 2.   Repeat the test procedure. If it is now possible to pressurize the valve port and hold pressure for a reasonable length of time, the load-check valve is suspect.

If the leakage remains unchanged after testing the valve with a good load-check valve, the spool is suspect.

If necessary, refer to the manufacturer's specifications for recommended leakage data.

3.   **Diagnostic Observation:** The MicroLeak analyzer fails to pressurize the "A" port passage of the directional control valve. Oil leaks profusely from the open ports in the valve.

Diagnosis: Leakage across the spool and/or the load check valve is excessive.

Since the load-check valve and valve spool are in parallel it will be necessary to conduct further testing to determine the root-cause:

Step 1.   Remove the load-check valve from the suspect port and substitute it for one from the opposite port or from an adjoining valve section. If necessary, a new load-check valve assembly can be used to help determine the root-cause.

Step 2.   Repeat the test procedure. If it is now possible to pressurize the valve port and hold pressure for a reasonable length of time, the load-check valve is suspect.

If the leakage remains unchanged after testing the valve with a good load-check valve, the spool is suspect.

If necessary, refer to the manufacturer's specifications for recommended leakage data.

Hydraulic Pilot Pressure
In-Circuit Test Procedure

What will this test procedure accomplish?

Directional control valve spools can be shifted with either hydraulic or air pressure applied to the ends of the main spool. Pilot pressure is supplied to the valve through pilot lines from a remote source; usually a joystick controller or a pilot valve.

Pilot-operated directional control valves offer some unique advantages:
    a. Larger valves can be mounted away from the operator's station keeping large hose and valve assemblies out of the immediate vicinity of the operator.
    b. Expensive hose assemblies are kept to a minimum.
    c. Smaller "joystick" valves provide effortless operation thus reducing operator fatigue.
    d. Proportional spool operation can be achieved with joystick control valves.
    e. Lower "in-cab" control pressures result in a safer operating environment.

Since a pilot-operated directional control valve usually utilizes lower operating pressures than the system's main pressure relief valve setting, a pressure reducing valve is used to reduce the pilot pressure if the pilot pressure source is in parallel with the system's main pressure relief valve or compensator setting.

A reduction in pilot pressure could cause a valve spool to malfunction or become erratic. An increase in pilot pressure beyond the valve manufacturer's recommendations could result in catastrophic pilot valve and/or main control valve end-cap failure.

The most common symptoms of problems associated with pilot pressure related
problems are:
    a. Directional control valve does not respond when the pilot valve is activated.
    b. Erratic actuator operation.

This test procedure will determine the pilot pressure at the end-cap inlet ports of a directional control valve.

 **-WARNING-** | **Do not work on or around hydraulic systems without wearing safety glasses which conform to ANSI Z87.1-1989 standard.**

YOU WILL NEED THE FOLLOWING DIAGNOSTIC EQUIPMENT TO CONDUCT THIS TEST PROCEDURE

CAUTION! The pressure and flow ratings of the diagnostic equipment which will be used to conduct this test procedure must be equal to, or greater than, the pressure and flow ratings of the system being tested. Refer to the system schematic for recommended pressure and flow data.

---

1. Pressure gauge   3. Pyrometer
2. Pressure gauge

PREPARATION

To conduct this test safely and accurately, refer to Figure 7-8 for correct placement of diagnostic equipment.

To record the test data, make a copy of the test worksheet on page 7-48 (Figure 7-9).

TEST PROCEDURE

Step 1. Shut the prime mover off.

Step 2. Lock the electrical system out or tag the keylock switch.

Step 3. Observe the system pressure gauge. Release any residual pressure trapped in the system by an accumulator, counterbalance or pilot-operated check valve, suspended load on an actuator, intensifier, or a pressurized reservoir.

NOTE: *If an end-cap appears to have been damaged by excessive pressure, do not reconnect the control pressure line to the replacement valve until the control pressure has been determined. In this case, connect the pressure gauges directly into the connection at the end of the pilot pressure transmission line and activate the pilot control valve.*

This procedure will provide a means of determining pilot pressure without compromising the end-caps. If adjusting the pressure reducing valve has no effect on the pilot pressure refer to the test procedure in chapter four (Page 4-35) for further diagnosis.

Step 4. Install pressure gauge (1) in parallel with the connector at the left end-cap control pressure port.

Step 5. Install pressure gauge (2) in parallel with the connector at the right end-cap control pressure port.

Step 6. Start the prime mover. Inspect the diagnostic equipment connectors for leaks.

1. Pressure gauge
2. Pressure gauge
3. Pyrometer

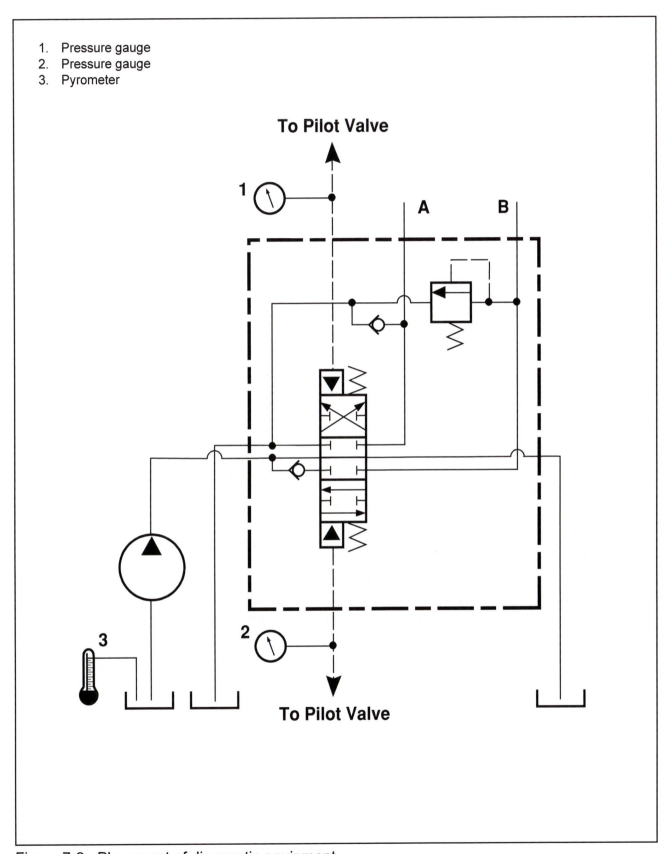

To Pilot Valve

To Pilot Valve

Figure 7-8   Placement of diagnostic equipment

Step 7.   Allow the system to warm up to approximately 130° F. (54° C.). Observe pyrometer (3).

Step 8.   Activate the pilot valve and direct pilot pressure to the left end-cap. Record on the test worksheet:
      a. The pressure indicated on pressure gauge (1).
      b. The pressure indicated on pressure gauge (2).

Step 9.   Activate the pilot valve and direct pilot pressure to the right end-cap. Record on the test worksheet:
      a. The pressure indicated on pressure gauge (2).
      b. The pressure indicated on pressure gauge (1).

Step 10. At the conclusion of this test procedure, remove the diagnostic equipment. Reconnect the transmission lines and tighten the connectors securely.

Step 11. Start the prime mover and inspect the connectors for leaks.

ANALYZING THE TEST RESULTS

1. **Diagnostic Observation:** Pressures indicated on pressure gauges (1) and (2) are nominal.

    Diagnosis: There is insufficient pilot pressure to shift the spool. There must be a pressure difference across the pilot ports of at least 70 PSI (4.8 bar) to shift the spool.

NOTE: *Pilot pressure may vary from one manufacturer to another. For specific pilot pressure data, refer to the manufacturer's specification.*

    The pilot pressure source and the pilot valve which controls the directional control valve are suspect.

2. **Diagnostic Observation:** The pressure indicated on pressure gauge (1) is not at least 70 PSI (4.8 bar) higher than the pressure indicated on pressure gauge (2).

    Diagnosis: The spool will not shift if there is insufficient pilot pressure and/or high pilot drain pressure.

    If pressure gauge (1) reading the pilot pressure indicates a pressure which is somewhat less than 70 PSI (4.8 bar), while pressure gauge (2) indicates a nominal pressure, the pilot pressure source is suspect.

    If pressure gauge (1) indicates a pressure which is somewhat less than 70 PSI (4.8 bar) while pressure gauge (2) indicates a pressure which is slightly less or equal to the pressure indicated on pressure gauge (1), there is evidence of restriction in the drain-line.

The restriction could be in the drain-line between the main directional control valve and the pilot valve. It could also be in the drain-line between the pilot valve and the reservoir.

To determine the root-cause install a pressure gauge in the next downstream connector and repeat the test procedure.

3.   **Diagnostic Observation:** Erratic actuator operation. Pressure gauges (1) and (2) indicate erratic pressures which appear to momentarily equalize.

Diagnosis: If the pressure indicated on pressure gauge (2) momentarily equals the pressure indicated on pressure gauge (1) the main valve spool will momentarily move to the center position. This action will cause the actuator to start and stop causing erratic actuator operation.

Pressure surges in the pilot valve's reservoir return-line will cause pressure transients in the pilot drain passage. This will cause the pressure on both sides of the main spool to momentarily equalize.

The primary causes of pressure surges in return-lines are return manifolds, heat-exchangers, and/or return filters.

If it is apparent that erratic actuator operation is caused by these components, the problem can be solved by re-routing the drain-line from the pilot valve directly back to the reservoir.

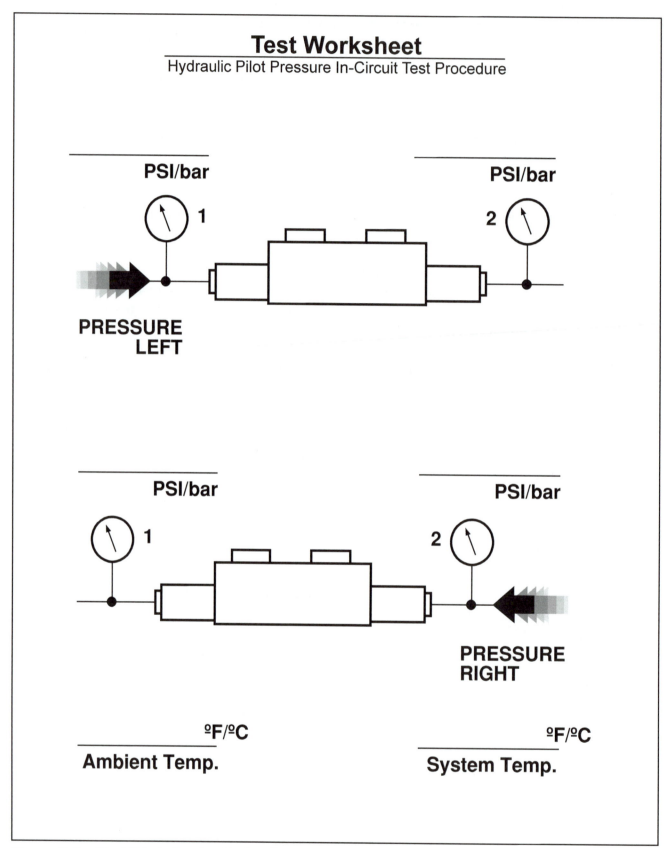

Figure 7-9   Test Worksheet - Hydraulic pilot pressure in-circuit test procedure

Pneumatic Pilot Pressure
In-Circuit Test Procedure

What will this test procedure accomplish?

Directional control valve spools can be shifted with either hydraulic or air pressure applied to the ends of the main spool. Pilot pressure is suppled to the valve through pilot lines from a remote source; usually a joystick controller or a pilot valve. Pilot-operated directional control valves offer some unique advantages:

    a. Larger valves can be mounted away from the operator's station keeping large hose and valve assemblies out of the immediate vicinity of the operator.
    b. Expensive hose assemblies are kept to a minimum.
    c. Smaller "joystick" valves provide effortless operation thus reducing operator fatigue.
    d. Proportional spool operation can be achieved with joystick control valves.
    e. Lower "in-cab" control pressures result in a safer operating environment.

Since a pilot-operated directional control valve usually utilizes lower operating pressures than the system's main pressure relief valve setting, a pressure reducing valve is used to reduce the pilot pressure if the pilot pressure source is in parallel with the main system's pressure relief valve.

A reduction in pilot pressure could cause the valve spool to malfunction or become erratic. An increase in pilot pressure beyond the valve manufacturer's recommendations could result in catastrophic failure of the pilot valve and/or the main control valve end-caps.

Cold temperature conditions could have an adverse effect on air-piloted directional control valves due to water vapor and condensation which tend to form ice in the pilot lines.

The most common symptoms of problems associated with pilot pressure related problems are:

    a. Directional control valve does not respond when the pilot valve is activated.
    b. Erratic actuator operation.

This test procedure will determine the pilot pressure at the end-cap inlet ports of a directional control valve.

 **-WARNING-** | Do not work on or around hydraulic systems without wearing safety glasses which conform to ANSI Z87.1-1989 standard.

YOU WILL NEED THE FOLLOWING DIAGNOSTIC EQUIPMENT
TO CONDUCT THIS TEST PROCEDURE

CAUTION! The pressure and flow ratings of the diagnostic equipment which will be used to conduct this test procedure must be equal to, or greater than, the pressure and flow ratings of the system being tested. Refer to the system schematic for recommended pressure and flow data.

1. Pressure gauge
2. Pressure gauge
3. Pyrometer

PREPARATION

To conduct this test safely and accurately, refer to Figure 7-10 for correct placement of diagnostic equipment.

To record the test data, make a copy of the test worksheet on page 7-53 (Figure 7-11).

TEST PROCEDURE

Step 1.   Shut the prime mover off.

Step 2.   Lock the electrical system out or tag the keylock switch.

Step 3.   Observe the system pressure gauge. Release any residual pressure trapped in the system by an accumulator, counterbalance or pilot-operated check valve, suspended load on an actuator, intensifier, or a pressurized reservoir.

Step 4.   Install pressure gauge (1) in parallel with the connector at the left end-cap control pressure port.

Step 5.   Install pressure gauge (2) in parallel with the connector at the right end-cap control pressure port.

Step 6.   Start the prime mover. Inspect the diagnostic equipment connectors for leaks.

Step 7.   Allow the system to warm up to approximately 130° F. (54° C.). Observe pyrometer (3).

Step 8.   Activate the pilot valve and direct pilot pressure to the left end-cap. Record on the test worksheet:
   a. The pressure indicated on pressure gauge (1).
   b. The pressure indicated on pressure gauge (2).

Step 9.   Activate the pilot valve and direct pilot pressure to the right end-cap. Record on the test worksheet:
   a. The pressure indicated on pressure gauge (2).
   b. The pressure indicated on pressure gauge (1).

1. Pressure gauge
2. Pressure gauge
3. Pyrometer

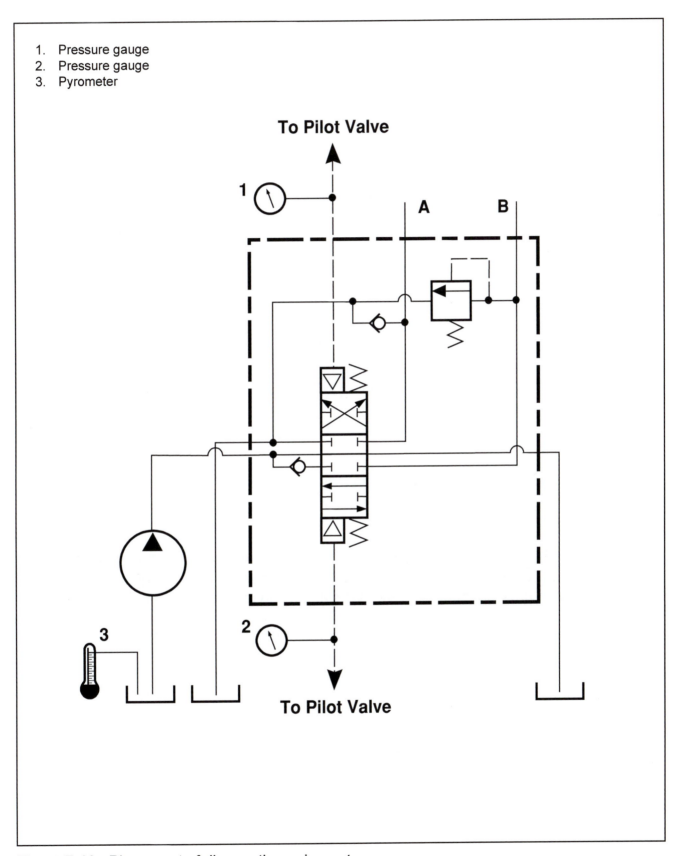

Figure 7-10   Placement of diagnostic equipment

Step 10. At the conclusion of this test procedure, remove the diagnostic equipment. Reconnect the transmission lines and tighten the connectors securely.

Step 11. Start the prime mover and inspect the connectors for leaks.

ANALYZING THE TEST RESULTS

1. **Diagnostic Observation:** Pressures indicated on pressure gauges (1) and (2) are nominal.

    Diagnosis: There is insufficient pilot pressure to shift the spool. There must be a pressure difference across the pilot ports of at least approximately 70 PSI (4.8 bar) to shift the spool.

NOTE: *Pilot pressure may vary from one manufacturer to another. For specific pilot pressure data, refer to the manufacturer's specification.*

    The pilot pressure source and the pilot valve which controls the directional control valve are suspect. Also check the air pressure regulator adjustment.

2. **Diagnostic Observation:** The pressure indicated on pressure gauge (1) is not at least 70 PSI (4.8 bar) higher than the pressure indicated on pressure gauge (2).

    Diagnosis: The spool will not shift if there is insufficient pilot pressure and/or high exhaust pressure.

    If pressure gauge (1), reading the pilot pressure, indicates a pressure which is somewhat less than 70 PSI (4.8 bar), while pressure gauge (2) indicates a nominal pressure, the pilot pressure source is suspect.

    If pressure gauge (1) indicates a pressure which is somewhat less than 70 PSI (4.8 bar) while pressure gauge (2) indicates a pressure which is slightly less or equal to the pressure indicated on pressure gauge (1), there is evidence of restriction in the exhaust line.

    To determine the root-cause install a pressure gauge in parallel with the exhaust port and repeat the test procedure.

3. **Diagnostic Observation:** Erratic actuator operation. Pressure gauges (1) and (2) indicate erratic pressures which appear to momentarily equalize.

    Diagnosis: If the pressure indicated on pressure gauge (2) momentarily equals the pressure indicated on pressure gauge (1), the main valve spool will momentarily move to the center position. This action will cause the actuator to start and stop resulting in erratic actuator operation.

    Pressure surges in the exhaust port will cause the pressure to momentarily equalize on both sides of the main spool.

Test Worksheet
Pneumatic Pilot Pressure In-Circuit Test Procedure

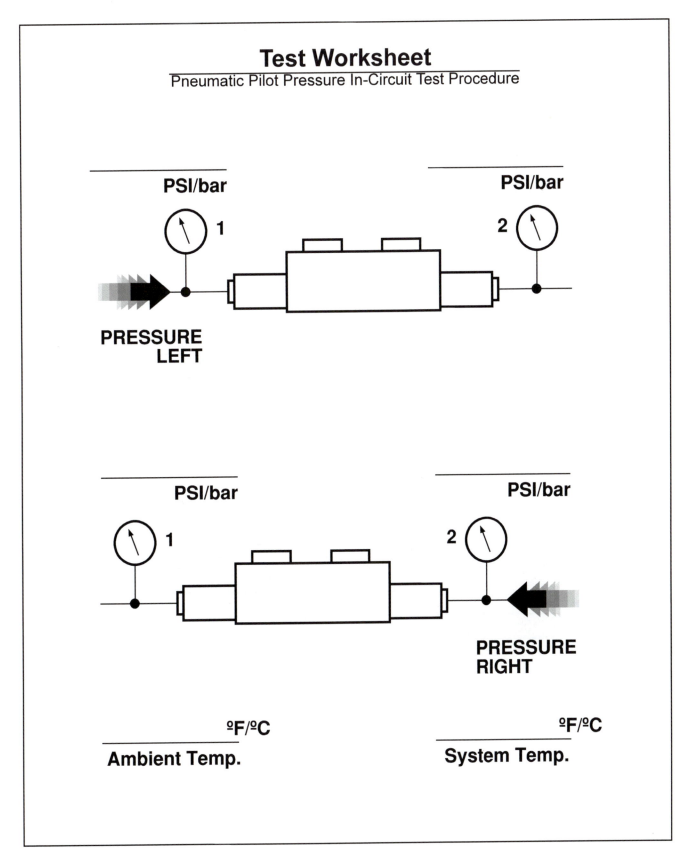

Figure 7-11   Test Worksheet - Pneumatic pilot pressure in-circuit test procedure

notes

chapter eight

Procedures for Testing Hydraulic Cylinders

Cylinder Seal/Bore "Dead-Head" Extend Test Procedure

What will this test procedure accomplish?

A hydraulic cylinder consists of a cylindrical body closed on each end. Inside the cylinder is a moveable piston attached to a piston rod. The piston has a seal on its circumference to prevent leakage across the piston.

The side of the piston to which the rod is attached is called the "live-end" or "rod-end" of the cylinder. The opposite side of the piston is called the "closed-end" or "blind-end" of the cylinder. A port is located at each end of the cylinder through which the oil can enter to move the piston.

Two types of seal systems are used in double-acting cylinders. A single bi-directional piston seal is sometimes used. It has the capability of sealing regardless of which direction the fluid enters the cylinder. The most common type is the uni-directional seal system. Two seals are installed on the piston back-to-back. One seals when the rod extends while the other seals during retract.

This test procedure is designed to check for cross-piston leakage. Cross-piston leakage can be caused by damage or wear in the internal cylinder tube wall or the piston seal system. It will also determine the condition of the live-end rod seal. It is not possible to isolate the problem to seal or bore damage by conducting this test. However, if there is an abnormal variation in cross-piston leakage the cylinder should be removed for repair.

Where the type of piston seal system is unknown it is recommended that the cylinder be tested in both the extend and retract positions.

The most common symptoms of problems associated with cross-piston leakage are:
   a. Cylinder fails to respond when directional control valve is activated.
   b. Reduced cylinder rod velocity in one or both directions of travel.
   c. Moderate to high increase in the operating temperature of the fluid.
   d. Cylinder rod speed reduction as the load increases.
   e. Actuator drifts when the directional control valve is in the neutral position.
   f. Cylinder rod extends regardless of which port the oil enters.

This test procedure will determine the amount of leakage across the piston of a double-acting cylinder.

-WARNING- **Do not work on or around hydraulic systems without wearing safety glasses which conform to ANSI Z87.1-1989 standard.**

YOU WILL NEED THE FOLLOWING DIAGNOSTIC EQUIPMENT TO CONDUCT THIS TEST PROCEDURE

CAUTION! The pressure and flow ratings of the diagnostic equipment which will be used to conduct this test procedure must be equal to, or greater than, the pressure and flow ratings of the system being tested. Refer to the system schematic for recommended pressure and flow data.

1.   Flow meter
2.   Pressure gauge
3.   Pyrometer

PREPARATION

To conduct this test safely and accurately, refer to Figure 8-1 for correct placement of diagnostic equipment.

To record the test data, make a copy of the test worksheet on page 8-6 (Figure 8-3).

TEST PROCEDURE

Step 1.   Shut the prime mover off.

Step 2.   Lock the electrical system out or tag the keylock switch.

Step 3.   Observe the system pressure gauge. Release any residual pressure trapped in the system by an accumulator, counterbalance or pilot-operated check valve, suspended load on an actuator, intensifier, or a pressurized reservoir.

Step 4.   Install flow meter (1) in series with the transmission line at the rod-end of the cylinder.

Step 5.   Install pressure gauge (2) in parallel with the connector at the closed-end of the cylinder.

Step 6.   Start the prime mover. Inspect the diagnostic equipment connectors for leaks.

Step 7.   Allow the system to warm up to approximately 130º F. (54º C.). Observe pyrometer (3).

Step 8.   Activate the directional control valve to the rod extend position. When the rod reaches the end of its stroke the pressure in the closed-end of the cylinder increases to the value of the main pressure relief valve or compensator setting.

1. Flow meter
2. Pressure gauge
3. Pyrometer

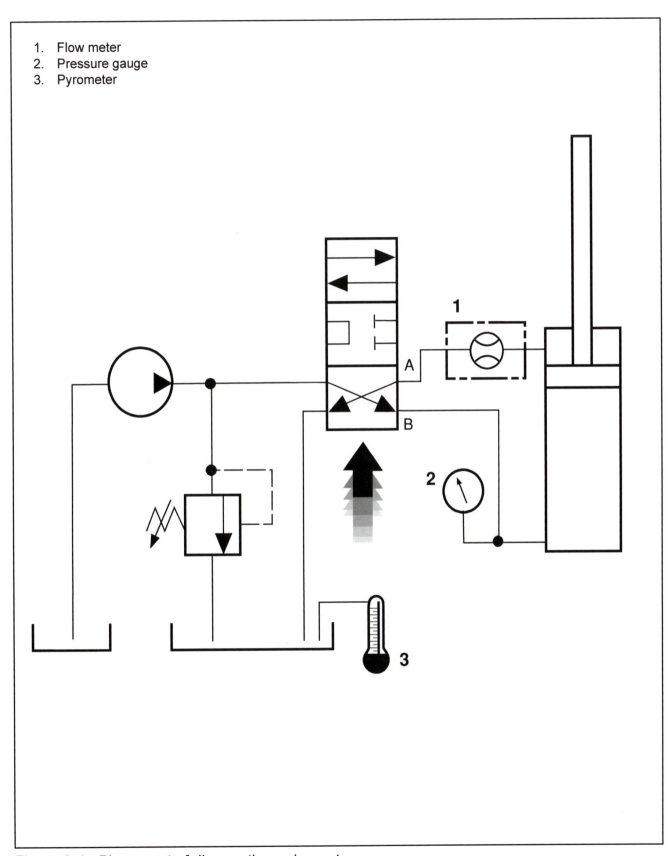

Figure 8-1   Placement of diagnostic equipment

Step 9.   While holding this position, record on the test worksheet:
a. Pressure indicated on pressure gauge (2).
b. Flow indicated on flow meter (1).

Step 10.   Release the directional control valve.

Step 11.   Shut the prime mover off and analyze the test results.

Step 12.   At the conclusion of this test procedure, remove the diagnostic equipment. Reconnect the transmission lines and tighten the connectors securely.

Step 13.   Start the prime mover and inspect the connectors for leaks.

ANALYZING THE TEST RESULTS

1.   **Diagnostic Observation**: Pressure gauge (2) indicates main pressure relief valve setting. Flow meter (1) indicates zero to marginal flow.

   Diagnosis: Leakage across the piston appears to be within design specification.

2.   **Diagnostic Observation:** Pressure gauge (2) indicates main pressure relief valve or compensator setting. Flow meter (1) indicates moderate to high flow.

   Diagnosis: Leakage across the piston appears to be excessive. If necessary, refer to the manufacturer's specification for recommended leakage data.

3.   **Diagnostic Observation:** Pressure gauge (2) indicates a pressure somewhat less than main pressure relief valve setting. Flow meter (1) indicates high flow.

   Diagnosis: Leakage across the piston is excessive. Service or replace the cylinder.

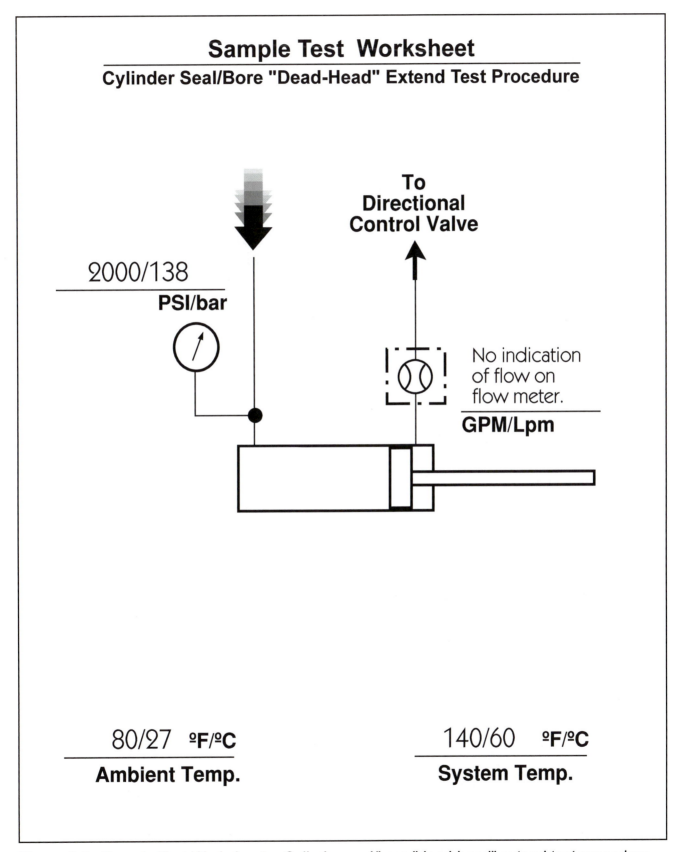

Figure 8-2   Sample Test Worksheet - Cylinder seal/bore "dead-head" extend test procedure

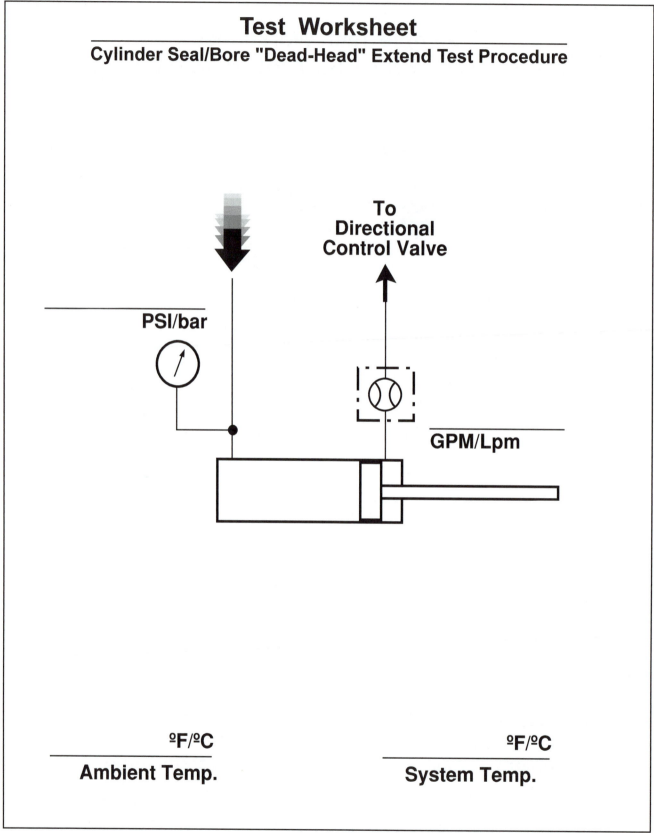

Figure 8-3   Test Worksheet - Cylinder seal/bore "dead-head" extend test procedure

Cylinder Seal/Bore "Dead-Head" Retract Test Procedure

What will this test procedure accomplish?

A hydraulic cylinder consists of a cylindrical body closed on each end. Inside the cylinder is a moveable piston attached to a piston rod. The piston has a seal on its circumference to prevent leakage across the piston.

The side of the piston to which the rod is attached is referred to as the "live-end" or "rod-end" of the cylinder. The opposite side of the piston is called the "closed-end" or "blind-end" of the cylinder. A port is located at each end of the cylinder through which the oil can enter to move the piston.

Two types of seal systems are used in double-acting cylinders. A single bi-directional piston seal is sometimes used. It has the capability of sealing regardless of which direction the fluid enters the cylinder. The most common type is the uni-directional seal system. Two seals are installed on the piston back-to-back. One seals when the rod extends while the other seals during retract.

This test procedure is designed to check for cross-piston leakage. Cross-piston leakage can be caused by damage or wear in the internal cylinder tube wall or the piston seal system. It will also determine the condition of the live-end rod seal. It is not possible to isolate the problem to seal or bore damage by conducting this test. However, if there is an abnormal variation in cross-piston leakage, the cylinder should be removed for repair.

Where the type of piston seal system is unknown, it is recommended that the cylinder be tested in both the extend and retract positions.

The most common symptoms of problems associated with cross-piston leakage are:
    a. Cylinder fails to respond when directional control valve is activated.
    b. Reduced cylinder rod velocity in one or both directions of travel.
    c. Moderate to high increase in the operating temperature of the fluid.
    d. Cylinder rod speed reduction as the load increases.
    e. Actuator drifts when the directional control valve is in the neutral position.
    f. Cylinder rod extends regardless of which port the oil enters.

This test procedure will determine the amount of leakage across the piston of a double-acting cylinder.

 **-WARNING-**   **Do not work on or around hydraulic systems without wearing safety glasses which conform to ANSI Z87.1-1989 standard.**

YOU WILL NEED THE FOLLOWING DIAGNOSTIC EQUIPMENT TO CONDUCT THIS TEST PROCEDURE

CAUTION! The pressure and flow ratings of the diagnostic equipment which will be used to conduct this test procedure must be equal to, or greater than, the pressure and flow ratings of the system being tested. Refer to the system schematic for recommended pressure and flow data.

---

1. Flow meter
2. Pressure gauge
3. Pyrometer

PREPARATION

To conduct this test safely and accurately, refer to Figure 8-4 for correct placement of diagnostic equipment.

To record the test data, make a copy of the test worksheet on page 8-12 (Figure 8-6).

TEST PROCEDURE

Step 1.   Shut the prime mover off.

Step 2.   Lock the electrical system out or tag the keylock switch.

Step 3.   Observe the system pressure gauge. Release any residual pressure trapped in the system by an accumulator, counterbalance or pilot-operated check valve, suspended load on an actuator, intensifier, or a pressurized reservoir.

Step 4.   Install flow meter (1) in series with the transmission line at the closed-end of the cylinder.

Step 5.   Install pressure gauge (2) in parallel with the connector at the rod-end of the cylinder.

Step 6.   Start the prime mover. Inspect the diagnostic equipment connectors for leaks.

Step 7.   Allow the system to warm up to approximately 130° F. (54° C.). Observe pyrometer (3).

Step 8.   Activate the directional control valve to the rod retract position. When the rod reaches the end of its stroke the pressure in the rod-end of the cylinder increases to the value of the main pressure relief valve setting.

Step 9.   While holding this position, record on the test worksheet :
   a. Flow indicated on flow-meter (1).
   b. Pressure indicated on pressure gauge (2).

1. Flow meter
2. Pressure gauge
3. Pyrometer

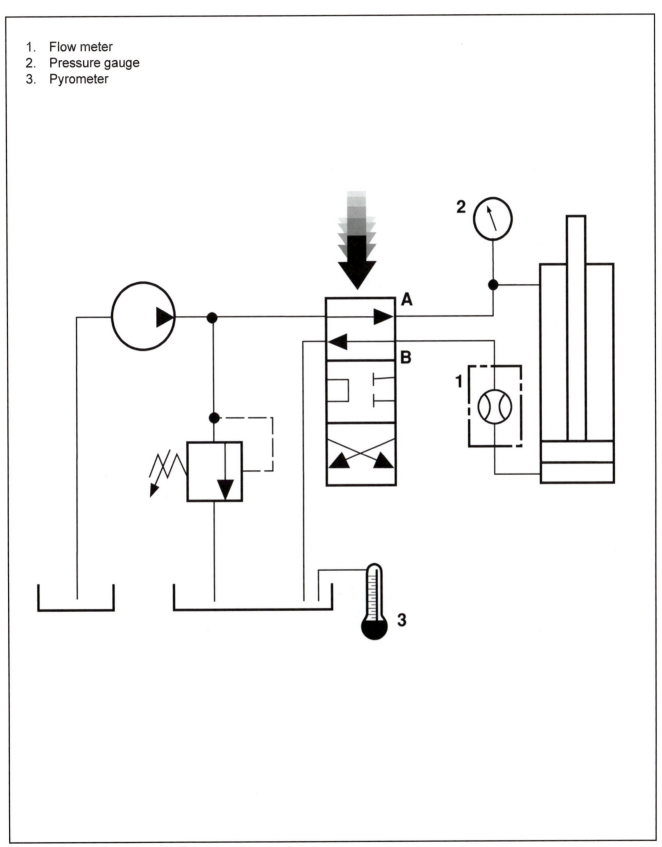

Figure 8-4   Placement of diagnostic equipment

Step 10. Release the directional control valve.

Step 11. Shut the prime mover off and analyze the test results.

Step 12. At the conclusion of this test procedure, remove the diagnostic equipment. Reconnect the transmission lines and tighten the connectors securely.

Step 13. Start the prime mover and inspect the connectors for leaks.

ANALYZING THE TEST RESULTS

1. **Diagnostic Observation**: Pressure gauge (2) indicates main pressure relief valve setting. Flow meter (1) indicates zero to marginal flow.

    Diagnosis: Leakage across the piston appears to be within design specification.

2. **Diagnostic Observation:** Pressure gauge (2) indicates main pressure relief valve setting. Flow meter (1) indicates moderate to high flow.

    Diagnosis: Leakage across the piston appears to be excessive. If necessary, refer to the manufacturer's specification for leakage data.

3. **Diagnostic Observation:** Pressure gauge (2) indicates a pressure somewhat less than main pressure relief valve setting. Flow meter (1) indicates high flow.

    Diagnosis: Leakage across the piston is excessive. Service or replace the cylinder.

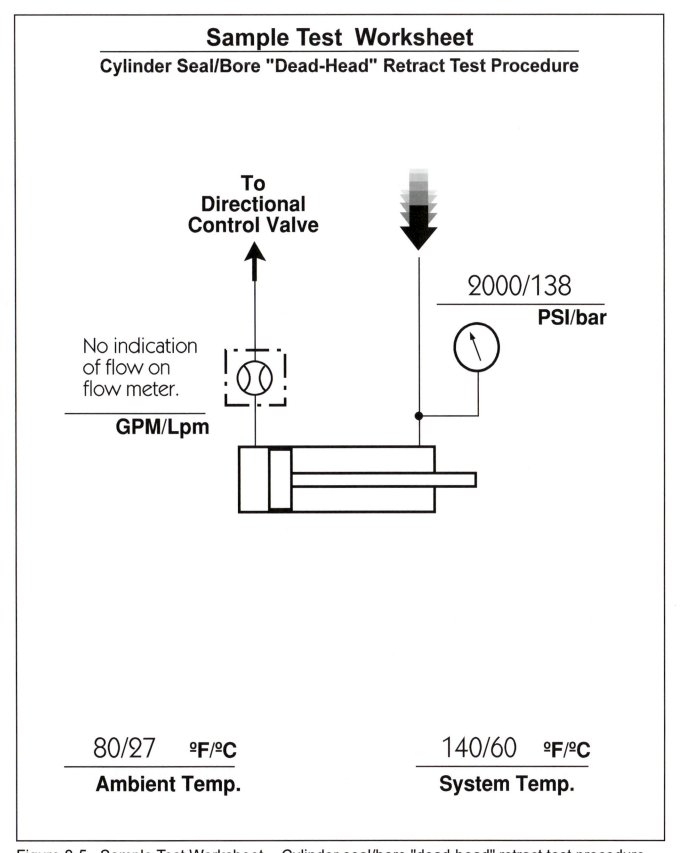

Figure 8-5   Sample Test Worksheet - Cylinder seal/bore "dead-head" retract test procedure

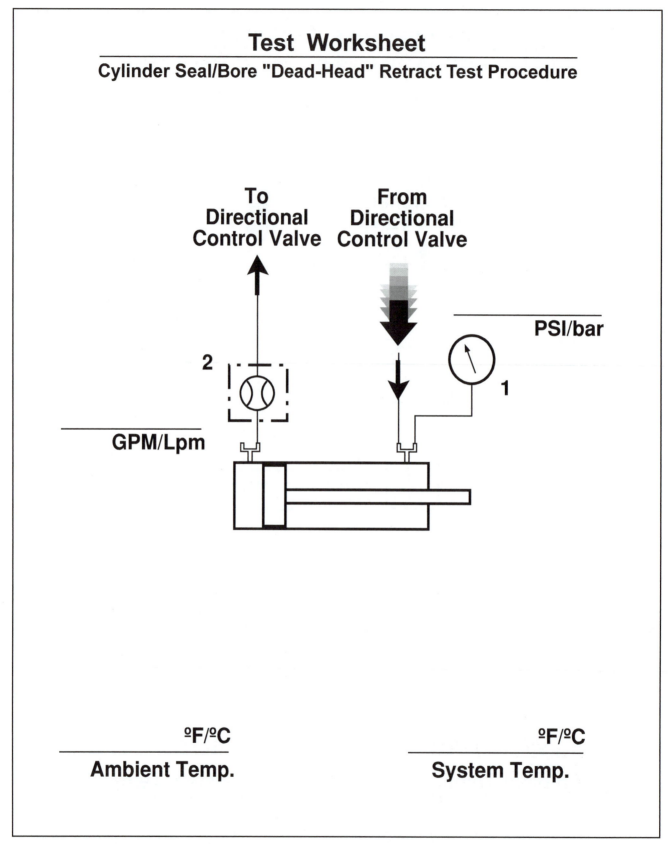

Figure 8-6   Test Worksheet - Cylinder seal/bore "dead-head" retract test procedure

Cylinder Intermediate Bore
Test Procedure

What will this test procedure accomplish?

The mechanical design parameters of industrial process and mobile equipment sometimes limits the amount of available cylinder stroke utilized during operation. When the piston in a hydraulic cylinder cycles at high frequency in an intermediate-stroke position, the cylinder tube wear tends to be localized. This can result in cylinder rod drifting caused by intermediate bore wear. The problem will usually go undetected in end-stroke testing.

Intermediate-stroke testing is an effective method of determining abnormal variations in cylinder wall/piston seal leakage in the intermediate-stroke position. Leakage in this area can also be caused by cylinder tube bulging from shock loads, or substandard tube machining, welding, or honing practices.

Typical causes of intermediate cylinder bore wear or damage are:
   a. Constant cycling in the intermediate-stroke position.
   b. Bent cylinder rod.
   c. Misalignment in cylinder mounting.
   d. Sub-standard cylinder mounting.
   e. Stop-tubes not installed or too short.
   f. External shock loads to the rod can cause tubes to bulge.
   g. Welding on the cylinder tube.
   h. Excessive force being applied to cylinder rod
      (force output of system exceeds rating of cylinder).

The most common symptoms of problems associated with intermediate cylinder bore damage or wear are:
   a. Load drifting in the intermediate position.
   b. Vertical load drifting in the intermediate position.
   c. Loss of precise positioning control in the intermediate position.
   d. Load "dropping" when cylinder is in the neutral (hold) position.
   e. Accumulator bleeding down rapidly in cylinder hold position in applications where an accumulator is utilized to make-up for normal leakage in neutral "extended hold" applications.
   f. Machine "wandering" in steering applications.

This test will determine the amount of leakage across the piston of a double-acting cylinder in the intermediate-stroke position.

-WARNING- Do not work on or around hydraulic systems without wearing safety glasses which conform to ANSI Z87.1-1989 standard.

YOU WILL NEED THE FOLLOWING DIAGNOSTIC EQUIPMENT TO CONDUCT THIS TEST PROCEDURE

CAUTION! The pressure and flow ratings of the diagnostic equipment which will be used to conduct this test procedure must be equal to, or greater than, the pressure and flow ratings of the system being tested. Refer to the system schematic for recommended pressure and flow data.

---

1. Flow meter
2. Pressure gauge
3. Pyrometer

PREPARATION

To conduct this test safely and accurately, refer to Figure 8-7 for correct placement of diagnostic equipment.

To record the test data, make a copy of the test worksheet on page 8-18 (Figure 8-9).

TEST PROCEDURE

Step 1.   Shut the prime mover off.

Step 2.   Lock the electrical system out or tag the keylock switch.

Step 3.   Observe the system pressure gauge. Release any residual pressure trapped in the system by an accumulator, counterbalance or pilot-operated check valve, suspended load on an actuator, intensifier, or a pressurized reservoir.

Step 4 .   Install flow meter (1) in series with the transmission line at the live-end of the cylinder.

Step 5.   Install pressure gauge (2) in parallel with the connector at the closed-end of the cylinder.

Step 6.   Start the prime mover. Inspect the diagnostic equipment connectors for leaks.

Step 7.   Allow the system to warm up to approximately 130° F. (54° C.). Observe pyrometer (3).

Step 8.   Adjust the pressure relief valve or compensator to its lowest pressure setting.

Step 9.   Extend the cylinder rod until it reaches the intermediate-stroke position. Using a suitable device block the cylinder in this position.

WARNING! DO NOT use cables or chains to stall an actuator as they are inclined to stretch under stress. Should either the cable or chain fail under stress, their behavior would be unpredictable.

---

1. Flow meter
2. Pressure gauge
3. Pyrometer

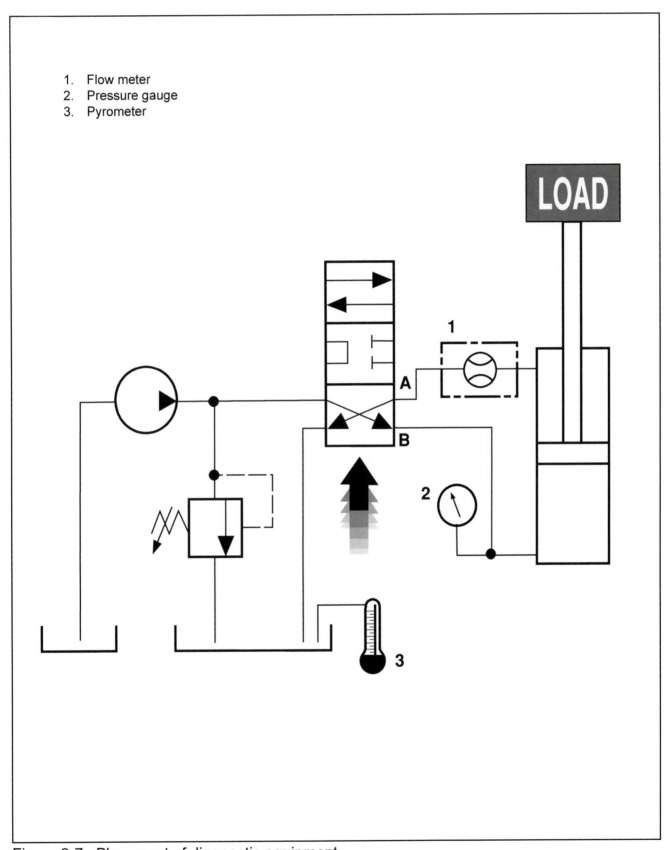

Figure 8-7   Placement of diagnostic equipment

Step 10. Activate the directional control valve. Select a position that will direct pressurized fluid to the side of the piston opposite the block. Hold the directional control valve in this position during the test procedure.

Step 11. On the test worksheet, record:
   a. Flow indicated on flow meter (1).
   b. Pressure indicated on pressure gauge (2).

NOTE: *It may be necessary to increase the pressure relief valve or compensator setting slightly (approximately 500 PSI (34.5 bar)) to conduct the test procedure.*

Step 12. Shut the prime mover off and analyze the test results.

Step 13. At the conclusion of this test procedure, remove the diagnostic equipment. Reconnect the transmission lines and tighten the connectors securely.

Step 14. Start the prime mover and inspect the connectors for leaks.

Step 15. Adjust the pressure relief valve or compensator back to its original setting. Refer to the manufacturer's specification for the recommended pressure relief valve setting.

ANALYZING THE TEST RESULTS

1. **Diagnostic Observation**: Pressure gauge (2) indicates main pressure relief valve setting. Flow meter (1) indicates zero to marginal flow.

   Diagnosis: Leakage across the piston, in the intermediate-stroke position, appears to be within design specification.

2. **Diagnostic Observation:** Pressure gauge (2) indicates main pressure relief valve setting. Flow meter (1) indicates moderate to high flow.

   Diagnosis: Leakage across the piston, in the intermediate-stroke position, appears to be excessive. If necessary, refer to the manufacturer's specification for leakage data.

3. **Diagnostic Observation:** Pressure gauge (2) indicates a pressure somewhat less than main pressure relief valve setting. Flow meter (1) indicates high flow.

   Diagnosis: Leakage across the piston, in the intermediate-stroke position, is excessive. Service or replace the cylinder.

Sample Test Worksheet
Cylinder Intermediate Bore Test Procedure

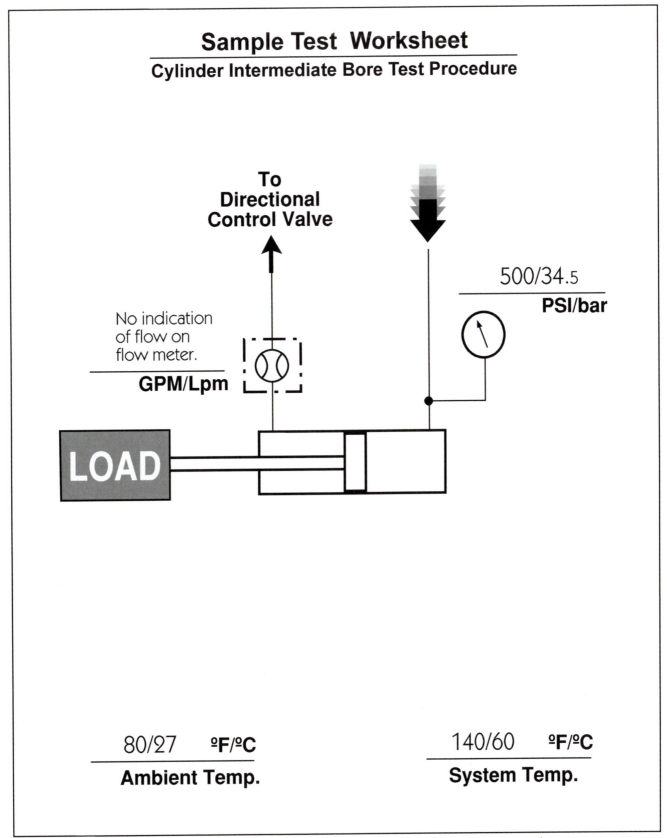

To
Directional
Control Valve

500/34.5
PSI/bar

No indication
of flow on
flow meter.
GPM/Lpm

LOAD

80/27   **ºF/ºC**
Ambient Temp.

140/60   **ºF/ºC**
System Temp.

Figure 8-8   Sample Test Worksheet - Cylinder intermediate bore test procedure

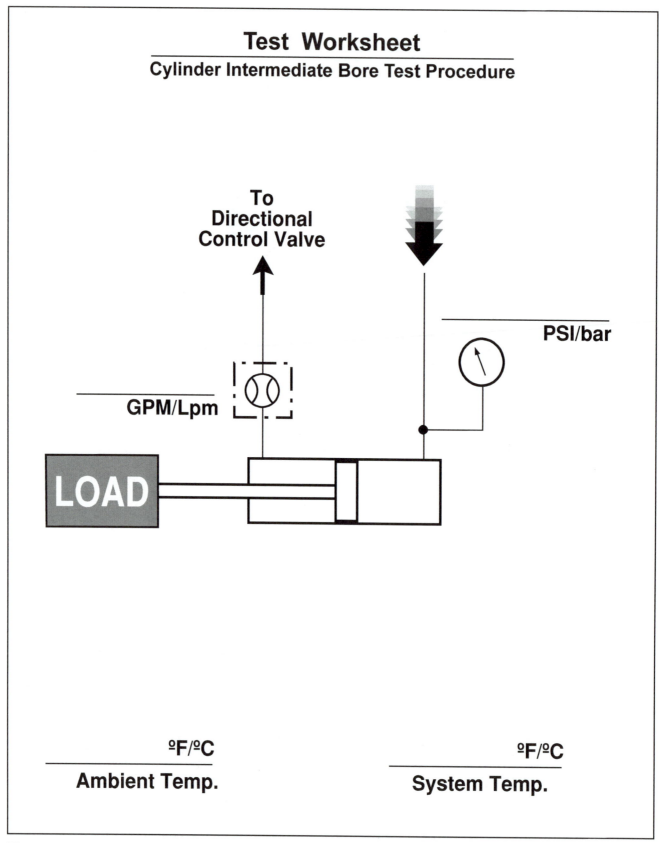

Figure 8-9   Test Worksheet - Cylinder intermediate bore test procedure

Cylinder Thru-Stroke
Test Procedure

What will this test procedure accomplish?

The mechanical design parameters of industrial process and mobile equipment sometimes limits the amount of available cylinder stroke utilized during operation. When the piston in a hydraulic cylinder cycles at high frequency in an intermediate-stroke position, the cylinder tube wear tends to be localized. This can result in cylinder rod drifting caused by intermediate bore wear. The problem will usually go undetected in end-stroke testing.

Thru-stroke testing is an effective method of determining abnormal variations in cylinder wall/piston seal leakage in the intermediate-stroke position. Leakage in this area can also be caused by cylinder tube bulging from shock loads, or substandard tube machining, welding, or honing practices.

Typical causes of intermediate cylinder bore wear or damage are:
 a. Constant cycling in the intermediate-stroke position.
 b. Bent cylinder rod.
 c. Misalignment in cylinder mounting.
 d. Sub-standard cylinder mounting.
 e. Stop-tubes not installed or too short.
 f. External shock loads to the rod can cause tubes to bulge.
 g. Welding on the cylinder tube.
 h. Excessive force being applied to cylinder rod
  (force output of system exceeds rating of cylinder).

The most common symptoms of problems associated with intermediate cylinder bore damage or wear are:
 a. Load drifting in the intermediate position.
 b. Vertical load drifting in the intermediate position.
 c. Loss of precise positioning control in the intermediate position.
 d. Load "dropping" when cylinder is in the neutral (hold) position.
 e. Accumulator bleeding down rapidly in cylinder "hold" position in applications where an accumulator is utilized to make-up for normal leakage in neutral "extended hold" applications.

This test procedure will determine if there is damage or wear on the inside diameter of a hydraulic cylinder.

 **-WARNING-**   **Do not work on or around hydraulic systems without wearing safety glasses which conform to ANSI Z87.1-1989 standard.**

YOU WILL NEED THE FOLLOWING DIAGNOSTIC EQUIPMENT TO CONDUCT THIS TEST PROCEDURE

CAUTION! The pressure and flow ratings of the diagnostic equipment which will be used to conduct this test procedure must be equal to, or greater than, the pressure and flow ratings of the system being tested. Refer to the system schematic for recommended pressure and flow data.

1. Needle valve
2. Pressure gauge
3. Pyrometer

PREPARATION

To conduct this test safely and accurately, refer to Figure 8-10 for correct placement of diagnostic equipment.

TEST PROCEDURE

Step 1.   Shut the prime mover off.

Step 2.   Lock the electrical system out or tag the keylock switch.

Step 3.   Observe the system pressure gauge. Release any residual pressure trapped in the system by accumulators, counterbalance or pilot-operated check valves, suspended loads on cylinders, intensifiers, or pressurized reservoirs.

Step 4 .   Install needle valve (1) in series with the transmission line at the rod-end of the cylinder.

Step 5.   Install pressure gauge (2) in parallel with the connector at the rod-end of the cylinder.

Step 6.   Start the prime mover. Inspect the diagnostic equipment connectors for leaks.

Step 7.   Allow the system to warm up to approximately 130° F. (54° C.). Observe pyrometer (3).

Step 8.   Open needle valve (1) fully (turn counter-clockwise).

Step 9.   Retract the cylinder rod fully.

Step 10.   Adjust the system's main pressure relief valve or compensator to its lowest setting.

Step 11.   Close needle valve (1) securely (turn clockwise).

1. Needle valve
2. Pressure gauge
3. Pyrometer

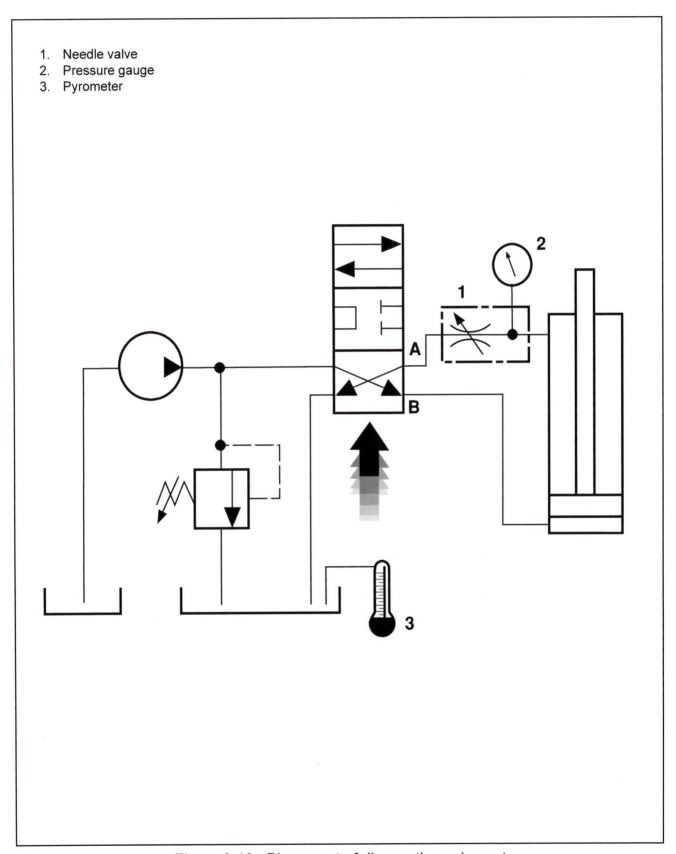

Figure 8-10   Placement of diagnostic equipment

Step 12. Activate the directional control valve and attempt to extend the cylinder rod. (The rod should not extend while the needle valve is closed).

Step 13. While maintaining pressure in the closed-end of the cylinder, adjust the system's main pressure relief valve or compensator setting to the pressure recommended by the manufacturer. Observe pressure gauge (2).

NOTE: *If the cylinder rod should "creep" out through its entire stroke while the needle valve is closed securely, suspect cylinder bore or piston seal damage. If the cylinder rod does not "creep", proceed with the next step.*

Step 14. Allow the cylinder rod to extend in short increments by opening and closing needle valve (1). (Each time the needle valve is closed, pause momentarily). Continue this procedure until the cylinder rod travels through its entire stroke.

NOTE: *If the cylinder rod extends while the needle valve is closed, suspect local cylinder bore damage. Observe how far the cylinder rod travels before it stops. This will determine the extent of the cylinder bore damage.*

Step 15. At the conclusion of this test procedure, remove the diagnostic equipment. Reconnect the transmission lines and tighten the connectors securely.

Step 16. Start the prime mover and inspect the connectors for leaks.

Step 17. Adjust the pressure relief valve or compensator back to it's original setting. Refer to the manufacturer's specification for the recommended pressure relief valve or compensator setting.

ANALYZING THE TEST RESULTS

1. **Diagnostic Observation**: With needle valve (1) securely closed and the pressure indicated on pressure gauge (2) at main pressure relief valve setting, the cylinder rod does not move.

   Diagnosis: The piston seal and local bore area appear to be in satisfactory operating condition.

2. **Diagnostic Observation:** With needle valve (1) securely closed and the pressure indicated on pressure gauge (2) at the main pressure relief valve setting, the cylinder rod extends slowly.

   Diagnosis: There is evidence of piston seal and/or bore damage. If necessary, refer to the manufacturer's specification for leakage data.

chapter nine

Procedures for Testing Hydraulic Motors

Internally Drained Motor
In-Circuit Test Procedure

What will this test procedure accomplish?

All hydraulic motors have component parts which move in relation to one another, separated by a small oil-filled clearance. These components are generally loaded toward one another by forces related to system pressure, surface area, and springs or seals.

Motors are very similar to pumps in construction and volumetric efficiency. However, there are some subtle but significant differences between pumps and motors. Pumps "push" on the fluid to generate flow. Pump input shaft rotation is typically uni-directional, and the inlet port is usually larger in diameter than the outlet port.

Motors, being actuators, are "pushed" by the fluid to develop continuous rotation or torque. Motor output shaft rotation is generally bi-directional, and the pressure ports are usually the same diameter.

Displacement is the amount of fluid required to turn the motor output shaft one revolution. Motor displacement determines output shaft speed relative to flow input, and output torque relative to the pressure drop across the motor.

Theoretically, a motor rotates at a speed equal to its displacement during each revolution. The actual speed is reduced because of internal leakage. As the pressure difference across a motor increases, leakage also increases causing a proportional decrease in motor shaft speed.

The amount of leakage is influenced by four factors:
1. Pressure difference across the clearances;
2. Oil viscosity;
3. Temperature; and
4. Size of clearance.

Internal leakage can either flow from the inlet port to the outlet port (internal drain), or from the inlet port into the motor case (external drain). If a motor is externally drained, a separate port, which is generally smaller in diameter than the pressure ports, is provided in the motor case. A case drain-line is connected to the drain port to transport the leakage directly back to the fluid reservoir.

Generally, the volumetric efficiency of a positive displacement hydraulic motor is between 80% and 97% depending upon design.

Since a motor is generally reversible, direction of rotation will determine which is the inlet port and which is the outlet port.

The volumetric efficiency or leakage of an internally drained motor can be determined by comparing the "no-load" motor shaft speed against the "full-load" speed while simultaneously monitoring the flow input with a flow meter. Some internally drained motors include: internal gear, external gear, gerotor, and balanced vane.

The most common symptoms of problems associated with excessive leakage across a motor are:
    a. A progressive loss of motor shaft speed as the fluid temperature increases.
    b. A moderate to high increase in the operating temperature of the fluid.
    c. Motor stalling at low speed.

This test procedure will determine the amount of leakage across the ports of an internally drained motor.

-WARNING- Do not work on or around hydraulic systems without wearing safety glasses which conform to ANSI Z87.1-1989 standard.

YOU WILL NEED THE FOLLOWING DIAGNOSTIC EQUIPMENT TO CONDUCT THIS TEST PROCEDURE

CAUTION! The pressure and flow ratings of the diagnostic equipment which will be used to conduct this test procedure must be equal to, or greater than, the pressure and flow ratings of the system being tested. Refer to the system schematic for recommended pressure and flow data.

    1.   Pressure gauge   3.   Tachometer
    2.   Flow meter   4.   Pyrometer

PREPARATION

To conduct this test safely and accurately, refer to Figure 9-1 for correct placement of diagnostic equipment.

To record the test data, make a copy of the test worksheet on page 9-8 (Figure 9-3).

TEST PROCEDURE

Step 1.   Shut the prime mover off.

Step 2.   Lock the electrical system out or tag the keylock switch.

Step 3.   Observe the system pressure gauge. Release any residual pressure trapped in the system by an accumulator, counterbalance or pilot-operated check valve, suspended load on an actuator, intensifier, or a pressurized reservoir.

1. Pressure gauge
2. Flow meter
3. Tachometer
4. Pyrometer

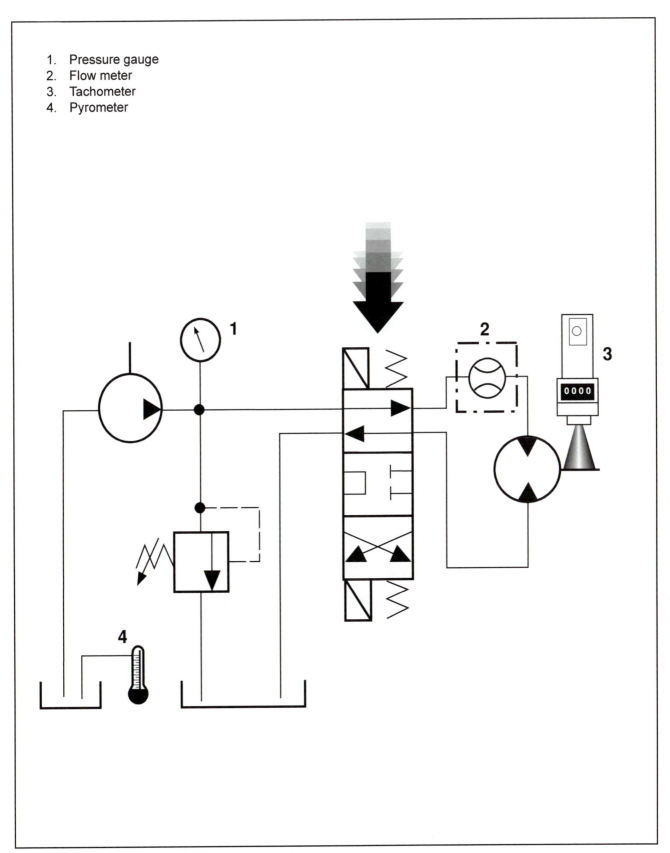

Figure 9-1 Placement of diagnostic equipment

Step 4.   Install flow meter (2) in series with the transmission line at the inlet port of the motor. The motor should be tested in forward and reverse if it is being used in a bi-directional rotation application.

Step 5.   Install pressure gauge (1) in parallel with the connector at the inlet port of flow meter (2).

Step 6.   Start the prime mover. Inspect the diagnostic equipment connectors for leaks.

NOTE: *If catastrophic failure is suspected, do not run the motor for longer than is absolutely necessary to determine its condition. If metal fragments are found in transmission lines, inlet or outlet ports, return line filter, or heat exchanger, DO NOT operate the motor. Replace the motor and conduct a post-catastrophic failure start-up procedure.*

Step 7.   Allow the system to warm up to approximately 130°F. (54°C.). Observe pyrometer (4).

NOTE: *For accurate test results, check and adjust (if necessary) the main pressure relief valve or compensator setting. Set to the pressure recommended by the manufacturer.*

Step 8.   Operate the prime mover at maximum speed. (Conduct Step 10 and Step 11 with the prime mover operating at maximum speed).

Step 9.   Record on the test worksheet the pump "no-load" input shaft speed indicated on tachometer (3).

NOTE: *If the tachometer indicates a pump "no-load" speed which is inconsistent with machine specification, adjust the speed accordingly and continue with the test procedure.*

Pump flow is proportional to input shaft speed. If there is a reduction in pump flow because of low prime mover speed, it will directly affect motor shaft speed. It is necessary to determine pump shaft speed prior to conducting an actuator speed test procedure.

Step 10.   Remove the load from the motor output shaft.

Step 11.   Activate the directional control valve and direct full pump flow through flow meter (2) into the motor. Record, on the test worksheet:
          a.   "no-load" pressure indicated on pressure gauge (1).
          b.   "no-load" flow into the motor indicated on flow meter (2).
          c.   "no-load" motor shaft speed indicated on tachometer (3).
          d.   Fluid temperature indicated on pyrometer (4).

Step 12.   Operate the machine through a normal "full-load" cycle. Record, on the test worksheet:
          a.   "full-load" pressure indicated on pressure gauge (1).
          b.   "full-load" flow into the motor indicated on flow meter (2).
          c.   "full-load" motor shaft speed indicated on tachometer (3).
          d.   "full-load" pump input shaft speed indicated on tachometer (3).
          e.   Fluid temperature indicated on pyrometer (4).

Step 13.   Shut the prime mover off and analyze the test results.

Step 14. At the conclusion of this test procedure, remove the diagnostic equipment. Reconnect the transmission lines and tighten the connectors securely.

Step 15. Start the prime mover and inspect the connectors for leaks.

ANALYZING THE TEST RESULTS

1. **Diagnostic Observation: "full-load" cycle:** Flow meter (2) indicates a nominal flow decrease as the pressure drop across the motor ports increases. The flow decrease is not less than the anticipated loss relative to the volumetric efficiency rating of the pump and a nominal decrease in prime mover speed.

    Tachometer (3) indicates a marginal decrease in motor speed. The speed decrease does not exceed the anticipated loss relative to the volumetric efficiency of the motor.

    The operating temperature of the fluid indicated on pyrometer (4) remains within design specification.

    Prime mover speed indicated on tachometer (3) remains within design specification.

    Diagnosis: Leakage across the motor appears to be within design specification.

2. **Diagnostic Observation: "full-load" cycle:** Flow meter (2) indicates a moderate to high flow decrease as the pressure drop across the hydraulic system increases.

    Tachometer (3) indicates a reduction in motor shaft speed which is proportional to the flow decrease.

    Pyrometer (4) indicates a moderate increase in the operating temperature of the fluid.

    Prime mover speed remains within design specification.

    Diagnosis: The loss in motor shaft speed is proportional to the decrease in flow into the motor. This indicates that the motor is not the cause of the problem.

    To determine the root-cause each component in the system will have to be individually tested.

3. **Diagnostic Observation: "full-load" cycle:** Flow meter (2) indicates a flow decrease which is proportional to the volumetric efficiency of the system.

    Tachometer (3) indicates a progressive decrease in motor shaft speed as the load at the motor output shaft increases. The speed decrease approaches, or is in excess of, 30% of motor no-load speed.

NOTE: *Component manufacturers rarely publish data on when the useful life of a component expires. The decision on when to remove a component is generally left to the discretion of the end-user.*

We conducted an extensive study on motor leakage versus useful life and it was determined that if the leakage approaches or exceeds approximately 30% of the theoretical (no-load) flow, at full-load (maximum rated pressure) it should be replaced.

Some motors will appear to operate normally above this figure, others will show problems well below this figure. However, the side effects of excessive leakage are detrimental to the well-being of the entire hydraulic system. It could cause a significant increase in the operating temperature of the fluid, and/or a loss in production.

A well managed, pro-active maintenance program will help establish the maximum leakage levels of your hydraulic components.

Pyrometer (4) indicates a progressive increase in the operating temperature of the fluid which does not appear to "level-off."

Prime mover speed remains within design specification.

Diagnosis: If the flow meter indicates that the flow into the motor remains within the volumetric efficiency of the system, and the tachometer indicates a speed decrease which causes an unacceptable loss in motor shaft speed, leakage across the motor is excessive. Service or replace the motor.

Leakage of this magnitude could also cause a significant increase in the operating temperature of the fluid.

If necessary, refer to the manufacturer's specification for recommended leakage data.

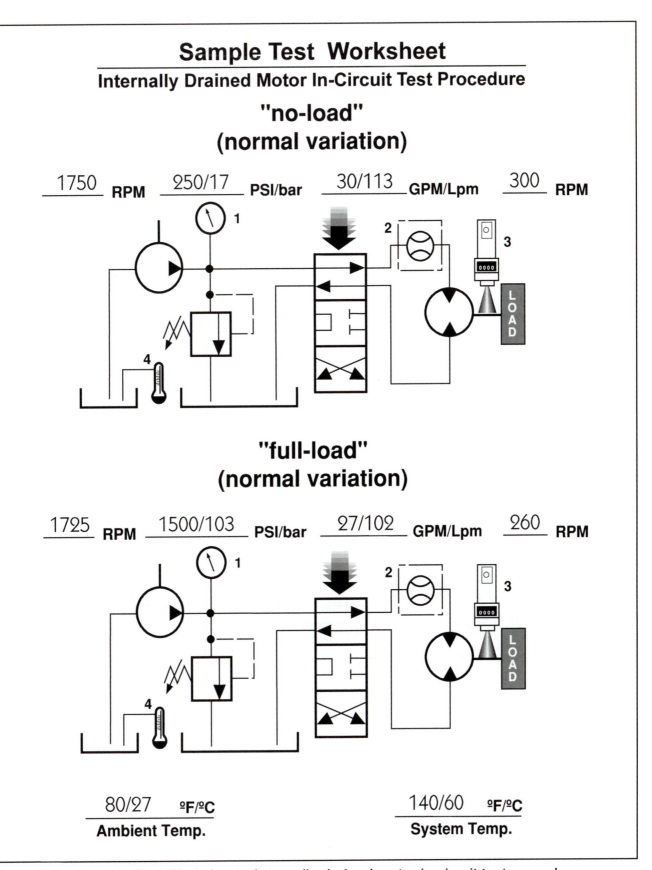

Figure 9-2   Sample Test Worksheet - Internally drained motor in-circuit test procedure

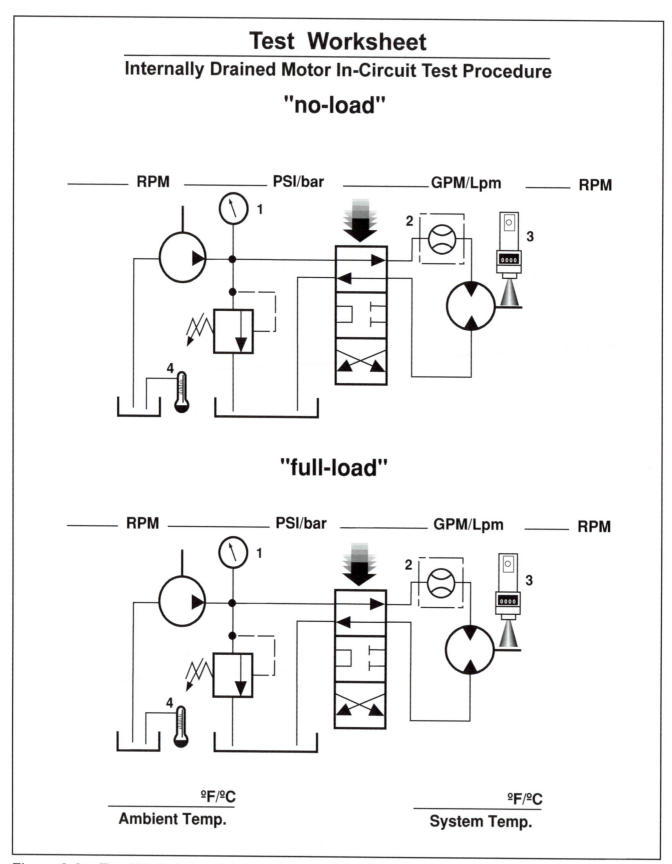

Figure 9-3   Test Worksheet - Internally drained motor in-circuit test procedure

Internally Drained Motor
Direct-Access Test Procedure

What will this test procedure accomplish?

All hydraulic motors have component parts which move in relation to one another, separated by a small oil-filled clearance. These components are generally loaded toward one another by forces related to system pressure, surface area, and springs or seals.

Motors are very similar to pumps in construction and volumetric efficiency. However, there are some subtle but significant differences between pumps and motors. Pumps "push" on the fluid to generate flow. Pump input shaft rotation is typically uni-directional, and the inlet port is usually larger in diameter than the outlet port.

Motors, being actuators, are "pushed" by the fluid to develop continuous rotation or torque. Motor output shaft rotation is generally bi-directional, and the pressure ports are usually the same diameter.

Displacement is the amount of fluid required to turn the motor output shaft one revolution. Motor displacement determines output shaft speed relative to flow input, and output torque relative to the pressure drop across the motor.

Theoretically, a motor rotates at a speed equal to its displacement during each revolution. The actual speed is reduced because of internal leakage. As the pressure difference across a motor increases, leakage also increases causing a proportional decrease in motor shaft speed.

The amount of leakage is influenced by four factors:
1. Pressure difference across the clearances;
2. Oil viscosity;
3. Temperature; and
4. Size of clearance.

Internal leakage can either flow from the inlet port to the outlet port (internally drained), or from the inlet port into the motor case (external drain). If a motor is externally drained, a separate port, which is generally smaller in diameter than the pressure ports, is provided in the motor case. A case drain-line is connected to the drain port to transport the leakage directly back to the fluid reservoir.

Generally, the volumetric efficiency of a positive displacement motor is between 80% and 97% depending upon the design.

Since a motor is generally reversible, direction of rotation will determine which is the inlet port and which is the outlet port.

The volumetric efficiency or leakage of an internally drained motor can be determined by comparing the "no-load" motor shaft speed against the "full-load" speed while simultaneously monitoring the flow input with a flow meter. Some internally drained motors include: internal gear, external gear, gerotor, and balanced vane.

The most common symptoms of problems associated with excessive leakage across a motor are:
    a. A progressive loss of motor speed as the fluid temperature increases.
    b. A moderate to high increase in the operating temperature of the fluid.
    c. Motor stalling at low speed.

This test procedure will determine the amount of leakage across the ports of an internally drained motor.

 **-WARNING-** Do not work on or around hydraulic systems without wearing safety glasses which conform to ANSI Z87.1-1989 standard.

YOU WILL NEED THE FOLLOWING DIAGNOSTIC EQUIPMENT
TO CONDUCT THIS TEST PROCEDURE

CAUTION! The pressure and flow ratings of the diagnostic equipment which will be used to conduct this test procedure must be equal to, or greater than, the pressure and flow ratings of the system being tested. Refer to the system schematic for recommended pressure and flow data.

1. Pressure gauge
2. Flow meter
3. Needle valve

4. Tachometer
5. Pyrometer

PREPARATION

To conduct this test safely and accurately, refer to Figure 9-4 for correct placement of diagnostic equipment.

To record the test data, make a copy of the test worksheet on page 9-16 (Figure 9-6).

TEST PROCEDURE

Step 1.   Shut the prime mover off.

Step 2.   Lock the electrical system out or tag the keylock switch.

1. Pressure gauge
2. Flow meter
3. Needle valve
4. Tachometer
5. Pyrometer

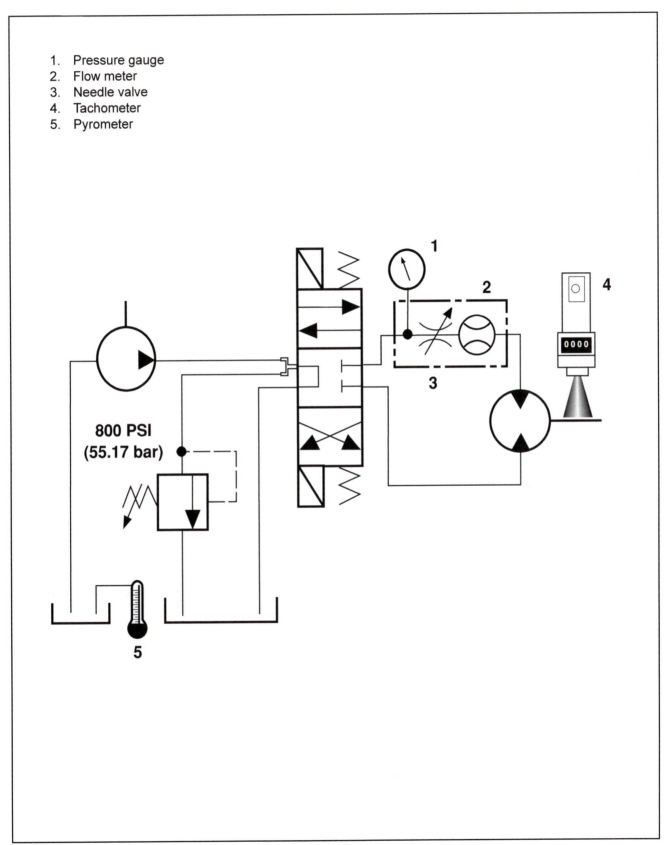

Figure 9-4   Placement of diagnostic equipment

Step 3.   Observe the system pressure gauge. Release any residual pressure trapped in the system by an accumulator, counterbalance or pilot-operated check valve, suspended load on an actuator, intensifier, or a pressurized reservoir.

Step 4.   Install flow meter (2) in series with the transmission line at the inlet port of the motor. The motor should be tested in forward and reverse if it is being used in a bi-directional rotation application.

Step 5.   Install needle valve (3) in series with the inlet port of flow meter (2). Needle valve (3) will be used to generate an "artificial" load at the inlet port of flow meter (2).

Step 6.   Install pressure gauge (1) in parallel with the connector at the inlet port of flow meter (2).

Step 7.   Open needle valve (3) fully (turn counter-clockwise).

Step 8.   Start the prime mover. Inspect the diagnostic equipment connectors for leaks.

NOTE: *If catastrophic failure is suspected, do not run the motor for longer than is absolutely necessary to determine its condition. If metal fragments are found in transmission lines, inlet or outlet ports, return line filter, or heat exchanger, DO NOT operate the motor. Replace the motor and conduct a post-catastrophic failure start-up procedure.*

Step 9.   Allow the system to warm up to approximately 130ºF. (54ºC.). Observe pyrometer (5).

NOTE: *For accurate test results, check and adjust (if necessary) the main pressure relief valve or compensator setting. Set to the pressure recommended by the manufacturer.*

Step 10.   Operate the prime mover at maximum speed. (Conduct Step 11 through Step 16 with the prime mover operating at maximum speed).

Step 11.   Record on the test worksheet, the pump "no-load" input shaft speed indicated on tachometer (4).

NOTE: *If the tachometer indicates a pump "no-load" speed which is inconsistent with machine specifications, adjust the speed accordingly and continue with the test procedure.*

Pump flow is proportional to input shaft speed. If there is a reduction in pump flow because of low prime mover speed, it will directly affect motor shaft speed. It is necessary to determine pump shaft speed prior to conducting actuator speed test procedures.

Step 12.   Remove the load from the motor output shaft.

Step 13.   Activate the directional control valve and direct full pump flow through flow meter (2), into the motor.

Step 14. Record on the test worksheet:
   a. "no-load" pressure indicated on pressure gauge (1).
   b. "no-load" flow into the motor indicated on flow meter (2).
   c. "no-load" motor shaft speed indicated on tachometer (4).
   d. Fluid temperature indicated on pyrometer (5).

Step 15. Gradually load the system by restricting the flow into the motor with needle valve (3) (turn clockwise). Load the system in 100 PSI (6.9 bar) increments.

Step 16. At each 100 PSI (6.9 bar) increment, record on the test worksheet:
   a. Pressure indicated on pressure gauge (1).
   b. Flow into the motor indicated on flow meter (2).
   c. Motor shaft speed indicated on tachometer (4).

Step 17. Stop the procedure when the value of the main pressure relief valve or compensator setting is reached.

Step 18. Record on the test worksheet the pump full-load speed indicated on tachometer (4).

Step 19. Open needle valve (3) fully (turn counter-clockwise).

Step 20. Shut the prime mover off and analyze the test results.

Step 21. At the conclusion of this test procedure, remove the diagnostic equipment. Reconnect the transmission lines and tighten the connectors securely.

Step 22. Start the prime mover and inspect the connectors for leaks.

ANALYZING THE TEST RESULTS

1. **Diagnostic Observation: "full-load" cycle:** Flow meter (2) indicates a nominal flow decrease as needle valve (3) creates an artificial load on the hydraulic system.

   The flow decrease does not exceed the anticipated loss relative to the volumetric efficiency rating of the pump, and a nominal decrease in prime mover speed.

   Tachometer (4) indicates a marginal decrease in motor shaft speed. The speed decrease does not exceed the anticipated loss relative to the volumetric efficiency rating of the pump and a nominal decrease in prime mover speed.

   The operating temperature of the fluid indicated on pyrometer (5) remains within design specification.

   Prime mover speed remains within design specification.

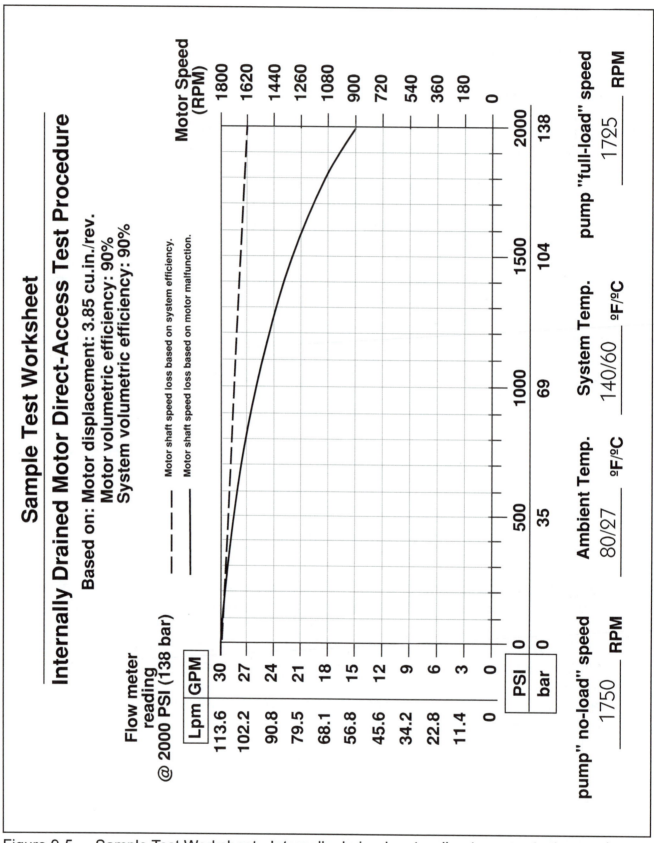

Figure 9-5   Sample Test Worksheet - Internally drained motor direct-access test procedure

Diagnosis: Leakage across the hydraulic system, excluding the motor, appears to be within design specification.

2. **Diagnostic Observation: "full-load" cycle:** Flow meter (2) indicates a moderate to high flow decrease as needle valve (3) creates an artificial load on the hydraulic system.

Tachometer (4) indicates a reduction in motor shaft speed which is proportional to the flow decrease.

Pyrometer (5) indicates a moderate increase in the operating temperature of the fluid.

Prime mover speed remains within design specification.

Diagnosis: The loss in motor shaft speed is proportional to the decrease in flow into the motor. This indicates that the motor is not the cause of the problem.

To determine the root-cause, each component in the system, upstream of the needle valve, will have to be individually tested.

3. **Diagnostic Observation: "full-load" cycle:** Flow meter (2) indicates a marginal flow decrease as needle valve (3) creates an artificial load on the hydraulic system. The flow decrease is not more than the anticipated loss relative to the volumetric efficiency rating of the pump, and a nominal decrease in prime mover speed.

During machine operation, tachometer (4) indicates a progressive decrease in motor shaft speed as the load at the motor output shaft increases during a normal load cycle. The speed decrease approaches, or exceeds, 30% of motor no-load speed (See **NOTE:** on page 9-6).

Pyrometer (5) indicates a progressive increase in the operating temperature of the fluid which does not appear to "level-off."

Prime mover speed remains within design specification.

Diagnosis: If the flow meter indicates that the flow into the motor remains within the volumetric efficiency of the system, and the tachometer indicates a speed decrease which causes an unacceptable loss in motor shaft speed, leakage across the motor is excessive. Service or replace the motor.

Leakage of this magnitude could also cause a significant increase in the operating temperature of the fluid.

If necessary, refer to the manufacturer's specification for recommended leakage data.

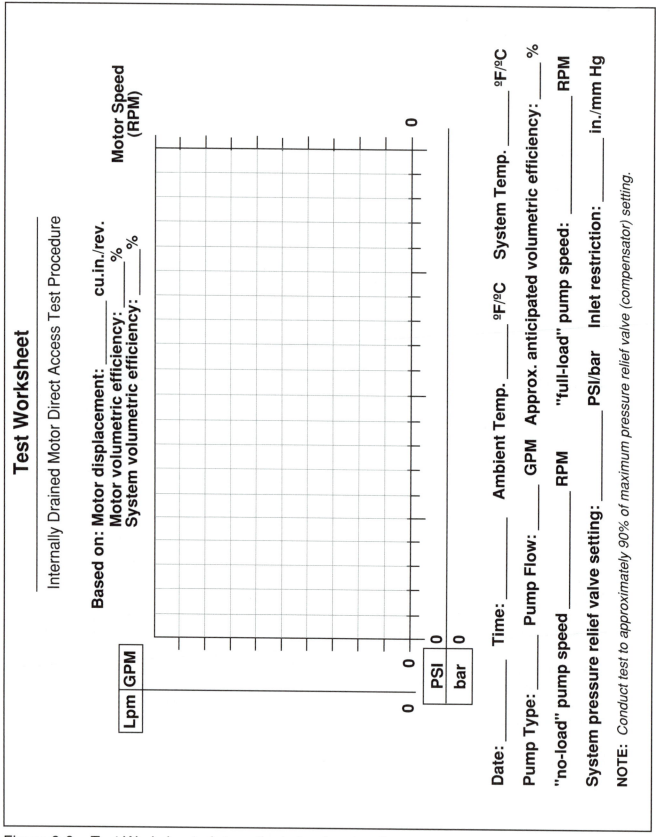

Figure 9-6   Test Worksheet - Internally drained motor direct-access test procedure

Externally Drained Motor
In-Circuit Test Procedure

What will this test procedure accomplish?

All hydraulic motors have component parts which move in relation to one another, separated by a small oil-filled clearance. These components are generally loaded toward one another by forces related to system pressure, surface area, and springs or seals.

Motors are very similar to pumps in construction and volumetric efficiency. However, there are some subtle but significant differences between pumps and motors. Pumps "push" on the fluid to generate flow. Pump input shaft rotation is typically uni-directional, and the inlet port is usually larger in diameter than the outlet port.

Motors, being actuators, are "pushed" by the fluid to develop continuous rotation or torque. Motor output shaft rotation is generally bi-directional, and the pressure ports are usually the same diameter.

Displacement is the amount of fluid required to turn the motor output shaft one revolution. Motor displacement determines output shaft speed relative to flow input, and output torque relative to the pressure drop across the motor.

Theoretically, a motor rotates at a speed equal to its displacement during each revolution. The actual speed is reduced because of internal leakage. As the pressure difference across a motor increases, leakage also increases causing a proportional decrease in motor shaft speed.

The amount of leakage is influenced by four factors:
1. Pressure difference across the clearances;
2. Oil viscosity;
3. Temperature; and
4. Size of clearance.

Internal leakage can either flow from the inlet port to the outlet port (internal drain), or from the inlet port into the motor case (external drain). If a motor is externally drained, a separate port, which is generally smaller in diameter than the pressure ports, is provided in the motor case. A case drain-line is connected to the drain port to transport the leakage directly back to the fluid reservoir.

Generally, the volumetric efficiency of a positive displacement motor is between 80% and 97% depending upon the design.

Since a motor is generally reversible, direction of rotation will determine which is the inlet port and which is the outlet port.

The volumetric efficiency or leakage of an externally drained motor can be determined by comparing the "no-load" motor shaft speed against the "full-load" speed while simultaneously monitoring case drain flow with a flow meter. Some externally drained motors include: axial piston, bent-axis, and radial piston.

The most common symptoms of problems associated with excessive motor leakage are:

    a. A progressive loss of motor shaft speed as the fluid temperature increases.
    b. A moderate to high increase in the operating temperature of the fluid.
    c. Motor stalling at low speed.

This test procedure will determine the amount of leakage across the ports of an externally drained motor.

 **-WARNING-** Do not work on or around hydraulic systems without wearing safety glasses which conform to ANSI Z87.1-1989 standard.

YOU WILL NEED THE FOLLOWING DIAGNOSTIC EQUIPMENT TO CONDUCT THIS TEST PROCEDURE

CAUTION! The pressure and flow ratings of the diagnostic equipment which will be used to conduct this test procedure must be equal to, or greater than, the pressure and flow ratings of the system being tested. Refer to the system schematic for recommended pressure and flow data.

    1.   Pressure gauge   3.   Flow meter
    2.   Tachometer   4.   Pyrometer

PREPARATION

To conduct this test safely and accurately, refer to Figure 9-7 for correct placement of diagnostic equipment.

To record the test data, make a copy of the test worksheet on page 9-23 (Figure 9-9).

TEST PROCEDURE

Step 1.   Shut the prime mover off.

Step 2.   Lock the electrical system out or tag the keylock switch.

Step 3.   Observe the system pressure gauge. Release any residual pressure trapped in the system by an accumulator, counterbalance or pilot-operated check valve, suspended load on an actuator, intensifier, or a pressurized reservoir.

1. Pressure gauge
2. Tachometer
3. Flow meter
4. Pyrometer

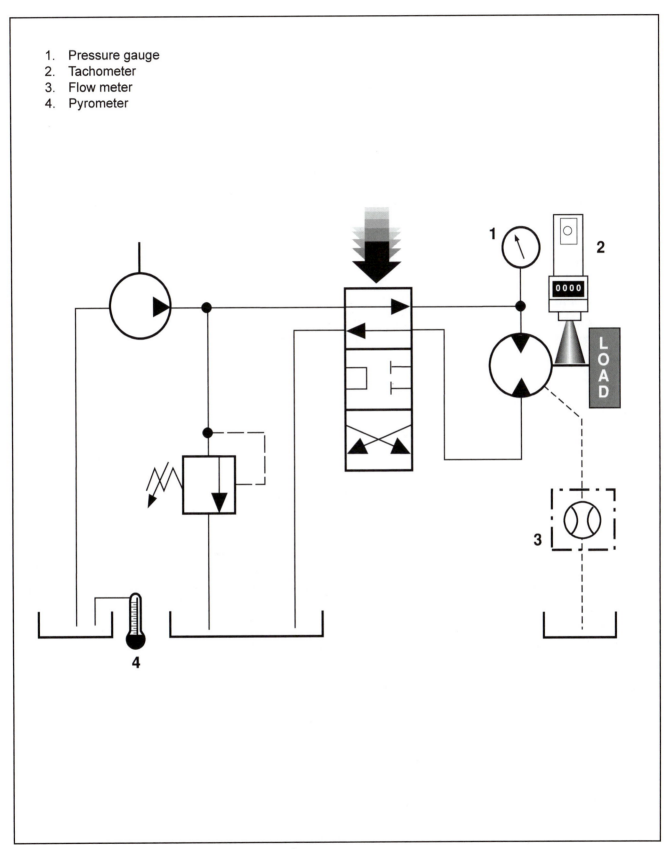

Figure 9-7   Placement of diagnostic equipment

Step 4.   Install flow meter (3) in series with the transmission line at the case drain port of the motor. The motor should be tested in forward and reverse if it is being used in a bi-directional rotation application.

Step 5.   Install pressure gauge (1) in parallel with the connector at the inlet port of the motor.

Step 6.   Start the prime mover. Inspect the diagnostic equipment connectors for leaks.

NOTE: *If catastrophic failure is suspected, do not run the motor for longer than is absolutely necessary to determine its condition. If metal fragments are found in transmission lines, inlet or outlet ports, return line filter, or heat exchanger, DO NOT operate the motor. Replace the motor and conduct a post-catastrophic failure start-up procedure.*

Step 7.   Allow the system to warm up to approximately 130ºF. (54ºC.). Observe pyrometer (4).

NOTE: *For accurate test results, check and adjust (if necessary), the main pressure relief valve or compensator setting. Set the relief valve or compensator to the pressure recommended by the manufacturer.*

Step 8.   Operate the prime mover at maximum speed. (Conduct Step 9 through Step 14 with the prime mover operating at maximum speed).

Step 9.   Record on the test worksheet (Figure 9-9), the pump "no-load" input shaft speed indicated on tachometer (2).

NOTE: *If the tachometer indicates a pump "no-load" speed which is inconsistent with machine specification, adjust the speed accordingly and continue with the test procedure.*

         Pump flow is proportional to input shaft speed. If there is a reduction in pump flow because of low prime mover speed, it will directly affect motor shaft speed. It is necessary to determine pump shaft speed prior to conducting an actuator speed test procedure.

Step 10.   Remove the load from the motor output shaft.

Step 11.   Activate the directional control valve and direct full pump flow into the motor.

Step 12.   Record on the test worksheet (Figure 9-9):
         a.   "no-load"pressure indicated on pressure gauge (1).
         b.   "no-load" motor shaft speed indicated on tachometer (2).
         c.   "no-load" case drain flow indicated on flow meter (3).
         d.   Fluid temperature indicated on pyrometer (4).

Step 13.   Operate the machine through a typical "full-load" cycle.

Step 14. Record on the test worksheet:
        a. "full-load"pressure indicated on pressure gauge (1).
        b. "full-load" motor shaft speed indicated on tachometer (2).
        c. "full-load" pump input shaft speed indicated on tachometer (2).
        d. "full-load"case drain flow indicated on flow meter (3).
        e. Fluid temperature indicated on pyrometer (4).

Step 15. Shut the prime mover off and analyze the test results.

Step 16. At the conclusion of this test procedure, remove the diagnostic equipment. Reconnect the transmission lines and tighten the connectors securely.

Step 17. Start the prime mover and inspect the connectors for leaks.

ANALYZING THE TEST RESULTS

1. **Diagnostic Observation: "full-load" cycle:** Flow meter (3) indicates a nominal flow increase as the pressure drop across the motor ports increases. The flow increase does exceed the anticipated flow increase relative to the volumetric efficiency of the motor.

    Motor shaft speed indicated on tachometer (2) remains within the volumetric efficiency rating of the motor.

    The operating temperature of the fluid indicated on pyrometer (4) remains within design specification.

    Prime mover speed remains within design specification.

    Diagnosis: Leakage across the motor appears to be within design specification.

2. **Diagnostic Observation: "full-load" cycle:** Flow meter (3) indicates a moderate to high increase in case drain flow as the pressure drop across the motor ports increases. The flow increase approaches 30% of the theoretical or "no-load" flow (See **NOTE:** on page 9-6).

    Pyrometer (4) indicates a moderate increase in the operating temperature of the fluid.

    Prime mover speed remains within design specification.

    Diagnosis: Leakage across the motor appears to be excessive. If there is an abnormal variation (reduction) in motor speed which causes an unacceptable production loss and/or elevated operating temperatures of the fluid, the motor should be replaced.

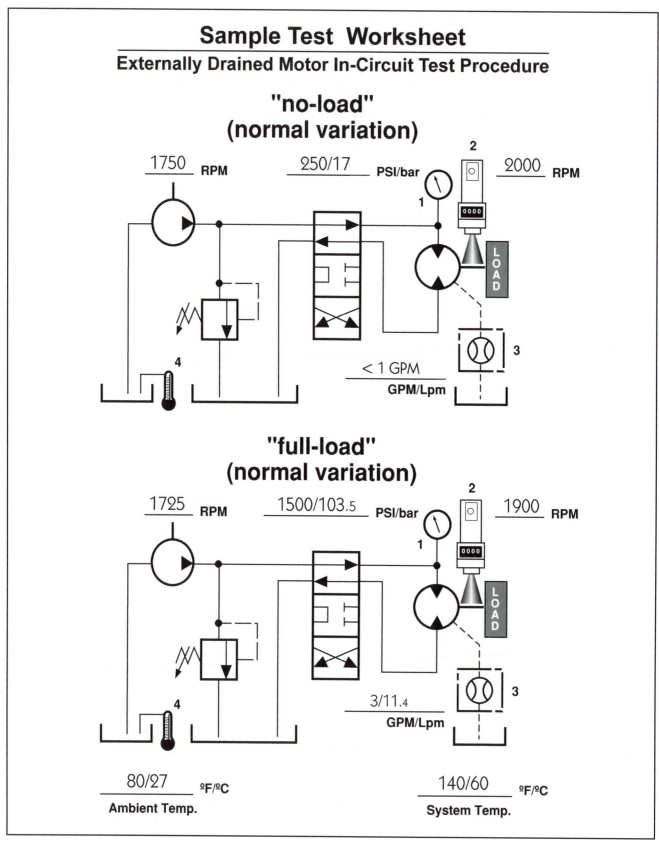

Figure 9-8   Sample Test Worksheet - Externally drained motor in-circuit test procedure

Test Worksheet

Externally Drained Motor In-Circuit Test Procedure

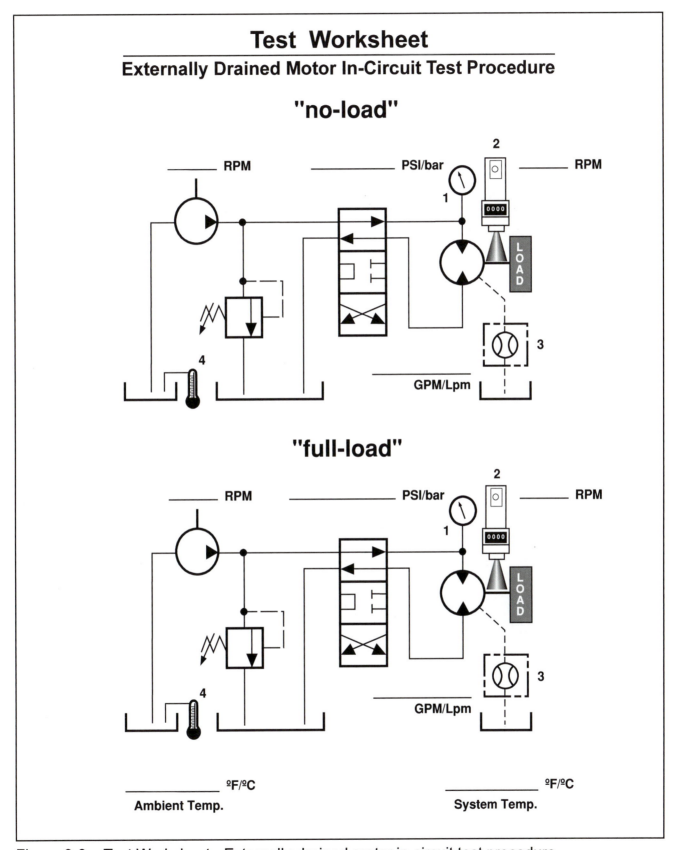

Figure 9-9   Test Worksheet - Externally drained motor in-circuit test procedure

3. **Diagnostic Observation: "full-load" cycle:** Flow meter (3) indicates a progressive flow increase as the pressure drop across the motor ports increases. The flow increase exceeds 30% of the theoretical or "no-load" flow (See **NOTE:** on page 9-6).

Pyrometer (4) indicates a progressive increase in the operating temperature of the fluid which does not appear to "level-off."

Prime mover speed remains within design specification.

Diagnosis: Leakage across the motor is excessive. A case drain flow increase of this magnitude will invariably cause an unacceptable loss in motor shaft speed. It could also cause a marked increase in the temperature of the fluid. If neglected, elevated fluid temperatures could lead to the degradation of the entire hydraulic system.

Externally Drained Motor
Direct-Access Test Procedure

What will this test procedure accomplish?

All hydraulic motors have component parts which move in relation to one another, separated by a small oil-filled clearance. These components are generally loaded toward one another by forces related to system pressure, surface area, and springs or seals.

Motors are very similar to pumps in construction and volumetric efficiency. However, there are some subtle, but significant differences between pumps and motors: Pumps "push" on the fluid to generate flow. Pump input shaft rotation is typically uni-directional, and the inlet port is usually larger in diameter than the outlet port.

Motors, being actuators, are "pushed" by the fluid to develop continuous rotation or torque. Motor output shaft rotation is generally bi-directional, and the pressure ports are usually the same diameter.

Displacement is the amount of fluid required to turn the motor output shaft one revolution. Motor displacement determines output shaft speed relative to flow input, and output torque relative to the pressure drop across the motor

Theoretically, a motor rotates at a speed equal to its displacement during each revolution. The actual speed is reduced because of internal leakage. As the pressure difference across a motor increases, leakage also increases causing a proportional decrease in motor shaft speed.

The amount of leakage is influenced by four factors:
1. Pressure difference across the clearances;
2. Oil viscosity;
3. Temperature; and
4. Size of clearance.

Internal leakage can either flow from the inlet port to the outlet port (internal drain), or from the inlet port into the motor case (external drain). If a motor is externally drained, a separate port, which is generally smaller in diameter than the pressure ports, is provided in the motor case. A case drain-line is connected to the drain port to transport the leakage directly back to the fluid reservoir.

Generally, the volumetric efficiency of a positive displacement motor is between 80% and 97% depending upon the design.

Since a motor is generally reversible, direction of rotation will determine which is the inlet port and which is the outlet port.

The volumetric efficiency or leakage of an externally drained motor can be determined by comparing the "no-load" motor shaft speed against the "full-load" speed, while simultaneously monitoring case drain flow with a flow meter. Some externally drained motors include: axial piston, bent-axis, and radial piston.

The most common symptoms of problems associated with excessive motor leakage are:
  a. A progressive loss of motor shaft speed as the fluid temperature increases.
  b. A moderate to high increase in the operating temperature of the fluid.
  c. Motor stalling at low speed.

This test procedure will determine the amount of leakage across the ports of an externally drained motor.

 **-WARNING-** Do not work on or around hydraulic systems without wearing safety glasses which conform to ANSI Z87.1-1989 standard.

YOU WILL NEED THE FOLLOWING DIAGNOSTIC EQUIPMENT TO CONDUCT THIS TEST PROCEDURE

CAUTION! The pressure and flow ratings of the diagnostic equipment which will be used to conduct this test procedure must be equal to, or greater than, the pressure and flow ratings of the system being tested. Refer to the system schematic for recommended pressure and flow data.

1.	Pressure gauge	4.	Flow meter
2.	Needle valve	5.	Pyrometer
3.	Tachometer		

PREPARATION

To conduct this test safely and accurately, refer to Figure 9-10 for correct placement of diagnostic equipment.
To record the test data, make a copy of the test worksheet on page 9-32 (Figure 9-12).

TEST PROCEDURE

Step 1. Shut the prime mover off.

Step 2. Lock the electrical system out or tag the keylock switch.

Step 3. Observe the system pressure gauge. Release any residual pressure trapped in the system by an accumulator, counterbalance or pilot-operated check valve, suspended load on an actuator, intensifier, or a pressurized reservoir.

1. Pressure gauge
2. Needle valve
3. Tachometer
4. Flow meter
5. Pyrometer

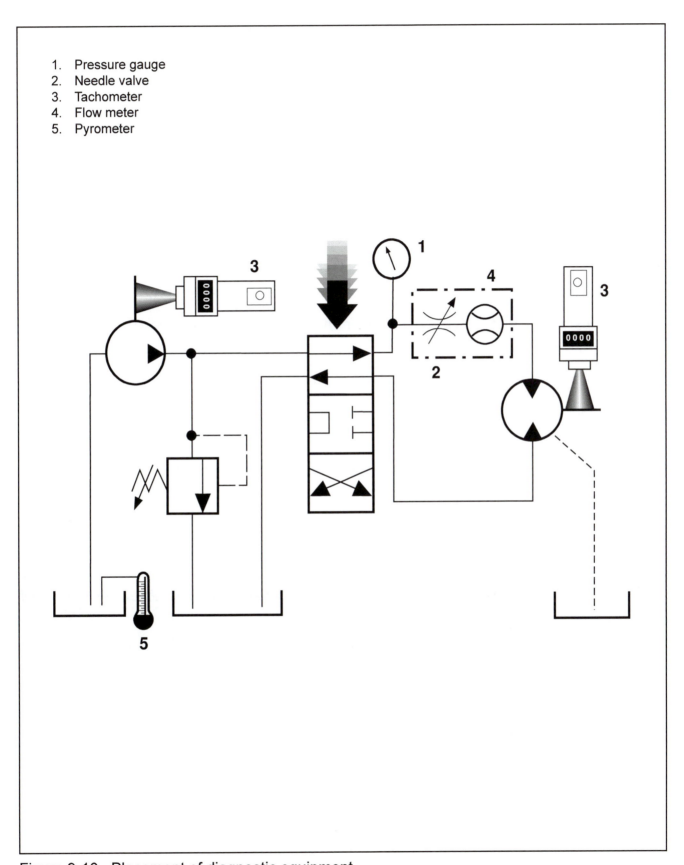

Figure 9-10   Placement of diagnostic equipment

Step 4.   Install flow meter (4) in series with the transmission line at the inlet port of the motor. The motor should be tested in forward and reverse if it is being used in a bi-directional rotation application.

Step 5.   Install needle valve (2) in series with the transmission line at the inlet port of the motor. Needle valve (2) will be used to generate an "artificial" load at the inlet port of the motor.

Step 6.   Install pressure gauge (1) in parallel with the connector at the inlet port of needle valve (2).

Step 7.   Open needle valve (2) fully (turn counter-clockwise).

Step 8.   Start the prime mover. Inspect the diagnostic equipment connectors for leaks.

NOTE: *If catastrophic motor failure is suspected, do not run the motor for longer than is absolutely necessary to determine its condition. If metal fragments are found in transmission lines, inlet or outlet ports, return line filter, or heat exchanger, DO NOT operate the motor. Replace the motor and conduct a post-catastrophic failure start-up procedure.*

Step 9.   Allow the system to warm up to approximately 130ºF. (54ºC.). Observe pyrometer (5).

NOTE: *For accurate test results, check and adjust (if necessary) the main pressure relief valve or compensator setting. Set to the pressure recommended by the manufacturer.*

Step 10.   Operate the prime mover at maximum speed. (Conduct Step 11 through Step 16 with the prime mover operating at maximum speed).

Step 11.   Record on the test worksheet, the pump "no-load" input shaft speed indicated on tachometer (3).

NOTE: *If the tachometer indicates a pump "no-load" speed which is inconsistent with machine specification, adjust the speed accordingly and continue with the test procedure.*

Pump flow is proportional to input shaft speed. If there is a reduction in pump flow because of low prime mover speed, it will directly affect motor shaft speed. It is necessary to determine pump shaft speed prior to conducting actuator speed test procedures.

Step 12.   Remove the load from the motor output shaft.

Step 13.   Activate the directional control valve and direct full pump flow into the motor.

Step 14.   Record on the test worksheet:
     a. "no-load" pressure indicated on pressure gauge (1).
     b. "no-load" motor shaft speed indicated on tachometer (3).
     c. "no-load" flow into the motor indicated on flow meter (4).
     d. Fluid temperature indicated on pyrometer (5).

Step 15. Gradually load the system by restricting the flow into the motor with needle valve (2) (turn clockwise). Load the system in 100 PSI (6.9 bar) increments.

Step 16. At each 100 PSI (6.9 bar) increment, record on the test worksheet:
   a. Pressure indicated on pressure gauge (1).
   b. Motor shaft speed indicated on tachometer (3).
   c. Flow into the motor indicated on flow meter (4).

Step 17. Stop the test procedure when the value of the main pressure relief valve is reached.

Step 18. Record on the test worksheet the pump full-load speed indicated on tachometer (3).

Step 19. Open needle valve (2) fully (turn counter-clockwise).

Step 20. Shut the prime mover off and analyze the test results.

Step 21. At the conclusion of this test procedure, remove the diagnostic equipment. Reconnect the transmission lines, and tighten the connectors securely.

Step 22. Start the prime mover and inspect the connectors for leaks.

ANALYZING THE TEST RESULTS

---

1. **Diagnostic Observation: "full-load" cycle:** Flow meter (4) indicates a nominal flow decrease as needle valve (2) creates an artificial load on the hydraulic system.

   The flow decrease does not exceed the anticipated loss relative to the volumetric efficiency rating of the pump and a nominal decrease in prime mover speed.

   Tachometer (3) indicates a marginal decrease in motor shaft speed. The speed decrease does not exceed the anticipated loss relative to the volumetric efficiency rating of the pump and a nominal decrease in prime mover speed.

   The operating temperature of the fluid indicated on pyrometer (5) remains within design specification.

   Prime mover speed remains within design specification.

   Diagnosis: Leakage across the hydraulic system, excluding the motor, appears to be within design specification.

2.   **Diagnostic Observation: "full-load" cycle:** Flow meter (4) indicates a moderate to high flow decrease as needle valve (2) creates an artificial load on the hydraulic system.

Tachometer (3) indicates a reduction in motor shaft speed which is proportional to the flow decrease.

Pyrometer (5) indicates a moderate increase in the operating temperature of the fluid.

Prime mover speed remains within design specification.

Diagnosis: The loss in motor shaft speed is proportional to the loss of flow into the motor. This indicates that the motor is not the cause of the problem.

To determine the root-cause, each component in the system, upstream of the needle valve, will have to be individually tested.

3.   **Diagnostic Observation: "full-load" cycle:** Flow meter (4) indicates a marginal flow decrease as needle valve (2) creates an artificial load on the hydraulic system. The flow decrease is not more than the anticipated loss relative to the volumetric efficiency rating of the pump, and a nominal decrease in prime mover speed.

During machine operation, tachometer (3) indicates a progressive decrease in motor shaft speed as the load at the motor output shaft increases during a normal load cycle. The speed decrease approaches, or exceeds, 30% of motor no-load speed (See **NOTE:** on page 9-6).

Pyrometer (5) indicates a progressive increase in the operating temperature of the fluid which does not appear to "level-off."

Prime mover speed remains within design specification.

Diagnosis: If the flow meter indicates that the flow into the motor remains within the volumetric efficiency of the system, and the tachometer indicates a speed decrease which causes an unacceptable loss in motor shaft speed, leakage across the motor is excessive. Service or replace the motor.

Leakage of this magnitude could also cause a significant increase in the operating temperature of the fluid.

If necessary, refer to the manufacturer's specification for recommended leakage data.

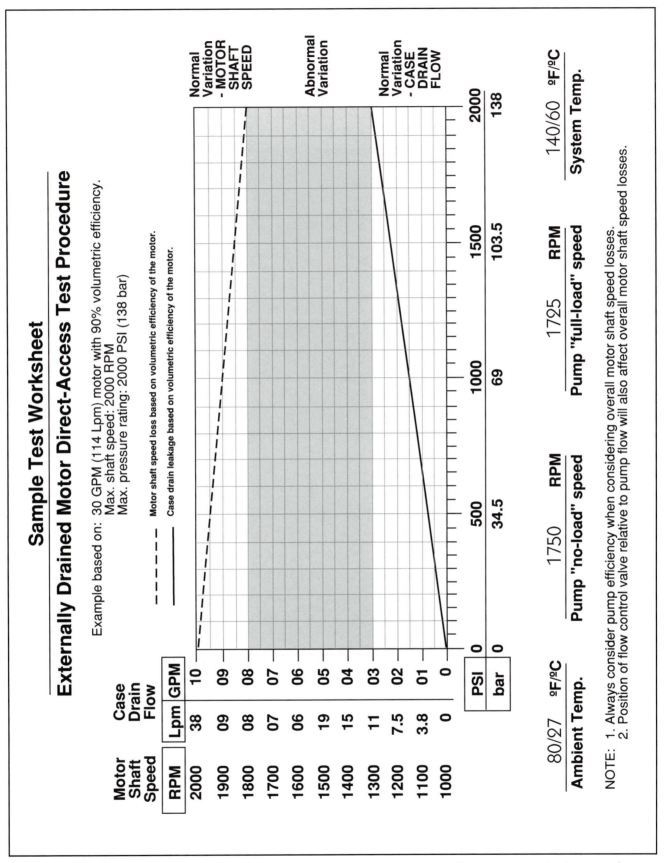

Figure 9-11   Sample Test Worksheet - Externally drained motor direct-access test procedure

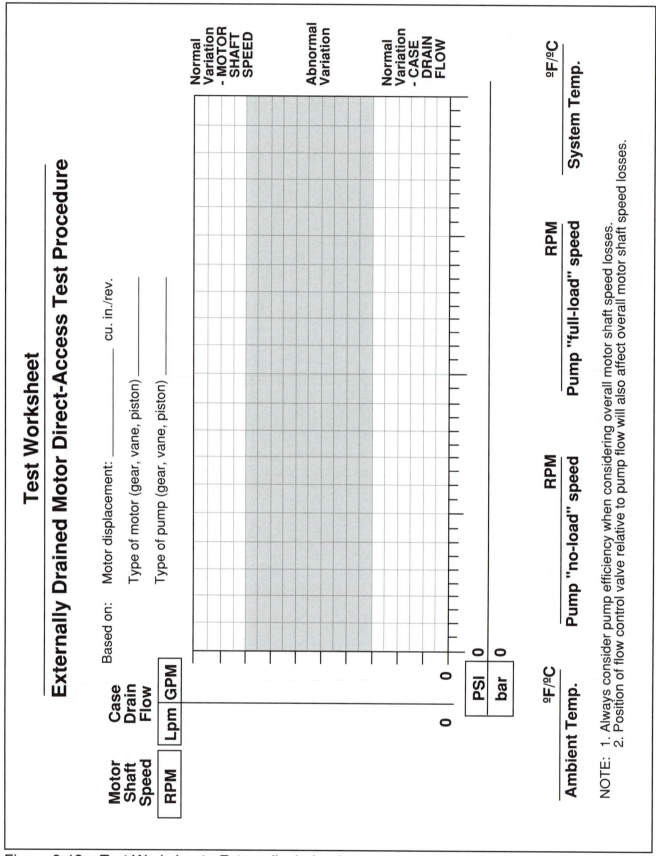

Figure 9-12   Test Worksheet - Externally drained motor direct-access test procedure

Motor Case Pressure Test Procedure

What will this test procedure accomplish?

All hydraulic motors have component parts which move in relation to one another, separated by a small oil-filled clearance. These components are generally loaded toward one another by forces related to system pressure, surface area, and springs or seals.

Motors are very similar to pumps in construction and volumetric efficiency. However, there are some subtle, but significant differences between pumps and motors: Pumps "push" on the fluid to generate flow. Pump input shaft rotation is typically uni-directional, and the inlet port is usually larger in diameter than the outlet port.

Motors, being actuators, are "pushed" by the fluid to develop continuous rotation or torque. Motor output shaft rotation is generally bi-directional, and the pressure ports are usually the same diameter.

Displacement is the amount of fluid required to turn the motor output shaft one revolution. Motor displacement determines output shaft speed relative to flow input and, output torque relative to the pressure drop across the motor.

Theoretically, a motor rotates at a speed equal to its displacement during each revolution. The actual speed is reduced because of internal leakage. As the pressure difference across a motor increases, leakage also increases causing a proportional decrease in motor shaft speed.

The amount of leakage is influenced by four factors:
1. Pressure difference across the clearances;
2. Oil viscosity;
3. Temperature; and
4. Size of clearance.

Internal leakage can either flow from the inlet port to the outlet port (internal drain), or from the inlet port into the motor case (external drain). If a motor is externally drained, a separate port, which is generally smaller in diameter than the pressure ports, is provided in the motor case. A case drain-line is connected to the drain port to transport the leakage directly back to the fluid reservoir.

The port in the motor case, and the external drain line are designed for relatively low flow rates depending on the volumetric efficiency of the motor. The motor housing is designed for relatively low pressures: from 15 PSI (1 bar) to 25 PSI (1.7 bar) with lip-type shaft seals, and up to approximately 40 PSI (2.8 bar) with mechanical shaft seals.

Any increase in internal leakage will result in increased flow into the case drain-line. This will cause the case pressure to increase because of higher flow resistance through the drain port in the motor housing, and the case drain-lines.

Excessive case pressure could result in shaft seal failure and motor case seal leakage. If the pressure in the motor case is high enough, it could literally split the motor case apart.

The most common symptoms of problems associated with excessive motor case pressure are:
    a. Shaft seal leakage.
    b. Catastrophic shaft seal failure.
    c. Motor case seal leakage.
    d. Moderate to high increase in the operating temperature of the fluid.

This test procedure will determine:
    a. Motor case pressure
    b. Pressure in the case drain-line
    c. If motor case pressure is excessive, why? Is it caused by:
        1. excessive internal leakage?
        2. restriction in the case drain-line?

-WARNING- **Do not work on or around hydraulic systems without wearing safety glasses which conform to ANSI Z87.1-1989 standard.**

YOU WILL NEED THE FOLLOWING DIAGNOSTIC EQUIPMENT TO CONDUCT THIS TEST PROCEDURE

CAUTION! The pressure and flow ratings of the diagnostic equipment which will be used to conduct this test procedure must be equal to, or greater than, the pressure and flow ratings of the system being tested. Refer to the system schematic for recommended pressure and flow data.

1.	Pressure gauge (case pressure)	4.	Pyrometer
2.	Pressure gauge (case pressure)		
3.	Pressure gauge (system pressure)		

PREPARATION

To conduct this test safely and accurately, refer to Figure 9-13 for correct placement of diagnostic equipment.
To record the test data, make a copy of the test worksheet on page 9-39 (Figure 9-15).

TEST PROCEDURE

Step 1. Shut the prime mover off.

Step 2. Lock the electrical system out or tag the keylock switch.

Step 3. Observe the system pressure gauge. Release any residual pressure trapped in the

1. Pressure gauge (case pressure)
2. Pressure gauge (case pressure)
3. Pressure gauge (system pressure)
4. Pyrometer

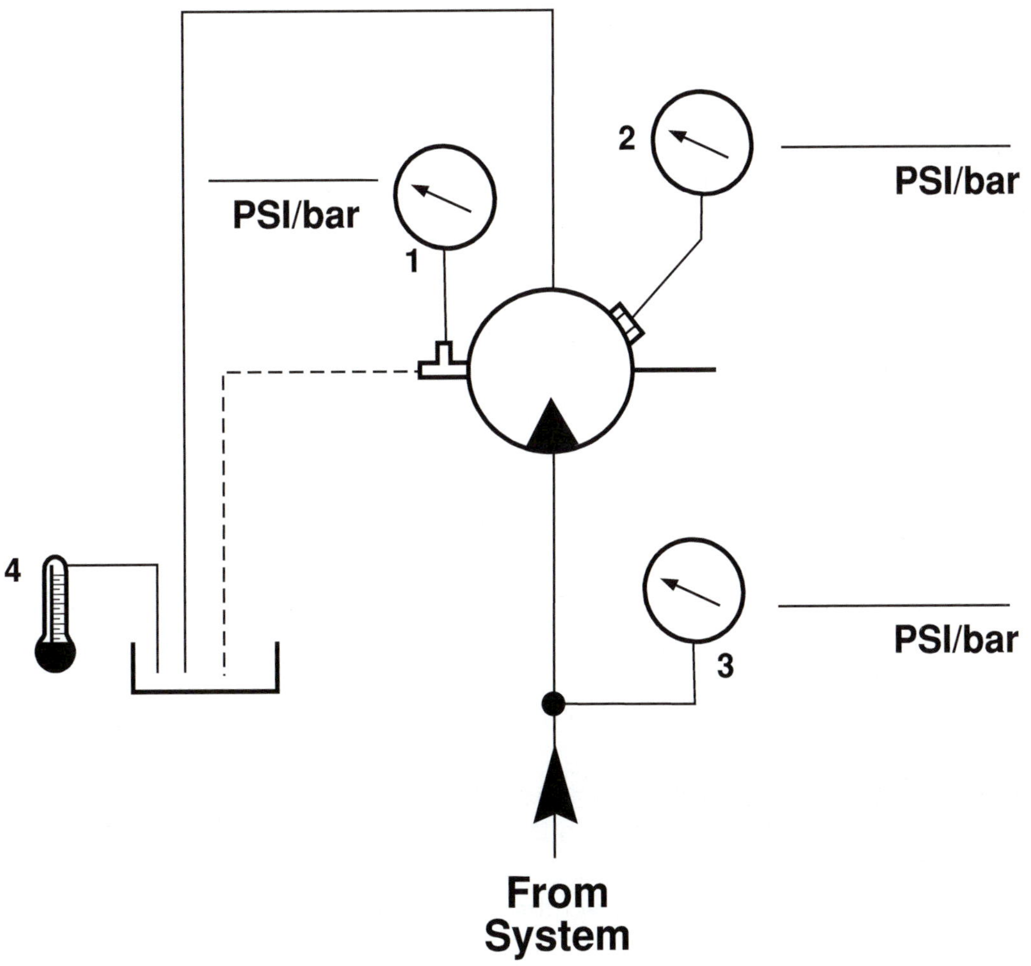

Figure 9-13   Placement of diagnostic equipment

system by an accumulator, counterbalance or pilot-operated check valve, suspended load on an actuator, intensifier, or a pressurized reservoir.

Step 4.   Install pressure gauge (1) in parallel with the connector at the case drain port of the motor.

NOTE: *Look around the motor housing for a secondary access port to the motor case. The secondary access port will usually be plugged. If necessary, refer to the manufacturer's specifications for guidance.*

Step 5.   Remove the plug from the secondary case drain port, and install pressure gauge (2) in series with the motor housing.

Step 6.   If the oil drains from the case while installing pressure gauge (2), fill the case with clean oil before starting.

Step 7.   Install pressure gauge (3) in parallel with the connector at the motor inlet port.

Step 8.   Start the prime mover.

Step 9.   Allow the system to warm up to approximately 130º F. (54 C.). Observe pyrometer (4).

Step 10.   Operate the machine through a "no-load" cycle. Record on the test worksheet:
      a.   Case drain-line pressure indicated on pressure gauge (1).
      b.   Case pressure indicated on pressure gauge (2).
      c.   System pressure indicated on pressure gauge (3).

Step 11.   Operate the machine through a "full-load" cycle. Record on the test worksheet:
      a.   Case drain-line pressure indicated on pressure gauge (1).
      b.   Case pressure indicated on pressure gauge (2).
      c.   System pressure indicated on pressure gauge (3).

Step 12.   At the conclusion of this test procedure, remove the diagnostic equipment. Reconnect the transmission lines, and tighten the connectors securely.

Step 13.   Start the prime mover and inspect the connectors for leaks.

ANALYZING THE TEST RESULTS

1.   **Diagnostic Observation: "no-load" cycle:** Pressure gauge (1) indicates a nominal pressure in the case drain-line. Pressure gauge (2) indicates a nominal case pressure. Pressure gauge (3) indicates a nominal system pressure.

The operating temperature of the fluid indicated on pyrometer (4) remains within design specification.

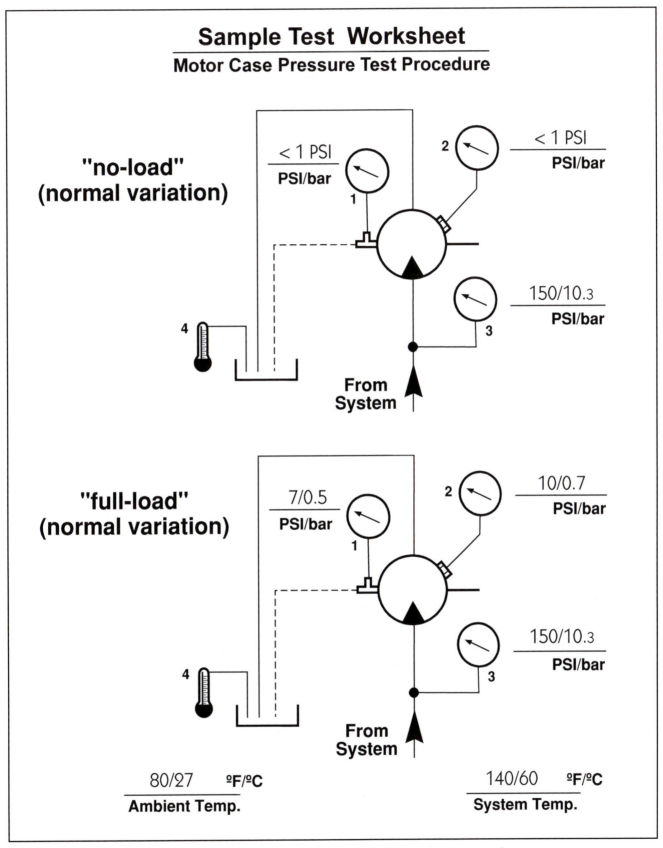

Sample Test Worksheet
Motor Case Pressure Test Procedure

"no-load"
(normal variation)

< 1 PSI
PSI/bar
1

2   < 1 PSI
PSI/bar

150/10.3
PSI/bar
3

4

From System

"full-load"
(normal variation)

7/0.5
PSI/bar
1

2   10/0.7
PSI/bar

150/10.3
PSI/bar
3

4

From System

80/27   ºF/ºC
Ambient Temp.

140/60   ºF/ºC
System Temp.

Figure 9-14   Sample Test Worksheet - Motor case pressure test procedure

Diagnosis: Motor case pressure appears to be within design specification.

2. **Diagnostic Observation: "full-load" cycle:** Pressure gauge (1) indicates a nominal pressure in the case drain-line. Pressure gauge (2) indicates a nominal pressure in the motor case. Pressure gauge (3) indicates maximum system pressure (Pressure relief valve or compensator setting).

   The operating temperature of the fluid indicated on pyrometer (4) remains within design specification.

   Diagnosis: Motor case pressure appears to be within design specification.

3. **Diagnostic Observation: "full-load" cycle:** Case pressure indicated on pressure gauge (2), progressively increases as the system pressure increases. It exceeds the maximum case pressure recommended by the motor manufacturer.

   Case drain-line pressure indicated on pressure gauge (1), also increases progressively. However, it does not increase to the pressure level recorded on gauge (2).

   Pyrometer (4) indicates a progressive increase in the operating temperature of the fluid which does not appear to "level-off."

   Diagnosis: There is evidence of excessive internal leakage in the motor. Excessive case pressure is usually caused by the reluctance of the internal leakage to flow through the port in the motor case and the case drain-line.

   Since the case drain port is in series with the case drain-line, pressure caused by resistance to flow in the case drain-line is added to the pressure in the case.

   If internal leakage is excessive, the resistance offered by the case drain port could cause the pressure in the motor case to exceed design specification.

   This will usually result in shaft seal leakage or housing seal leakage.

4. **Diagnostic Observation: "full-load" cycle:** Case drain-line pressure indicated on pressure gauge (1), progressively increases as the system pressure increases. It increases to a level which exceeds the maximum case pressure recommended by the motor manufacturer.

   Case pressure indicated on pressure gauge (2), increases in proportion to the pressure increase indicated on gauge (1). It too increases to a level which exceeds the maximum case pressure recommended by the motor manufacturer.

   Pressure gauge (3) indicates the pressure relief valve or pump compensator setting.

   There is a moderate to high increase in the operating temperature of the fluid indicated on pyrometer (4).

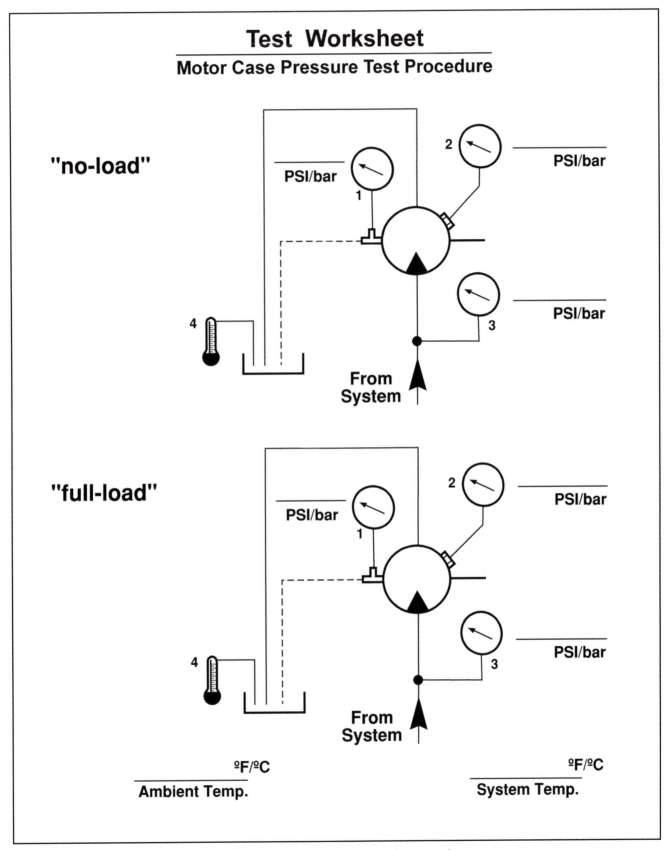

Figure 9-15   Test Worksheet - Motor case pressure test procedure

Diagnosis: There is evidence of excessive resistance in the case drain-line. Case drain-line pressure gauge (1) will generally record only the pressure caused by the resistance to flow in the case drain-line.

To determine if excessive case drain-line pressure is caused by high internal leakage, excessive case drain-line resistance, or a combination of both, it may be necessary to conduct a case drain flow test.

The following steps can be used to determine the root-cause with a pressure gauge:

Step 1.   Install case drain-line pressure gauge (1) in parallel with the next downstream connector and repeat the "full-load" test procedure.

Step 2.   If pressure gauge (1) indicates a reduction in pressure, the resistance is in the line upstream of the pressure gauge.

Step 3.   If there is no change in pressure, move pressure gauge (1) to the next downstream connector and repeat the test procedure.

Step 4.   Continue moving pressure gauge (1) through the case drain-line progressively until the resistance is isolated.

NOTE: *If the high case drain-line resistance is caused by a return line filter, manifold and/or cooler, the case drain-line will have to be re-routed so the fluid flows directly back to the reservoir. Submit to application engineering for design review.*

Motor Variable Displacement Control
Test Procedure

What will this test procedure accomplish?

Variable displacement motors provide the advantage of variable torque and speed capability while the pump flow remains constant. Maximum torque is achieved by stroking the swashplate in the motor to its maximum angle, while maximum speed is achieved by stroking the swashplate to its minimum angle.

If for any reason, the swashplate fails to move from the maximum speed/minimum torque position to the maximum torque/minimum speed position, system pressure will compensate by increasing to meet the load demand. However, before maximum torque/minimum speed is reached the oil will usually pass over the system's main pressure relief valve.

Usually the machine will appear to lose power as the load (pressure) increases, associated with a possible overheating condition.

The most common symptoms of problems associated with variable displacement motor control are:
    a. Moderate to high increase in the operating temperature of the fluid.
    b. Lack of power when the actuator is under load.
    c. Actuator stalls when the machine is under load.

This test procedure will determine if the motor swashplate angle is changing to meet the torque and speed demands of the machine.

Several different methods are used to change swashplate angle:
    a. Cable operated from the operations console- manual.
    b. Mechanically activated by an actuator (automatic or manual) sensing system.
    c. Proportional or servo controller connected to the swashplate control mechanism. The actuator is normally operated by sensing system pressure or by pilot pressure.

 **-WARNING-** Do not work on or around hydraulic systems without wearing safety glasses which conform to ANSI Z87.1-1989 standard.

YOU WILL NEED THE FOLLOWING DIAGNOSTIC EQUIPMENT TO CONDUCT THIS TEST PROCEDURE

CAUTION! The pressure and flow ratings of the diagnostic equipment which will be used to conduct this test procedure must be equal to, or greater than, the pressure and flow ratings of the system being tested. Refer to the system schematic for recommended pressure and flow data.

1.   Pressure gauge   2.   Pyrometer

PREPARATION

To conduct this test safely and accurately, refer to Figure 9-16 for correct placement of diagnostic equipment.

To record the test data, make a copy of the test worksheet on page 9-45 (Figure 9-17).

TEST PROCEDURE

Step 1.   Shut the prime mover off.

Step 2.   Lock the electrical system out or tag the keylock switch.

Step 3.   Observe the system pressure gauge. Release any residual pressure trapped in the system by an accumulator, counterbalance or pilot-operated check valve, suspended load on an actuator, intensifier, or a pressurized reservoir.

Step 4.   Install pressure gauge (1) in parallel with the connector at the outlet port of the pump.

Step 5.   Start the prime mover. Inspect the diagnostic equipment connectors for leaks.

Step 6.   Allow the system to warm up to approximately 130º F. (54º C.) Observe pyrometer (2).

NOTE: *For accurate test results, check and adjust (if necessary), the pressure relief valve or pump compensator setting. Set according to the manufacturer's specification.*

Step 7.   Engage the high torque/low speed position. Record on the test worksheet the pressure indicated on pressure gauge (1).

Step 8.   Operate the machine through a "full-load" cycle. Record on the test worksheet the pressure indicated on pressure gauge (1).

Step 9.   At the conclusion of this test procedure, remove the diagnostic equipment. Reconnect the transmission lines, and tighten the connectors securely.

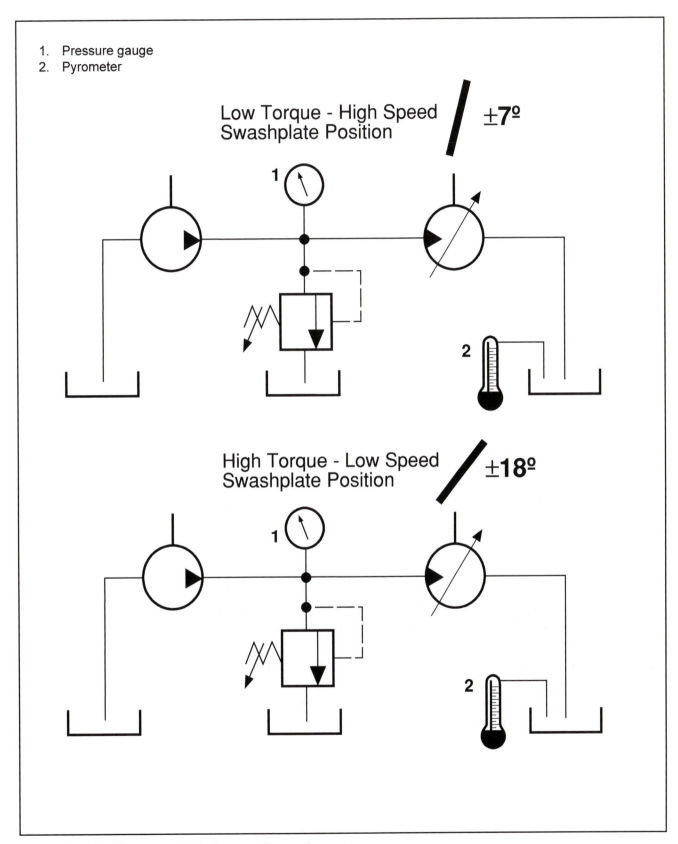

Figure 9-16   Placement of diagnostic equipment

Step 10.   Start the prime mover and inspect the connectors for leaks.

ANALYZING THE TEST RESULTS

1.   **Diagnostic Observation: "no-load" cycle:** High torque/low speed position engaged. No-load on the machine. The pressure indicated on pressure gauge (1) is low to moderate.

When the low torque/high speed position is selected, the pressure indicated on pressure gauge (1) increases. It does not exceed the value of the main pressure relief valve setting.

Diagnosis: The high speed/low torque mechanism appears to be in satisfactory operating condition.

2.   **Diagnostic Observation: "full-load" cycle:** High torque/low speed position engaged. Full-load on the machine. The pressure indicated on pressure gauge (1), is moderate to high. It does not exceed the value of the main pressure relief valve.

Diagnosis: The high speed/low torque mechanism appears to be in satisfactory operating condition.

3.   **Diagnostic Observation: "no-load" cycle:** High torque/low speed position engaged. No-load on the machine. The pressure indicated on pressure gauge (1) approaches or exceeds the value of the main pressure relief valve setting.

Diagnosis: It appears that the high torque/low speed control mechanism is not changing to meet the load demands of the machine.

When the motor displacement changes to increase motor shaft speed, there is a proportional decrease in torque output. The only way the system can respond to maintain the same torque output is for the system pressure to increase.

Generally, the main pressure relief valve setting is insufficient to allow the system to operate at maximum torque in the minimum displacement position. Thus, the fluid is forced to flow over the pressure relief valve. This problem could result in a significant increase in the operating temperature of the fluid.

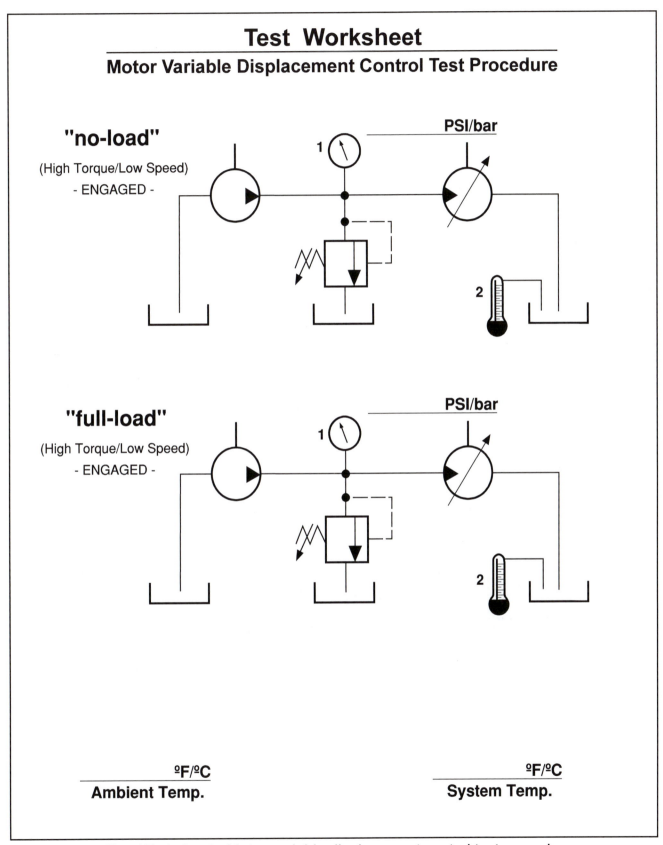

Figure 9-17   Test Worksheet - Motor variable displacement control test procedure

notes

Parallel Dual Motor Application
In-Circuit Test Procedure

What will this test procedure accomplish?

All hydraulic motors have component parts which move in relation to one another, separated by a small oil-filled clearance. These components are generally loaded toward one another by forces related to system pressure, surface area, and springs or seals.

Motors are very similar to pumps in construction and volumetric efficiency. However, there are some subtle, but significant differences between pumps and motors: Pumps "push" on the fluid to generate flow. Pump input shaft rotation is typically uni-directional, and the inlet port is usually larger in diameter than the outlet port.

Motors, being actuators, are "pushed" by the fluid to develop continuous rotation or torque. Motor output shaft rotation is generally bi-directional, and the pressure ports are usually the same diameter.

Displacement is the amount of fluid required to turn the motor output shaft one revolution. Motor displacement determines output shaft speed relative to flow input, and output torque relative to the pressure drop across the motor.

Theoretically, a motor rotates at a speed equal to its displacement during each revolution. The actual speed is reduced because of internal leakage. As the pressure difference across a motor increases, leakage also increases causing a proportional decrease in motor shaft speed.

The amount of leakage is influenced by four factors:
1. Pressure difference across the clearances;
2. Oil viscosity;
3. Temperature; and
4. Size of clearance.

Internal leakage can either flow from the inlet port to the outlet port (internal drain), or from the inlet port into the motor case (external drain). If a motor is externally drained, a separate port, which is generally smaller in diameter than the pressure ports, is provided in the motor case. A case drain-line is connected to the drain port to transport the leakage directly back to the fluid reservoir.

Generally, the volumetric efficiency of a positive displacement hydraulic motor is between 80% and 97% depending upon the design.

If motor shaft speed remains within 30% (rule-of-thumb) of "no-load" speed, under "full-load" conditions, motor performance is probably acceptable.

Since a motor is generally reversible, direction of rotation will determine which is the inlet port and which is the outlet port.

The volumetric efficiency or leakage of an internally drained motor is determined by comparing the "no-load" motor shaft speed against the "full-load" speed, while simultaneously monitoring the flow input with a flow meter. Some internally drained motors include: internal gear, external gear, gerotor, and balanced vane.

The most common symptoms of problems associated with excessive motor leakage are:
a. A progressive loss of motor shaft speed as the fluid temperature increases.
b. A moderate to high increase in the operating temperature of the fluid.
c. Motor stalling at low speed.

This test procedure will determine the amount of leakage across the ports of each motor. It will isolate a motor which may be leaking excessively.

 **-WARNING-** Do not work on or around hydraulic systems without wearing safety glasses which conform to ANSI Z87.1-1989 standard.

YOU WILL NEED THE FOLLOWING DIAGNOSTIC EQUIPMENT TO CONDUCT THIS TEST PROCEDURE

CAUTION! The pressure and flow ratings of the diagnostic equipment which will be used to conduct this test procedure must be equal to, or greater than, the pressure and flow ratings of the system being tested. Refer to the system schematic for recommended pressure and flow data.

1. Flow Meter
2. Pressure Gauge
3. Pyrometer
4. Tachometer

PREPARATION - Step 1

To conduct this test safely and accurately, refer to Figure 9-18 for correct placement of diagnostic equipment.
To record the test data, make a copy of the test worksheet on page 9-54 (Figure 9-20).

TEST PROCEDURE - Step 1

Step 1. Shut the prime mover off.

Step 2. Lock the electrical system out or tag the keylock switch.

Step 3. Observe the system pressure gauge. Release any residual pressure trapped in the system by an accumulator, counterbalance or pilot-operated check valve, suspended load on an actuator, intensifier, or a pressurized reservoir.

1. Flow meter
2. Pressure gauge
3. Pyrometer
4. Tachometer

Step 1

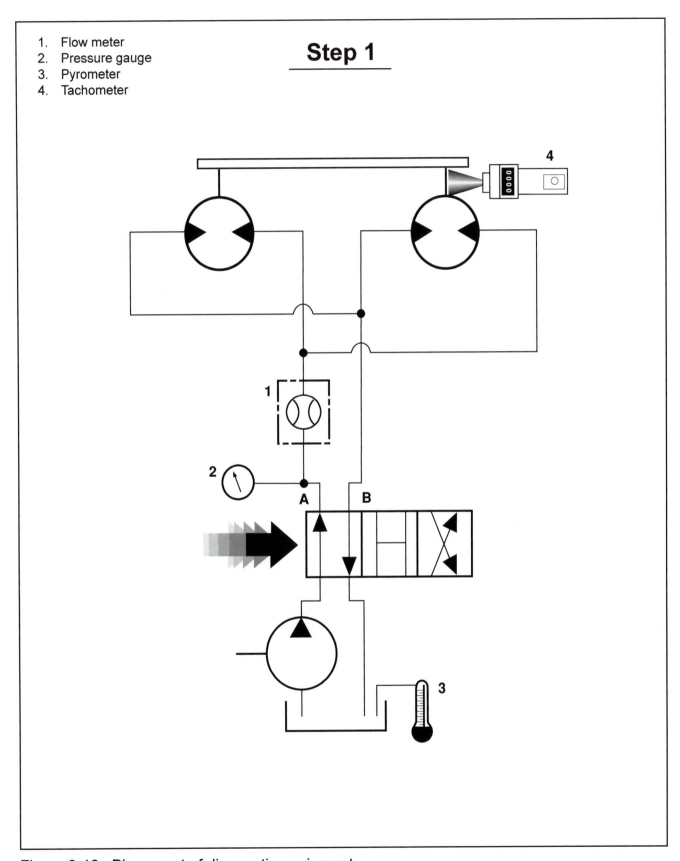

Figure 9-18   Placement of diagnostic equipment

Step 4.   Install flow meter (1) in series with the transmission line at the "A" port of the directional control valve.

NOTE: *The flow meter must be installed in the transmission line before it divides in parallel to each motor.*

Step 5.   Install pressure gauge (2) in parallel with the connector at the inlet port of flow meter (1).

Step 6.   Start the prime mover. Inspect the diagnostic equipment connectors for leaks.

NOTE: *If catastrophic motor failure is suspected, do not run the motor for longer than is absolutely necessary to determine its condition. If metal fragments are found in transmission lines, inlet or outlet ports, return line filter, or heat exchanger, DO NOT operate the motor. Replace the motor and conduct a post-catastrophic failure start-up procedure.*

Step 7.   Allow the system to warm up to approximately 130° F. (54° C.). Observe pyrometer (3).

Step 8.   Record on the test worksheet, the pump "no-load" input shaft speed indicated on tachometer (4). (Conduct Step 10. through Step 13. with the prime mover operating at maximum speed).

NOTE: *If the tachometer indicates a pump "no-load" speed which is inconsistent with machine specifications, adjust the speed accordingly and continue with the test procedure.*

Pump flow is proportional to input shaft speed. If there is a reduction in pump flow because of low prime mover speed, it will directly affect motor shaft speed. It is necessary to determine pump shaft speed prior to conducting actuator speed test procedures.

Step 9.   Remove the load from the motors.

Step 10.   Operate the machine through a "no-load" cycle.

Step 11.   Record on the test worksheet:
          a. Pump flow indicated on flow meter (1).
          b. "no-load" system pressure indicated on pressure gauge (2).
          c. "no-load" motor shaft speeds indicated on tachometer (4).

Step 12.   Operate the machine through a "full load" cycle.

Step 13.   Record on the test worksheet:
          a. Pump flow indicated on flow meter (1).
          b. "full-load" system pressure indicated on pressure gauge (2).
          c. "full-load" motor shaft speeds indicated on tachometer (4).

ANALYZING THE TEST RESULTS - Step 1

1. **Diagnostic Observation: "full-load" cycle:** Flow meter (1) indicates a nominal flow decrease as the system pressure increases.

   The flow decrease does not exceed the anticipated loss relative to the volumetric efficiency rating of the pump, and a nominal decrease in prime mover speed.

   Tachometer (4) indicates a marginal decrease in the speed of the motors. The speed decrease does not exceed the anticipated loss relative to the volumetric efficiency of the motors.

   The operating temperature of the fluid indicated on pyrometer (3) remains within design specification.

   Prime mover speed remains within design specification.

   Diagnosis: Leakage across the motors appears to be within design specification.

2. **Diagnostic Observation: "full-load" cycle:** Flow meter (1) indicates a progressive flow decrease as the load on the motors increases. The flow decrease approaches, but does not exceed, 30% of no-load pump flow (See **NOTE:** on page 9-6).

   Tachometer (4) indicates a reduction in the speed of the motors which is proportional to the flow decrease.

   Pyrometer (3) indicates a moderate to high increase in the operating temperature of the fluid.

   Prime mover speed remains within design specification.

   Diagnosis: The loss in the shaft speed of the motors is proportional to the flow loss indicated on the flow meter. This indicates that the problem lies within the system. All the components on the inlet side of the flow meter will have to be individually tested to determine the root-cause.

3. **Diagnostic Observation: "full-load" cycle:** Flow meter (1) indicates a flow decrease as the system pressure increases. The flow decrease does not exceed the volumetric efficiency rating of the pump.

   Tachometer (4) indicates a progressive decrease in motor shaft speed as the load on the motors increases. The speed decrease exceeds 30% of the no-load speed of the motors (See **NOTE:** on page 9-6).

   Pyrometer (3) indicates a progressive increase in the operating temperature of the fluid which does not appear to "level-off."

Prime mover speed remains within design specification.

Diagnosis: The flow meter indicates that the flow into the motors remains within the volumetric efficiency of the system.

The tachometer indicates that the decrease in the speed of the motors exceeds the volumetric efficiency rating of the motors. The speed decrease is unacceptable by normal operating standards. Leakage across the motors is excessive.

NOTE: *Motors connected in parallel, will normally wear out uniformly. However, excessive leakage could be caused by one motor. It is recommended that each motor be tested independently to isolate the problem.*

TEST PROCEDURE - Step 2

To isolate the problem to a specific motor, proceed as follows:

Refer to the test worksheet on page 9-56 (Figure 9-22) for correct placement of diagnostic equipment and to record the test data.

Step 1.   Install flow meter (1) in series with the inlet port of the first motor.

Step 2.   Repeat Step 6. through Step 13 (Page 9-50).

NOTE: *Conduct this test procedure at the inlet port of each motor in parallel.*

ANALYZING THE TEST RESULTS - Step 2
---

1.   **Diagnostic Observation: "full-load" cycle:** If the output shafts of the motors are mechanically connected and the displacement of the motors is equal, the pump flow should be divided equally between the two motors.

More flow is inclined to flow through the motor which may be damaged or worn due to the larger clearances.

Flow meter (1) connected in series with the inlet port of motor "A," indicates a flow which is greater than half the pump flow. The flow exceeds the volumetric efficiency rating of the motor.

Flow meter (1) connected in series with the inlet port of motor "B," indicates a flow which is less than half the pump flow.

Diagnosis: Motor "A" is receiving more flow than motor "B." It appears that leakage across motor "A" is excessive.

Sample Test Worksheet - Step 1
Parallel Dual Motor Application In-Circuit Test Procedure

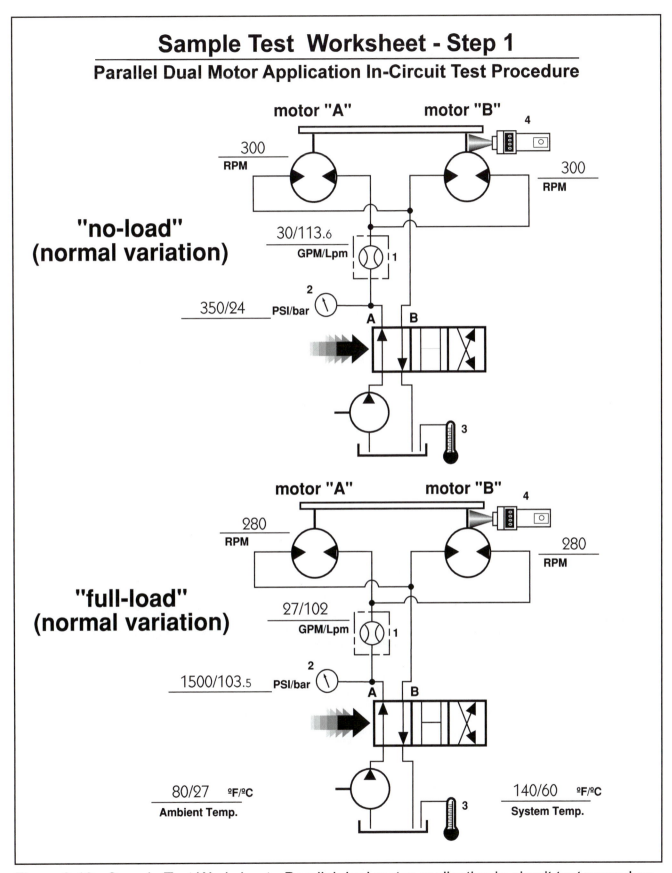

Figure 9-19   Sample Test Worksheet - Parallel dual motor application in-circuit test procedure

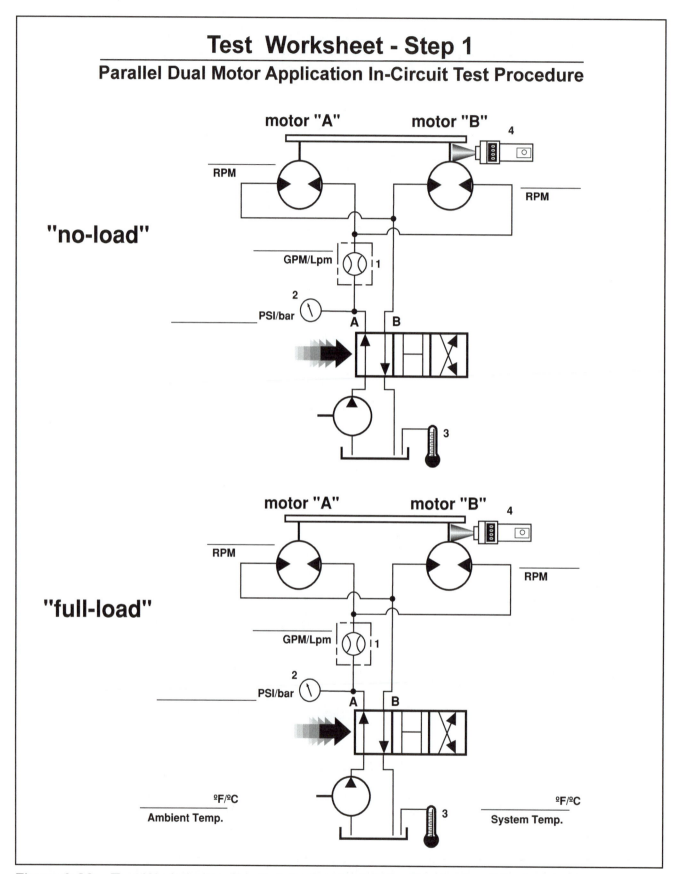

Test Worksheet - Step 1
Parallel Dual Motor Application In-Circuit Test Procedure

Figure 9-20   Test Worksheet - Parallel dual motor application in-circuit test procedure

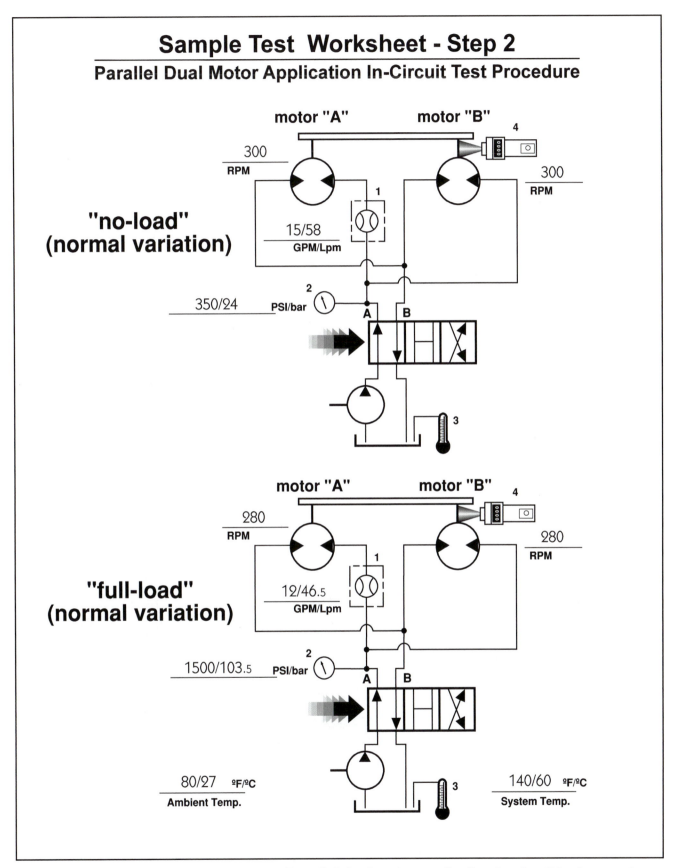

Sample Test Worksheet - Step 2
Parallel Dual Motor Application In-Circuit Test Procedure

motor "A" motor "B"

**"no-load"
(normal variation)**

300
RPM

300
RPM

15/58
GPM/Lpm

350/24 PSI/bar

motor "A" motor "B"

**"full-load"
(normal variation)**

280
RPM

280
RPM

12/46.5
GPM/Lpm

1500/103.5 PSI/bar

80/27 ºF/ºC
Ambient Temp.

140/60 ºF/ºC
System Temp.

Figure 9-21   Sample Test Worksheet - Parallel dual motor application in-circuit test procedure

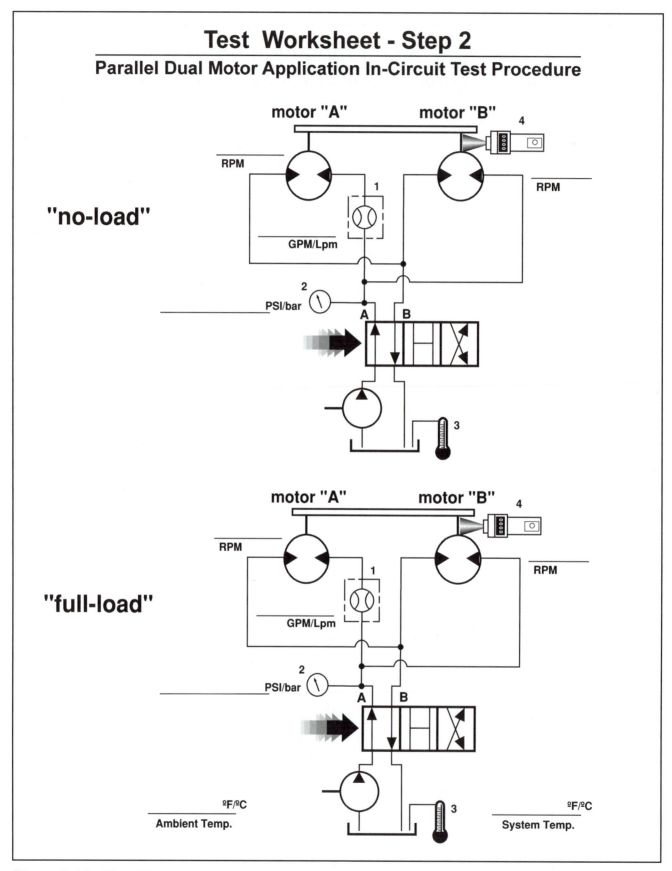

Figure 9-22   Test Worksheet - Parallel dual motor application in-circuit test procedure

Parallel Dual Motor Application
Direct-Access Test Procedure

What will this test procedure accomplish?

All hydraulic motors have component parts which move in relation to one another, separated by a small oil-filled clearance. These components are generally loaded toward one another by forces related to system pressure, surface area, and springs or seals.

Motors are very similar to pumps in construction and volumetric efficiency. However, there are some subtle, but significant differences between pumps and motors: Pumps "push" on the fluid to generate flow. Pump input shaft rotation is typically uni-directional, and the inlet port is usually larger in diameter than the outlet port.

Motors, being actuators, are "pushed" by the fluid to develop continuous rotation or torque. Motor output shaft rotation is generally bi-directional, and the pressure ports are usually the same diameter.

Displacement is the amount of fluid required to turn the motor output shaft one revolution. Motor displacement determines output shaft speed relative to flow input, and output torque relative to the pressure drop across the motor.

Theoretically, a motor rotates at a speed equal to its displacement during each revolution. The actual speed is reduced because of internal leakage. As the pressure difference across a motor increases, leakage also increases causing a proportional decrease in motor shaft speed.

The amount of leakage is influenced by four factors:
1. Pressure difference across the clearances;
2. Oil viscosity;
3. Temperature; and
4. Size of clearance.

Internal leakage can either flow from the inlet port to the outlet port (internal drain), or from the inlet port into the motor case (external drain). If a motor is externally drained, a separate port, which is generally smaller in diameter than the pressure ports, is provided in the motor case. A case drain-line is connected to the drain port to transport the leakage directly back to the fluid reservoir.

Generally, the volumetric efficiency of a positive displacement hydraulic motor is between 80% and 97% depending upon the design.

If motor shaft speed remains within 30% (rule-of-thumb) of "no-load" speed, under "full-load" conditions, motor performance is probably acceptable.

Since a motor is generally reversible, direction of rotation will determine which is the inlet port and which is the outlet port.

The volumetric efficiency or leakage of an internally drained motor is determined by comparing the "no-load" motor shaft speed against the "full-load" speed, while simultaneously monitoring the flow input with a flow meter. Some internally drained motors include: internal gear, external gear, gerotor, and balanced vane.

The most common symptoms of problems associated with excessive motor leakage are:
- a. A progressive loss of motor shaft speed as the fluid temperature increases.
- b. A moderate to high increase in the operating temperature of the fluid.
- c. Motor stalling at low speed.

This test procedure will determine the amount of leakage across the ports of each motor. It will isolate the motor which may be leaking excessively.

 **-WARNING-** Do not work on or around hydraulic systems without wearing safety glasses which conform to ANSI Z87.1-1989 standard.

YOU WILL NEED THE FOLLOWING DIAGNOSTIC EQUIPMENT TO CONDUCT THIS TEST PROCEDURE

CAUTION! The pressure and flow ratings of the diagnostic equipment which will be used to conduct this test procedure must be equal to, or greater than, the pressure and flow ratings of the system being tested. Refer to the system schematic for recommended pressure and flow data.

1.	Tachometer	4.	Pressure gauge
2.	Flow meter	5.	Pyrometer
3.	Needle valve		

PREPARATION - Step 1

To conduct this test safely and accurately, refer to Figure 9-23 for correct placement of diagnostic equipment.
To record the test data, make a copy of the test worksheet on page 9-63 (Figure 9-25).

TEST PROCEDURE - Step 1

Step 1.   Shut the prime mover off.

Step 2.   Lock the electrical system out or tag the keylock switch.

Step 3.   Observe the system pressure gauge. Release any residual pressure trapped in the system by an accumulator, counterbalance or pilot-operated check valve, suspended load on an actuator, intensifier, or a pressurized reservoir.

1. Tachometer
2. Flow meter
3. Needle valve
4. Pressure gauge
5. Pyrometer

Step 1

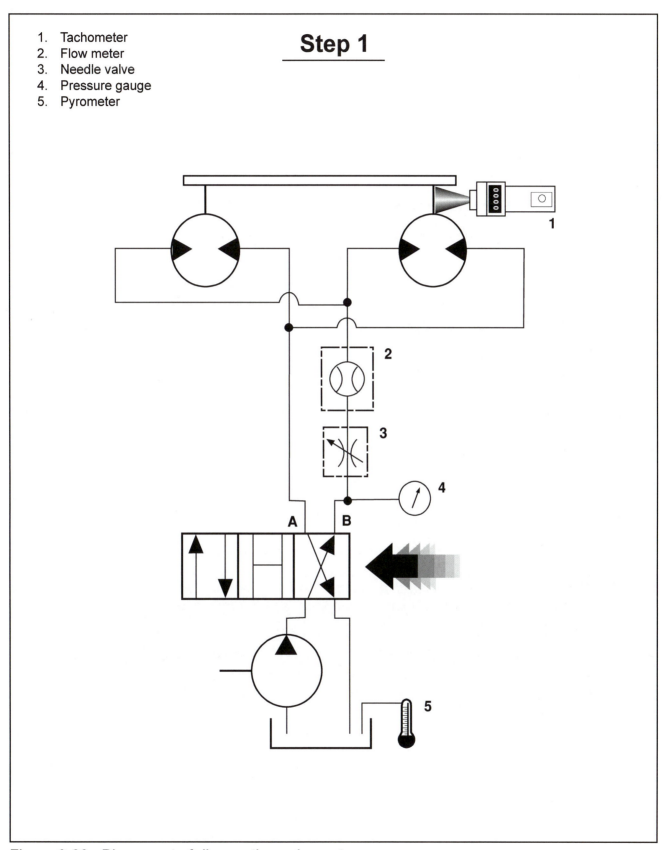

Figure 9-23   Placement of diagnostic equipment

Step 4.   Install flow meter (2) in series with the transmission line at the "B" port of the directional control valve.

NOTE: *The flow meter must be installed in the transmission line before it divides in parallel to each motor.*

Step 5.   Install needle valve (3) in series with the transmission line at the inlet port of flow meter (2).

Step 6.   Install pressure gauge (4) in parallel with the connector at the inlet port of the needle valve.

Step 7.   Open needle valve (3) fully (turn counter-clockwise).

Step 8.   Start the prime mover. Inspect the diagnostic equipment connectors for leaks.

NOTE: *If catastrophic motor failure is suspected, do not run the motor for longer than is absolutely necessary to determine its condition. If metal fragments are found in transmission lines, inlet or outlet ports, return line filter, or heat exchanger, DO NOT operate the motor. Replace the motor and conduct a post-catastrophic failure start-up procedure.*

Step 9.   Allow the system to warm up to approximately 130º F. (54º C.). Observe pyrometer (5).

Step 10.   Remove the load from the motors.

NOTE: *For accurate test results, check and adjust (if necessary) the main pressure relief valve or compensator setting. Set the relief valve or compensator to the pressure recommended by the manufacturer.*

Step 11.   Operate the prime mover at maximum speed. Record on the test worksheet the pump "no-load" input shaft speed indicated on tachometer (1). (Conduct Step 13. through Step 16. with the prime mover operating at maximum speed).

NOTE: *If the tachometer indicates a pump "no-load" speed which is inconsistent with machine specifications, adjust the speed accordingly and continue with the test procedure.*

Pump flow is proportional to input shaft speed. If there is a reduction in pump flow because of low prime mover speed, it will directly affect motor shaft speed. It is necessary to determine pump shaft speed prior to conducting actuator speed test procedures.

Step 12.   Remove the load from the motor output shaft.

Step 13.   Activate the directional control valve and direct full pump flow into the motors. Record on the test worksheet:
    a. "no-load" motor shaft speed indicated on tachometer (3).
    b. "no-load" system pressure indicated on pressure gauge (4).
    c. Fluid temperature indicated on pyrometer (5).

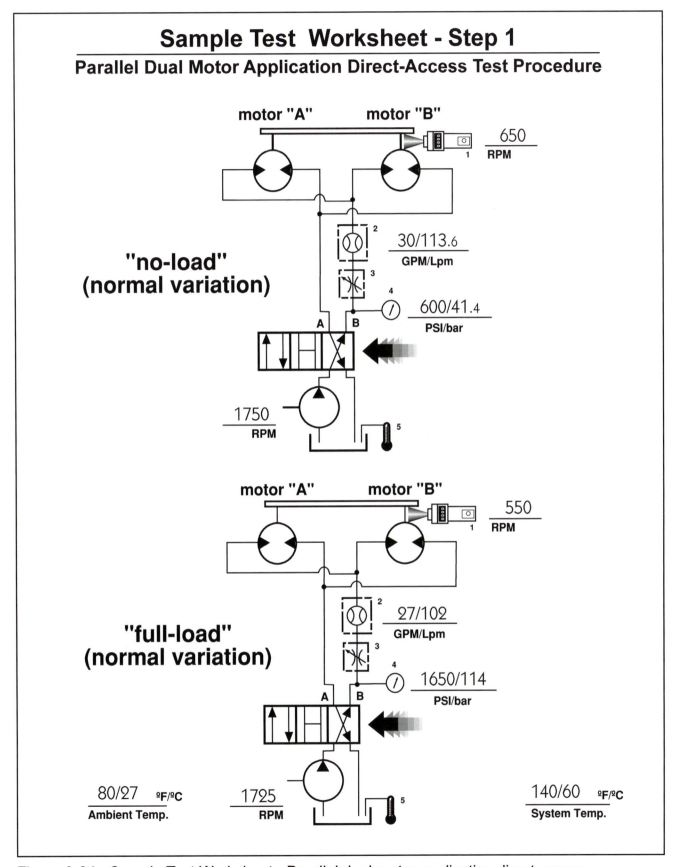

Sample Test Worksheet - Step 1
Parallel Dual Motor Application Direct-Access Test Procedure

Figure 9-24   Sample Test Worksheet - Parallel dual motor application direct-access
test procedure

Step 14. Gradually load the system by restricting the flow into the motors with needle valve (2) (turn clockwise).

Step 15. When the pressure reaches approximately 80% of the pressure relief valve setting, record on the test worksheet:
a. Pressure indicated on pressure gauge (4).
b. Flow indicated on flow meter (2).

ANALYZING THE TEST RESULTS - Step 1

---

1. **Diagnostic Observation: "full-load" cycle:** The flow meter indicates a nominal flow decrease as the needle valve creates an artificial load on the hydraulic system.

The flow decrease does not exceed the anticipated loss relative to the volumetric efficiency rating of the pump, and a nominal decrease in prime mover speed.

The tachometer indicates a decrease in the speed of the motors which is in proportion to the marginal loss of flow indicated on the flow meter.

The operating temperature of the fluid indicated on the pyrometer remains within design specification.

Prime mover speed remains within design specification.

Diagnosis: Leakage across the hydraulic system, excluding the motors, appears to be within design specification.

2. **Diagnostic Observation: "full-load" cycle:** The flow meter indicates a moderate to high flow decrease as the needle valve creates an artificial load on the hydraulic system.

The tachometer indicates a reduction in the speed of the motors which is proportional to the flow decrease.

The pyrometer indicates a moderate increase in the operating temperature of the fluid.

Prime mover speed remains within design specification.

Diagnosis: The loss in the speed of the motors is proportional to the decrease in flow into the motors. This indicates that the motors are not the cause of the problem.

To determine the root-cause, each component in the system, upstream of the needle valve, will have to be individually tested.

3. **Diagnostic Observation: "full-load" cycle:** The flow meter indicates a marginal flow decrease as the needle valve creates an artificial load on the hydraulic system. The flow

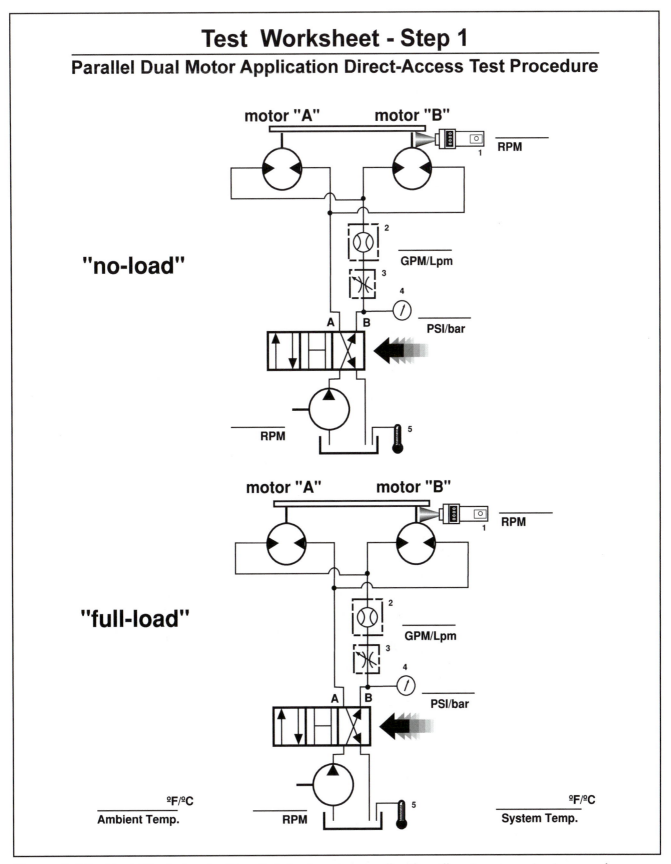

Test Worksheet - Step 1

Parallel Dual Motor Application Direct-Access Test Procedure

Figure 9-25   Test Worksheet - Parallel dual motor application direct-access test procedure

Step 2

1. Tachometer
2. Flow meter
3. Needle valve
4. Pressure gauge
5. Pyrometer

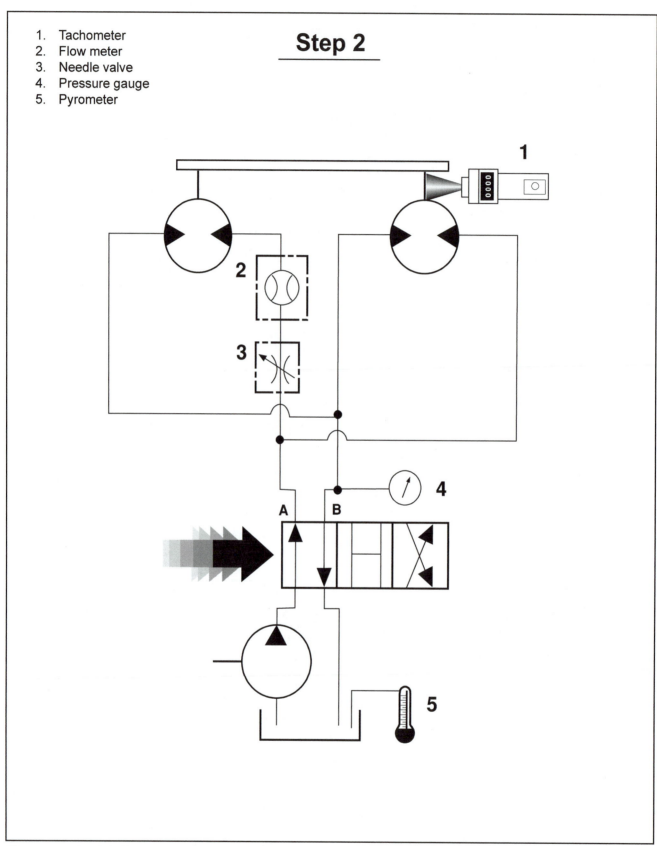

Figure 9-26   Placement of diagnostic equipment

decrease is not more than the anticipated loss relative to the volumetric efficiency rating of the pump, and a nominal decrease in prime mover speed.

During a normal machine load cycle, the tachometer indicates a progressive decrease in the speed of the motors as the load on their output shafts increases. The speed decrease approaches, or exceeds, 30% of the no-load speed of the motors (See **NOTE:** on page 9-6).

The pyrometer indicates a progressive increase in the operating temperature of the fluid which does not appear to "level-off."

Prime mover speed remains within design specification.

Diagnosis: If the flow meter indicates that the flow into the motors remains within the volumetric efficiency of the system, and the tachometer indicates a speed decrease which causes an unacceptable loss in motor shaft speed, leakage across the motors is excessive. Service or replace the motors.

Leakage of this magnitude could also cause a significant increase in the operating temperature of the fluid.

If necessary, refer to the manufacturer's specification for recommended leakage data.

NOTE: *Motors connected in parallel, will normally wear out uniformly. However, excessive leakage could be caused by one motor. It is recommended that each motor be tested independently to isolate the problem.*

TEST PROCEDURE - Step 2

To isolate the problem to a specific motor, proceed as follows:

Refer to Figure 9-26 for correct placement of diagnostic equipment. To record the test data, make a copy of the test worksheet on page 9-68 (Figure 9-28).

Step 1.   Remove flow meter (2), needle valve (3), and pressure gauge (4).

Step 2.   Install flow meter (2) and needle valve (3) in series with the transmission line at the inlet port of the first motor.

Step 3.   Install pressure gauge (4) in parallel with the connector at the inlet port of needle valve (3).

NOTE: *If the motors are connected in parallel, and are the same displacement, the flow recorded in step 14 (pump flow) should be equally divided between the number of motors in the circuit.*

Step 4.   Start the prime mover. Inspect the diagnostic equipment connectors for leaks.

Step 5.   Operate the motors at maximum speed.

Step 6.   Create an "artificial" load in the system by closing needle valve (3) gradually. Build pressure, in 100 PSI (6.9 bar) increments, up to approximately 80% of relief valve setting.

NOTE: *Conduct this test procedure at the inlet port of each motor in parallel.*

ANALYZING THE TEST RESULTS - Step 2

1.   **Diagnostic Observation: "full-load" cycle:** If the output shafts of the motors are mechanically connected, and the displacement of the motors is identical, the pump flow should be divided equally between the motors.

More flow will be inclined to flow through a motor which may be damaged or worn due to the larger clearances.

Flow meter (1), connected in series with the inlet port of motor "A," indicates a flow which is greater than half the pump flow. The flow exceeds the volumetric efficiency of the motor.

Flow meter (1), connected in series with the inlet port of motor "B," indicates a flow which is less than half the pump flow.

Diagnosis: Motor "A" is receiving more flow than motor "B." Thus, motor "A" is the cause of the problem.

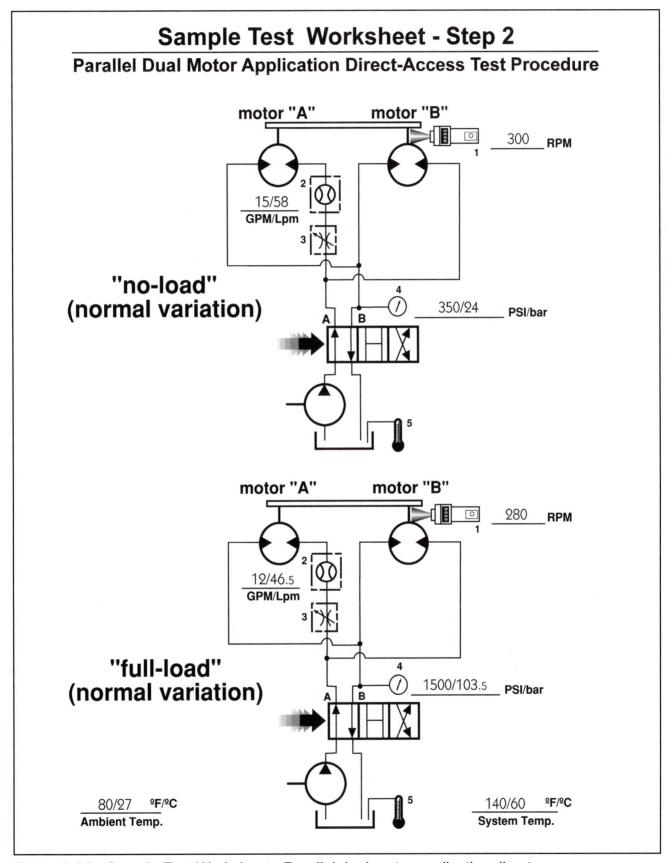

Sample Test Worksheet - Step 2
Parallel Dual Motor Application Direct-Access Test Procedure

Figure 9-27   Sample Test Worksheet - Parallel dual motor application direct-access
        test procedure

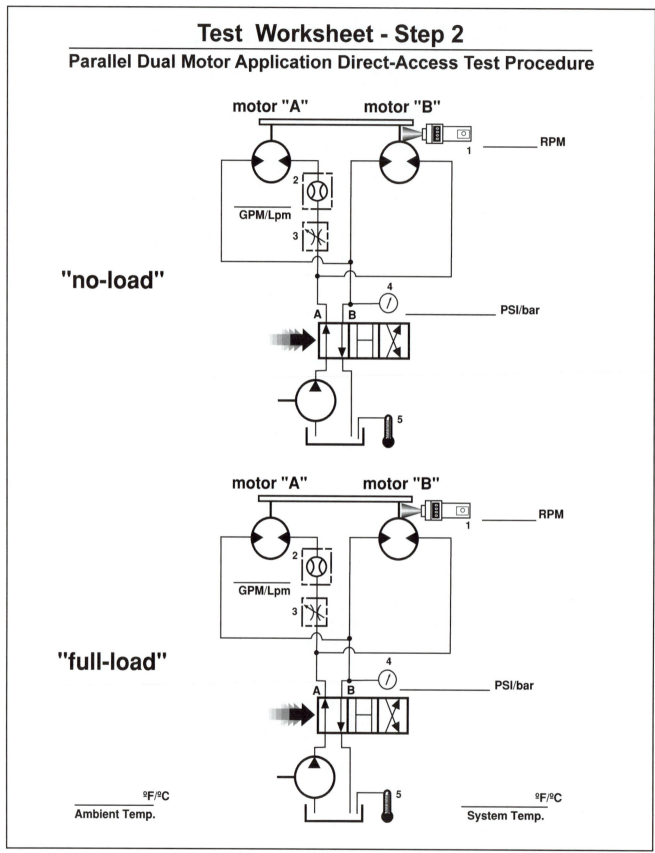

Test Worksheet - Step 2

Parallel Dual Motor Application Direct-Access Test Procedure

Figure 9-28   Test Worksheet - Parallel dual motor application direct-access test procedure

chapter ten

Procedures for Testing Cartridge Valves

Cartridge Valve Test Procedure

What will this test procedure accomplish?

Cartridge valves are "miniature" screw-in valves which offer the same control functions found in conventional valves. They are found in both mobile and industrial applications.

Screw-in cartridge valves have a wide variety of mounting options which include: in-line, direct mount, sandwich mount, subplate mount, and custom manifold block mount.

Because cartridge valves are so compact and have extremely low leakage rates they are highly susceptible to contamination related failures.

Due to the wide variety of mounting configurations available, troubleshooting cartridge valves can be challenging. Before attempting to troubleshoot cartridge valves there are certain factors that need to be understood:

    a. Cartridge valve manufacturers make several standard cartridge sizes to accomodate different flow rates. Each size is designed to screw into a standard cavity.

    b. A variety of physically identical cartridges with different control functions will fit into the same cavity. However, "physical" interchangeability of cartridges does not necessarily mean "functional" interchangeability.

    c. Physical interchangeability due to cavity design may be a standard feature on a specific manufacturers product range. However, caution must be used when changing from one brand of cartridge valve to another, as cavities are not necessarily identical. Always refer to the manufacturer's product literature before retrofitting cartridge valves.

    d. Most cartridge valves can be cleaned without altering pressure or flow settings. Most manufacturers caution against field disassembly and recommend cartridge replacement only.

This test procedure will determine:

    a. How much fluid leaks across the ports of a normally-closed (flow path closed in the "inactive" position) cartridge valve,

    b. How to isolate a cartridge valve malfunction in a multiple cartridge manifold application.

-WARNING- **Do not work on or around hydraulic systems without wearing safety glasses which conform to ANSI Z87.1-1989 standard.**

YOU WILL NEED THE FOLLOWING DIAGNOSTIC EQUIPMENT
TO CONDUCT THIS TEST PROCEDURE

CAUTION! The pressure and flow ratings of the diagnostic equipment which will be used to conduct this test procedure must be equal to, or greater than, the pressure and flow ratings of the system being tested. Refer to the system schematic for recommended pressure and flow data.

1.	Cartridge	3.	Test body
2.	Pressure gauge	4.	MicroLeak analyzer

PREPARATION

To conduct this test safely and accurately, refer to Figure 10-1 for correct placement of diagnostic equipment.

NOTE: a. *To test in-line and direct mounted cartridge valves, refer to chapter four- "Procedures for Testing Pressure Control Valves".*

b. *If the valves are sandwich or subplate mounted, refer to chapter five- "Procedures for Testing Industrial Directional Control Valves-".*

c. *This test procedure is primarily aimed at testing complex valve manifold assemblies utilizing two or more cartridge valves which are connected in parallel.*

TEST PROCEDURE

Step 1.   Shut the prime mover off.

Step 2.   Lock the electrical system out or tag the keylock switch.

Step 3.   Observe the system pressure gauge. Release any residual pressure trapped in the system by an accumulator, counterbalance or pilot-operated check valve, suspended load on an actuator, intensifier, or a pressurized reservoir.

Step 4.   Determine, from the circuit schematic, specifically which cartridge valve(s) in the circuit control the function(s) of the actuator(s).

Step 5.   Make a note of each valve, specifically manufacturer and part number. If the information is not on the circuit schematic refer to the bill of materials. The part number will provide information on the cavity size for the cartridge.

Step 6.   Obtain a single, line-mounted body to suit the cartridge valve to be tested.

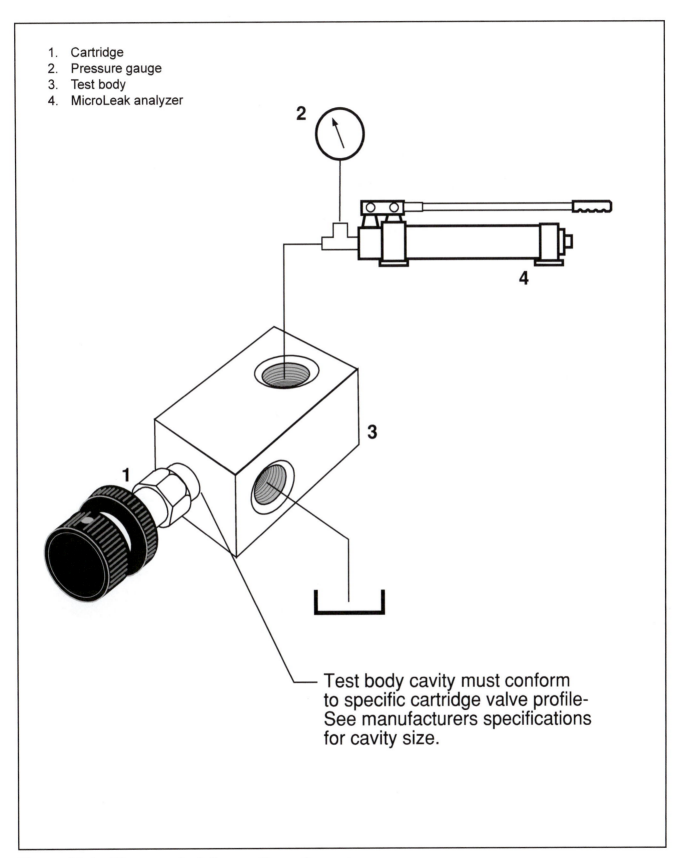

1. Cartridge
2. Pressure gauge
3. Test body
4. MicroLeak analyzer

Test body cavity must conform to specific cartridge valve profile- See manufacturers specifications for cavity size.

Figure 10-1   Placement of diagnostic equipment

Step 7.　Remove the suspect cartridge(s) one at a time from the circuit and screw them into the test body.

NOTE:　*Determine from the circuit diagram what type of valve is being tested. Refer to the index in the front of this manual for the test procedure applicable to the type of valve being tested. Conduct the test accordingly.*

Step 8.　Test each valve for proper operation until the suspect valve is located.

Step 9.　Clean or replace the cartridge and adjust to manufacturer's specifications.

chapter eleven

Procedures for Testing Flow Control Valves

Needle Valve Needle/Seat
Direct-Access Test Procedure

What will this test procedure accomplish?

A needle valve is an adjustable orifice. It is generally used in applications where it is desirable to adjust the speed of an actuator.

A needle valve consists of a valve body with an inlet and outlet port. The ports can be opposite one another or in some cases are 90° apart. The fluid passing through the valve passes through an opening which is the seat for a cone-shaped tip which is connected to the end of a threaded rod. The opposite end of the threaded rod is attached to an externally adjustable knob.

Turning the knob moves the rod through a threaded hole in the valve body. This changes the position of the cone in relation to its seat, which in turn adjusts the flow rate through the valve.

The adjustment sensitivity of a needle valve is determined by the shape of the cone and the thread dimension of the rod.

Fine contaminants in a high velocity stream of fluid could eventually erode the seat and cone. This will cause the needle valve to lose its adjustment sensitivity. In applications where a closed needle valve is used to hold a load, seat and cone damage could cause a load to drift.

The most common symptoms of problems associated with flow control valves are:
  a. Loss of control sensitivity.
  b. Loss of fine metering control.
  c. Actuator drifting in load-holding applications.

This test procedure will determine the amount of leakage across the ports of a needle valve when it is in the "closed" position.

 **-WARNING-** **Do not work on or around hydraulic systems without wearing safety glasses which conform to ANSI Z87.1-1989 standard.**

YOU WILL NEED THE FOLLOWING DIAGNOSTIC EQUIPMENT
TO CONDUCT THIS TEST PROCEDURE

CAUTION! The pressure and flow ratings of the diagnostic equipment which will be used to conduct this test procedure must be equal to, or greater than, the pressure and flow ratings of the system being tested. Refer to the system schematic for recommended pressure and flow data.

---

1.   MicroLeak analyzer   2.   Pressure gauge

PREPARATION

To conduct this test safely and accurately, refer to Figure 11-1 for correct placement of diagnostic equipment.

TEST PROCEDURE

Step 1.   Shut the prime mover off.

Step 2.   Lock the electrical system out or tag the keylock switch.

Step 3.   Observe the system pressure gauge. Release any residual pressure trapped in the system by an accumulator, counterbalance or pilot-operated check valve, suspended load on an actuator, intensifier, or a pressurized reservoir.

Step 4.   Disconnect the transmission line from the inlet port of the needle valve.

NOTE: *The needle valve can be removed from the system for this test procedure.*

Step 5.   Connect MicroLeak analyzer (1) in series with the inlet port of the needle valve.

Step 6.   Install pressure gauge (2) in parallel with the connector at the outlet port of MicroLeak analyzer(1).

Step 7.   Close the needle valve securely (turn clockwise).

Step 8.   Gradually pressurize the inlet port of the needle valve with MicroLeak analyzer (1).

Step 9.   Stop and maintain pressure when the value of the system's main pressure relief valve setting is reached.

1. MicroLeak analyzer
2. Pressure gauge

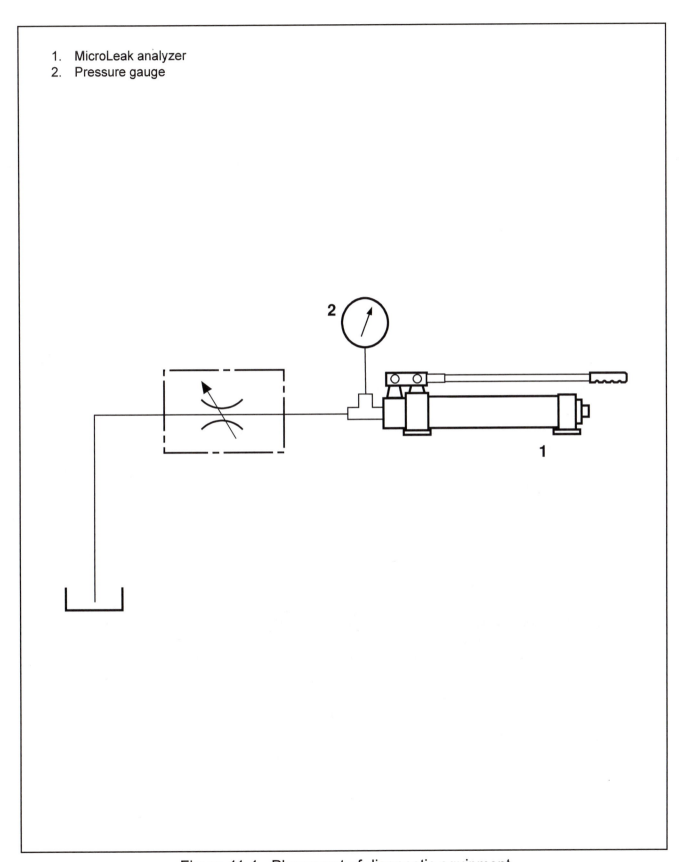

Figure 11-1   Placement of diagnostic equipment

NOTE: *Oil should now be trapped between MicroLeak analyzer (1) and the inlet port of the needle valve.*

Step 10. Observe pressure gauge (2). The pressure should "hold" for a reasonable length of time.

Step 11. Open the pressure release valve on MicroLeak analyzer (1) and release the pressure between MicroLeak analyzer (1) and the needle valve.

ANALYZING THE TEST RESULTS

1. **Diagnostic Observation:** The pressure indicated on pressure gauge (2) holds steady at the value of the system's main pressure relief valve setting.

   There may be some minor leakage indicated by a very gradual pressure loss.

   NOTE: *There is no general rule-of-thumb regarding the volumetric efficiency of needle valves. If it is necessary to determine accurate leakage rates, refer to the valve manufacturer's specifications.*

   Diagnosis: Leakage across the needle valve appears to be within design specification.

2. **Diagnostic Observation:** The pressure indicated on pressure gauge (2) attempts to increase when the MicroLeak analyzer is stroked. However, between pumping strokes the pressure drops rapidly.

   If the outlet port of the needle valve is open, a steady stream of oil can be seen pouring from the valve.

   The MicroLeak analyzer fails to pressurize the inlet port of the needle valve to the value of the system's main pressure relief valve setting.

   Diagnosis: Leakage across the needle valve appears to be excessive. Although this amount of leakage may appear to be insignificant, it could cause flow control problems and/or a marked increase in the operating temperature of the fluid.

   If necessary, refer to the manufacturer's specifications for recommended leakage data.

3. **Diagnostic Observation:** The MicroLeak analyzer fails to pressurize the inlet port of the needle valve. Oil leaks profusely from the outlet port of the needle valve.

   Diagnosis: Leakage across the needle valve is excessive. Replace the needle valve.

Flow Control Valve Needle/Seat/Reverse Check Direct-Access Test Procedure

What will this test procedure accomplish?

A flow control valve is basically a needle valve with an integral reverse flow check valve. The check valve is positioned in parallel with the needle valve. It is used in applications where it is desirable to adjust the speed of an actuator in one direction, while allowing free-flow in the opposite direction.

A flow control valve consists of a valve body with an inlet and outlet port. The ports can be opposite one another or in some cases are 90° apart.

The fluid passing through the valve in the "controlled flow" direction holds the check valve poppet closed. It passes, in parallel, through an opening which is the seat for a cone-shaped tip which is connected to the end of a threaded rod. The opposite end of the threaded rod is attached to an externally-adjustable knob.

Turning the knob moves the rod through a threaded hole in the valve body. This changes the position of the cone in relation to its seat which in turn adjusts the flow rate through the valve.

The adjustment sensitivity of a flow control valve is determined by the shape of the cone, and the thread dimension of the rod.

In the reverse flow direction, the flow control valve restricts the flow. However, the integral check valve, which is positioned in parallel with the needle valve, offers a lower resistance than the needle valve and allows unrestricted flow to return through the valve.

Fine contaminants in a high velocity stream of fluid could eventually erode the seat and cone in the needle valve assembly or the seat and poppet assembly in the check valve. This could cause the needle valve to loose its adjustment sensitivity. In applications where a flow control valve is used to hold a load, wear in the valve may cause the load to drift.

The most common symptoms of problems associated with flow control valves are:
    a. Loss of control sensitivity.
    b. Loss of fine metering control.
    c. Actuator drifting in load-holding applications.

This test procedure will determine the amount of leakage across the ports of a flow control valve when it is in the "closed" position.

 **-WARNING-** Do not work on or around hydraulic systems without wearing safety glasses which conform to ANSI Z87.1-1989 standard.

YOU WILL NEED THE FOLLOWING DIAGNOSTIC EQUIPMENT
TO CONDUCT THIS TEST PROCEDURE

CAUTION! The pressure and flow ratings of the diagnostic equipment which will be used to conduct this test procedure must be equal to, or greater than, the pressure and flow ratings of the system being tested. Refer to the system schematic for recommended pressure and flow data.

---

1. MicroLeak analyzer   2. Pressure gauge

PREPARATION

To conduct this test safely and accurately, refer to Figure 11-2 for correct placement of diagnostic equipment.

TEST PROCEDURE

Step 1.   Shut the prime mover off.

Step 2.   Lock the electrical system out or tag the keylock switch.

Step 3.   Observe the system pressure gauge. Release any residual pressure trapped in the system by an accumulator, counterbalance or pilot-operated check valve, suspended load on an actuator, intensifier, or a pressurized reservoir.

NOTE: *The flow control valve can be removed from the system for this test procedure.*

Step 4.   Connect MicroLeak analyzer (1) in series with the inlet (controlled flow) port of the flow control valve.

Step 5.   Install pressure gauge (2) in parallel with the connector at the outlet port of MicroLeak analyzer (1).

Step 6.   Close the flow control valve securely (turn clockwise).

Step 7.   Gradually pressurize the inlet (controlled flow) port of the flow control valve with Micro Leak analyzer (1).

Step 8.   Stop and maintain pressure when the value of the system's main pressure relief valve or compensator setting is reached.

NOTE: *Oil should now be trapped between MicroLeak analyzer (1) and the inlet (controlled flow) port of the flow control valve.*

1. MicroLeak analyzer
2. Pressure gauge

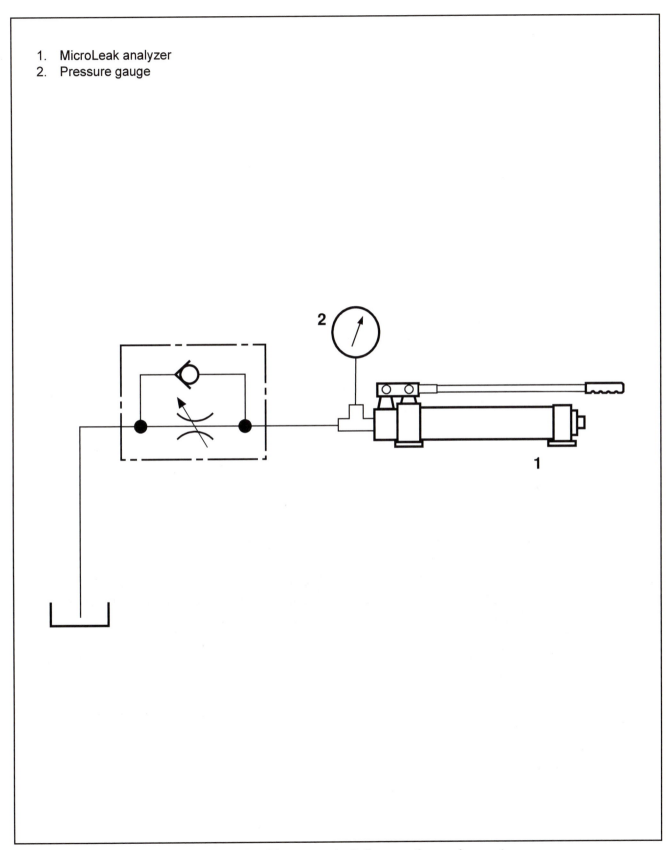

Figure 11-2   Placement of diagnostic equipment

Step 9.   Observe pressure gauge (2). The pressure should "hold" for a reasonable length of time.

Step 10.   Open the pressure release valve on MicroLeak analyzer (1) and release the pressure between MicroLeak analyzer (1) and the flow control valve.

ANALYZING THE TEST RESULTS

1.   **Diagnostic Observation:** The pressure indicated on pressure gauge (2) holds steady at the value of the system's main pressure relief valve setting.

There may be some minor leakage indicated by a very gradual pressure loss.

NOTE: *There is no general rule-of-thumb regarding the volumetric efficiency of flow control valves. If it is necessary to determine accurate leakage rates refer to the valve manufacturer's specifications.*

Diagnosis: Leakage across the flow control valve appears to be within design specification.

2.   **Diagnostic Observation:** The pressure indicated on pressure gauge (2) attempts to increase when the MicroLeak analyzer is stroked. However, between pumping strokes the pressure drops rapidly.

If the outlet port of the flow control valve is open, a steady stream of oil can be seen pouring from the valve.

The MicroLeak analyzer fails to pressurize the inlet port of the flow control valve to the value of the system's main pressure relief valve setting.

Diagnosis: Leakage across the flow control valve appears to be excessive. Although this amount of leakage may appear to be insignificant, it could cause flow control problems and/or a marked increase in the operating temperature of the fluid.

If necessary, refer to the manufacturer's specifications for recommended leakage data.

3.   **Diagnostic Observation:** The MicroLeak analyzer fails to pressurize the inlet port of the flow control valve. Oil leaks profusely from the outlet port of the flow control valve.

Diagnosis: Leakage across the flow control valve is excessive. Cross-port leakage can be caused by needle/seat damage or wear. It could also be caused by reverse flow check valve damage or wear. Service or replace the flow control valve.

chapter twelve

Procedures for Testing Check Valves

Check Valve
Direct-Access Test Procedure

What will this test procedure accomplish?

A check valve consists of a steel or brass body with an inlet and outlet port. Inside the body is an active poppet which is biased toward the inlet port by a mechanical spring force.

The position of the poppet relative to the inlet and outlet ports allows fluid to flow through the valve in one direction only while "checking" or blocking flow in the opposite direction.

Fluid flowing into the inlet port of the check valve pushes the poppet away from its seat. The fluid then passes around the sides of the poppet and exhausts through the outlet port of the valve.

If the fluid attempts to flow in the reverse direction, the poppet is pushed against its seat blocking the outlet port.

Check valves generally have extremely low leakage rates.

The most common symptoms of problems associated with check valves are:
    a. Premature system "bleed-down" in accumulator applications.
    b. Control problems with valves utilizing check valves for reverse flow or by-pass functions, e.g. sequence valves, counterbalance valves, flow control valves, etc.
    c. Leakage across check valves used as low pressure relief valves, filter by-pass, and heat exchanger by-pass valves.

This test procedure will determine the amount of leakage across the blocked-flow port of a check valve.

 **-WARNING-** Do not work on or around hydraulic systems without wearing safety glasses which conform to ANSI Z87.1-1989 standard.

YOU WILL NEED THE FOLLOWING DIAGNOSTIC EQUIPMENT TO CONDUCT THIS TEST PROCEDURE

CAUTION! The pressure and flow ratings of the diagnostic equipment which will be used to conduct this test procedure must be equal to, or greater than, the pressure and flow ratings of the system being tested. Refer to the system schematic for recommended pressure and flow data.

---

1.   MicroLeak analyzer   2.   Pressure gauge

PREPARATION

To conduct this test safely and accurately, refer to Figure 12-1 for correct placement of diagnostic equipment.

TEST PROCEDURE

Step 1.   Shut the prime mover off.

Step 2.   Lock the electrical system out or tag the keylock switch.

Step 3.   Observe the system pressure gauge. Release any residual pressure trapped in the system by an accumulator, counterbalance or pilot-operated check valve, suspended load on an actuator, intensifier, or a pressurized reservoir.

NOTE: *The check valve can be removed from the system for this test procedure.*

Step 4.   Connect MicroLeak analyzer (1) in series with the transmission line at the outlet (blocked flow) port of the check valve.

Step 5.   Install pressure gauge (2) in parallel with the connector at the outlet port of MicroLeak analyzer (1).

Step 6.   Gradually pressurize the outlet (blocked flow) port of the check valve with MicroLeak analyzer (1).

Step 7.   Stop and maintain pressure when the the value of the system's main pressure relief valve or compensator setting is reached.

NOTE: *Oil should now be trapped between MicroLeak analyzer(1) and the outlet port (blocked flow) of the check valve.*

1. MicroLeak analyzer
2. Pressure gauge

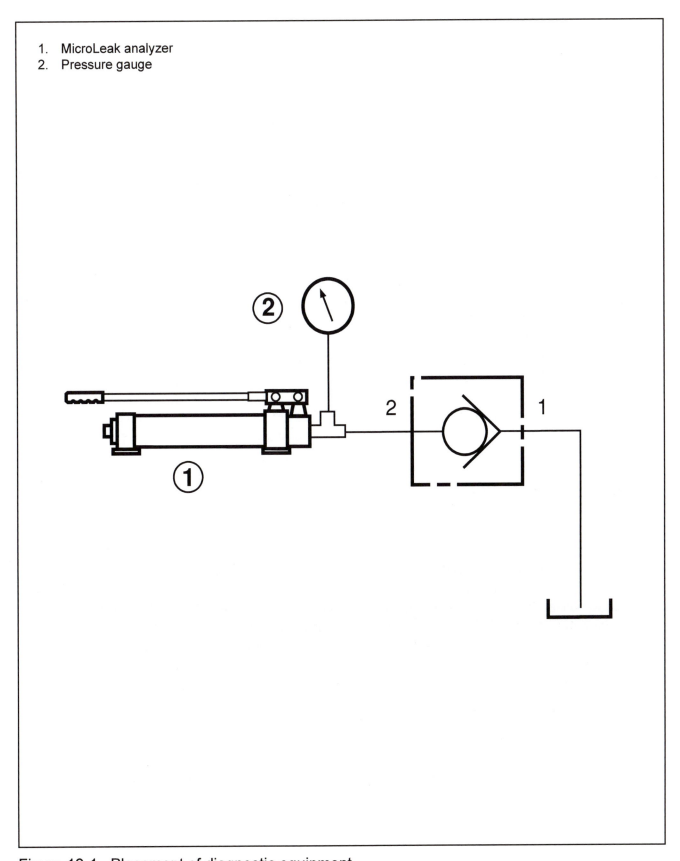

Figure 12-1   Placement of diagnostic equipment

Step 8. Observe pressure gauge (2). The pressure should hold for a reasonable length of time.

Step 9. Open the pressure release valve on MicroLeak analyzer (1) and release the pressure between MicroLeak analyzer (1) and the check valve.

ANALYZING THE TEST RESULTS

1. **Diagnostic Observation:** The pressure indicated on pressure gauge (2) holds steady at the value of the system's main pressure relief valve or compensator setting.

   There may be some minor leakage indicated by a very gradual pressure loss.

NOTE: *There is no general rule-of-thumb regarding the volumetric efficiency of check valves. If it is necessary to determine accurate leakage rates, refer to the valve manufacturer's specifications.*

   Diagnosis: Leakage across the blocked-flow port of the check valve appears to be within design specification.

2. **Diagnostic Observation:** The pressure indicated on pressure gauge (2) attempts to increase when the MicroLeak analyzer is stroked. However, between pumping strokes the pressure drops rapidly.

   If the outlet port of the check valve is open, a steady stream of oil can be seen pouring from the valve.

   The MicroLeak analyzer fails to pressurize the inlet port of the check valve to the value of the system's main pressure relief valve or compensator setting.

   Diagnosis: Leakage across the blocked-flow port of the check valve appears to be excessive. Although this amount of leakage may appear to be insignificant, it could cause control problems and/or a marked increase in the operating temperature of the fluid.

   If necessary, refer to the manufacturer's specifications for recommended leakage data.

3. **Diagnostic Observation:** The MicroLeak analyzer fails to pressurize the outlet port of the check valve. Oil leaks profusely from the valve's inlet port.

   Diagnosis: Leakage across the blocked-flow port of the check valve is excessive. Replace the check valve.

Pilot-Operated Check Valve - Poppet/Seat
Direct-Access Test Procedure

What will this test procedure accomplish?

A pilot-to-open check valve, more commonly known as a pilot-operated check valve, is basically a check valve with an integrated, single-acting, spring-return cylinder attached to one end.

It consists of a body with an inlet and outlet port. Inside the body is an active poppet which is biased toward the inlet port by a mechanical spring force.

The single-acting, spring-return cylinder consists of a cylindrical body closed at each end. Inside the cylinder is a moveable piston which is attached to a rod. The rod, which is located on the same axis as the check valve, protrudes from one end of the cylinder and faces the check valve.

A pilot pressure port is drilled into the opposite end of the cylinder. When pilot pressure acting against the back of the piston reaches a pre-determined level, it forces the piston to move. This causes the rod, which is attached to the cylinder, to come into physical contact with the check valve poppet and force it open.

As long as pilot pressure is present behind the piston, the check valve will remain open. If pilot pressure should drop below a pre-determined level, the bias spring on the rod side of the piston will force the piston to retract, moving it away from the check valve poppet.

The check valve poppet bias spring will immediately force the check valve poppet against its seat blocking the flow path through the valve.

The most common symptoms of problems associated with pilot-operated check valve poppet/seat leakage are:
  a. Actuator drifting in pressure manifold applications.
  b. Actuator drifting in vertical load "hold" applications.

Typical causes of pilot operated check valve deterioration:
  a. Contamination in the oil.
  b. High cycling rates deteriorate seats and accelerates fatigue in components.
  c. Overheating leads to seat deterioration.

This test procedure will determine the amount of leakage across the blocked-flow port of a pilot-operated check valve.

 **-WARNING-** | Do not work on or around hydraulic systems without wearing safety glasses which conform to ANSI Z87.1-1989 standard.

YOU WILL NEED THE FOLLOWING DIAGNOSTIC EQUIPMENT
TO CONDUCT THIS TEST PROCEDURE

CAUTION! The pressure and flow ratings of the diagnostic equipment which will be used to conduct this test procedure must be equal to, or greater than, the pressure and flow ratings of the system being tested. Refer to the system schematic for recommended pressure and flow data.

---

1.   MicroLeak analyzer   2.   Pressure gauge

PREPARATION

To conduct this test safely and accurately, refer to Figure 12-2 for correct placement of diagnostic equipment.

TEST PROCEDURE

Step 1.   Shut the prime mover off.

Step 2.   Lock the electrical system out or tag the keylock switch.

Step 3.   Observe the system pressure gauge. Release any residual pressure trapped in the system by an accumulator, counterbalance or pilot-operated check valve, suspended load on an actuator, intensifier, or a pressurized reservoir.

CAUTION! Even though the system pressure gauge indicates zero pressure, there could be significant pressure trapped between the pilot-operated check valve and the actuator. Follow the manufacturer's recommended "bleed-down" procedure before removing the valve.

Step 4.   Disconnect the transmission line from the outlet (blocked-flow) port of the pilot-operated check valve.

NOTE: *The pilot-operated check valve can be removed from the system for this test procedure.*

Step 5.   Connect MicroLeak analyzer (1) in series with the transmission line at the outlet (blocked flow) port of the pilot-operated check valve.

Step 6.   Install pressure gauge (2) in parallel with the connector at the outlet port of MicroLeak analyzer (1).

Step 7.   Gradually pressurize the outlet (blocked-flow) port of the pilot-operated check valve with MicroLeak analyzer (1).

1. MicroLeak analyzer
2. Pressure gauge

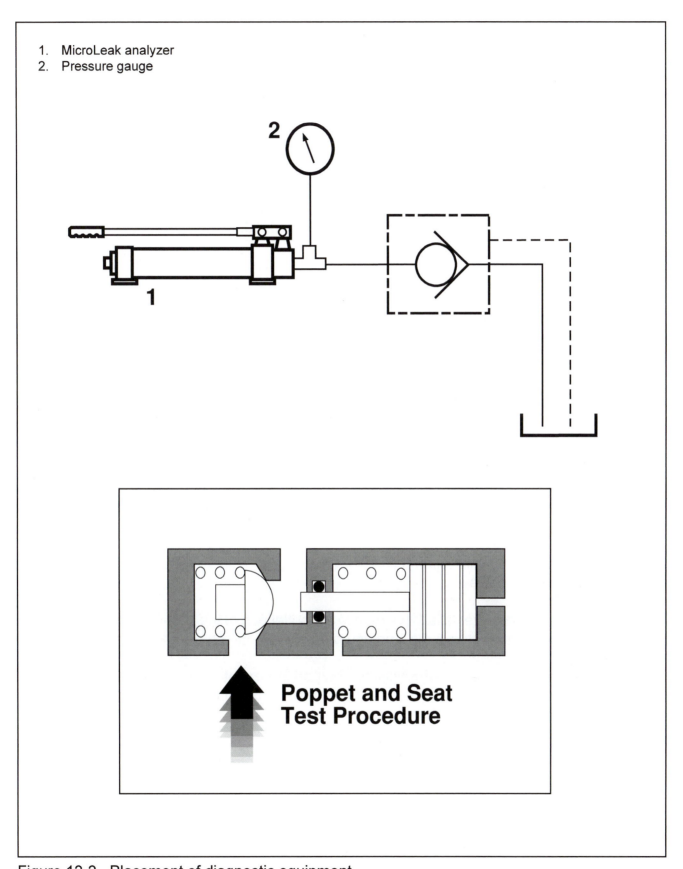

Figure 12-2   Placement of diagnostic equipment

Step 8. Stop and maintain pressure when the value of the system's main pressure relief valve or compensator setting is reached.

NOTE: *Oil should now be trapped between MicroLeak analyzer (1) and the outlet (blocked flow) port of the pilot-operated check valve.*

Step 9. Observe pressure gauge (2). The pressure should hold for a reasonable length of time.

Step 10. Open the pressure release valve on MicroLeak analyzer (1) and release the pressure between MicroLeak analyzer (1) and the pilot-operated check valve.

ANALYZING THE TEST RESULTS

---

1. **Diagnostic Observation:** The pressure indicated on pressure gauge (2) holds steady at the value of the system's main pressure relief valve or compensator setting.

    There may be some minor leakage indicated by a very gradual pressure loss.

NOTE: *There is no general rule-of-thumb regarding the volumetric efficiency of check valves. If it is necessary to determine accurate leakage rates, refer to the valve manufacturer's specifications.*

    Diagnosis: Leakage across the blocked-flow port of the pilot-operated check valve appears to be within design specification.

2. **Diagnostic Observation:** The pressure indicated on pressure gauge (2) attempts to increase when the MicroLeak analyzer is stroked. However, between pumping strokes the pressure drops rapidly.

    If the inlet port of the pilot-operated check valve is open a steady stream of oil can be seen pouring from the valve.

    The MicroLeak analyzer fails to pressurize the outlet port of the pilot-operated check valve to the value of the system's main pressure relief valve or compensator setting.

    Diagnosis: Leakage across the blocked-flow port of the pilot-operated check valve appears to be excessive. Although this amount of leakage may appear to be insignificant, it could cause actuator drifting problems.

    If necessary, refer to the manufacturer's specifications for recommended leakage data.

3. **Diagnostic Observation:** The MicroLeak analyzer fails to pressurize the outlet port of the pilot-operated check valve. Oil leaks profusely from the valve's inlet port.

    Diagnosis: Leakage across the blocked-flow port of the pilot-operated check valve is excessive. Replace the pilot-operated check valve.

Pilot-Operated Check Valve--Pilot Piston Seal
Direct Access Test Procedure

What will this test procedure accomplish?

A pilot-to-open check valve, more commonly known as a pilot-operated check valve, is basically a check valve with an integrated, single-acting, spring-return cylinder attached to one end.

It consists of a body with an inlet and outlet port. Inside the body is an active poppet which is biased toward the inlet port by a mechanical spring force.

The single-acting, spring-return cylinder consists of a cylindrical body closed at each end. Inside the cylinder is a moveable piston which is attached to a rod. The rod, which is located on the same axis as the check valve, protrudes from one end of the cylinder and faces the check valve.

A pilot pressure port is drilled into the opposite end of the cylinder. When pilot pressure acting against the back of the piston reaches a pre-determined level, it forces the piston to move. This causes the rod, which is attached to the cylinder, to come into physical contact with the check valve poppet and force it open.

As long as pilot pressure is present behind the piston, the check valve will remain open. If pilot pressure should drop below a pre-determined level, the bias spring on the rod side of the piston will force the piston to retract, moving it away from the check valve poppet.

The check valve poppet bias spring will immediately force the check valve poppet against its seat blocking the flow path through the valve.

The most common symptoms of problems associated with pilot pressure in pilot-operated check valve applications are:
   a. Actuator appears to "lock-up" when directional control valve is activated.
   b. Vertical load drifting when directional control valve is in the neutral "hold" position-linear actuator application.
   c. Horizontal load drifting (in rod extend direction only) when prime mover is operating. This problem is common to pressure manifold systems.

This test procedure will determine if:
   a. There is sufficient pilot pressure to open the pilot-operated check valve.
   b. Pilot pressure builds when the directional control valve is in the neutral position.
   c. Pilot pressure is maintained at the pilot pressure inlet port when the directional control valve is in the neutral position.

 **-WARNING-** | **Do not work on or around hydraulic systems without wearing safety glasses which conform to ANSI Z87.1-1989 standard.**

YOU WILL NEED THE FOLLOWING DIAGNOSTIC EQUIPMENT
TO CONDUCT THIS TEST PROCEDURE

CAUTION! The pressure and flow ratings of the diagnostic equipment which will be used to conduct this test procedure must be equal to, or greater than, the pressure and flow ratings of the system being tested. Refer to the system schematic for recommended pressure and flow data.

---

1.   MicroLeak analyzer   2.   Pressure gauge

PREPARATION

To conduct this test safely and accurately, refer to Figure 12-3 for correct placement of diagnostic equipment.

TEST PROCEDURE

Step 1.   Shut the prime mover off.

Step 2.   Lock the electrical system out or tag the keylock switch.

Step 3.   Observe the system pressure gauge. Release any residual pressure trapped in the system by an accumulator, counterbalance or pilot-operated check valve, suspended load on an actuator, intensifier, or a pressurized reservoir.

CAUTION! Even though the system pressure gauge indicates zero pressure, there could be significant pressure trapped between the pilot-operated check valve and the actuator. Follow the manufacturer's recommended "bleed-down" procedure before removing the valve.

Step 4.   Disconnect the transmission line from the outlet (blocked-flow) port of the pilot-operated check valve.

NOTE: *The pilot-operated check valve can be removed from the system for this test procedure.*

Step 5.   Connect MicroLeak analyzer (1) in series with the transmission line at the pilot pressure port of the pilot-operated check valve.

Step 6.   Connect pressure gauge (2) in parallel with the connector at the outlet port of MicroLeak analyzer (1).

Step 7.   Gradually pressurize the pilot pressure port of the pilot-operated check valve with MicroLeak analyzer (1).

1. MicroLeak analyzer
2. Pressure gauge

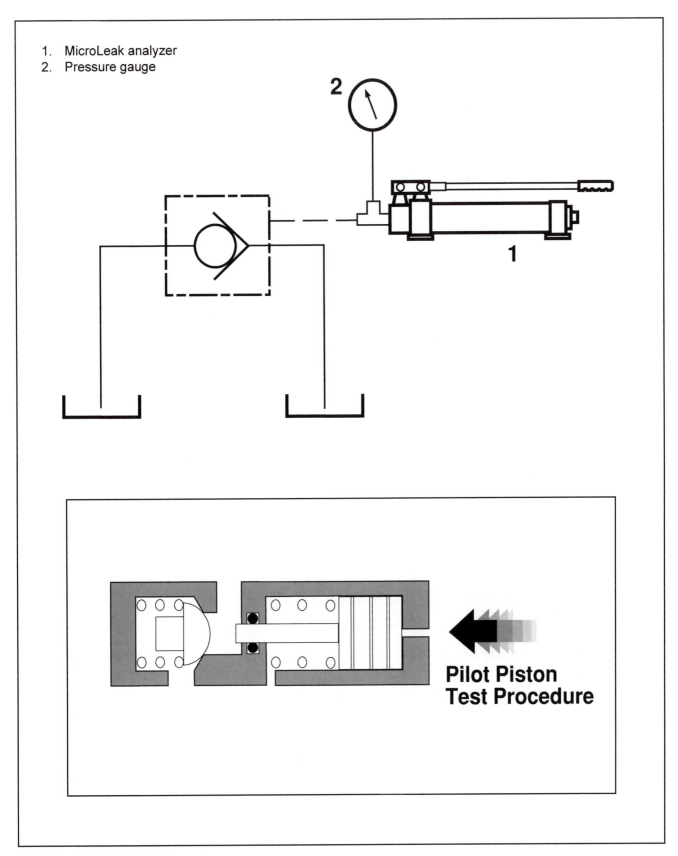

Figure 12-3   Placement of diagnostic equipment

Step 8.   Stop and maintain pressure when the value of the system's main pressure relief valve or compensator setting is reached.

Step 9.   Observe pressure gauge (2). The pressure should hold for a reasonable length of time.

Step 10. Open the pressure release valve on MicroLeak analyzer (1) and release the pressure between the MicroLeak analyzer (1) and the pilot pressure port.

ANALYZING THE TEST RESULTS

1.   **Diagnostic Observation:** The pressure indicated on pressure gauge (2) holds steady at the value of the system's main pressure relief valve or compensator setting.

There may be some minor leakage indicated by a very gradual pressure loss.

NOTE: *There is no general rule-of-thumb regarding the volumetric efficiency of the pilot piston in pilot-operated check valves. If it is necessary to determine accurate leakage rates, refer to the valve manufacturer's specifications.*

Diagnosis: Leakage across the pilot piston appears to be within design specification.

2.   **Diagnostic Observation:** The pressure indicated on pressure gauge (2) attempts to increase when the MicroLeak analyzer is stroked. However, between pumping strokes the pressure drops rapidly.

If the inlet and outlet ports of the pilot-operated check valve are open a steady stream of oil can be seen pouring from the valve.

The MicroLeak analyzer fails to pressurize the pilot pressure port of the pilot-operated check valve.

Diagnosis: Leakage across the pilot piston appears to be excessive. Although this amount of leakage may appear to be insignificant, it could cause either the pilot-operated check valve to operate erratically, or, total valve malfunction.

If necessary, refer to the manufacturer's specifications for recommended leakage data.

3.   **Diagnostic Observation:** The MicroLeak analyzer fails to pressurize the pilot port of the pilot-operated check valve. Oil leaks profusely from the valve's inlet and outlet ports.

Diagnosis: Leakage across the pilot piston is excessive. Replace the pilot-operated check valve.

Pilot-Operated Check Valve - Pilot Pressure
In-Circuit Test Procedure

What will this test procedure accomplish?

A pilot-to-open check valve, more commonly known as a pilot-operated check valve, is basically a check valve with an integrated, single-acting, spring-return cylinder attached to one end.

It consists of a body with an inlet and outlet port. Inside the body is an active poppet which is biased toward the inlet port by a mechanical spring force.

The single-acting, spring-return cylinder consists of a cylindrical body closed at each end. Inside the cylinder is a moveable piston which is attached to a rod. The rod, which is located on the same axis as the check valve, protrudes from one end of the cylinder and faces the check valve.

A pilot pressure port is drilled into the opposite end of the cylinder. When pilot pressure acting against the back of the piston reaches a pre-determined level, it forces the piston to move. This causes the rod, which is attached to the cylinder, to come into physical contact with the check valve poppet and force it open.

As long as pilot pressure is present behind the piston, the check valve will remain open. If pilot pressure should drop below a pre-determined level, the bias spring on the rod side of the piston will force the piston to retract, moving it away from the check valve poppet.

The check valve poppet bias spring will immediately force the check valve poppet against its seat blocking the flow path through the valve.

The most common symptoms of problems associated with pilot pressure in pilot-operated check valve applications are:
    a. Actuator appears to "lock-up" when directional control valve is activated.
    b. Vertical load drifting when directional control valve is in the neutral "hold" position - linear actuator application.
    c. Horizontal load drifting (in rod extend direction only) when prime mover is operating. This problem is common to pressure manifold systems.

This test procedure will determine if:
    a. There is sufficient pilot pressure to open the pilot-operated check valve.
    b. Pilot pressure builds when the directional control valve is in the neutral position.
    c. Pilot pressure is maintained at the pilot pressure inlet port when the directional control valve is in the neutral position.

-WARNING-   **Do not work on or around hydraulic systems without wearing safety glasses which conform to ANSI Z87.1-1989 standard.**

YOU WILL NEED THE FOLLOWING DIAGNOSTIC EQUIPMENT
TO CONDUCT THIS TEST PROCEDURE

CAUTION! The pressure and flow ratings of the diagnostic equipment which will be used to conduct this test procedure must be equal to, or greater than, the pressure and flow ratings of the system being tested. Refer to the system schematic for recommended pressure and flow data.

---

1.   Pressure gauge

PREPARATION

To conduct this test safely and accurately, refer to Figure 12-4 for correct placement of diagnostic equipment.

To record the test data, make a copy of the test worksheet on page 12-18 (Figure 12-5).

TEST PROCEDURE

Step 1.   Shut the prime mover off.

Step 2.   Lock the electrical system out or tag the keylock switch.

Step 3.   Observe the system pressure gauge. Release any residual pressure trapped in the system by an accumulator, counterbalance or pilot-operated check valve, suspended load on an actuator, intensifier, or a pressurized reservoir.

NOTE: *Determine from the circuit schematic if there is a pilot-operated check valve in one or both actuator ports.*

To test a pilot-operated check valve which is controlling the live-end of an actuator, pump flow must be directed to the closed-end of the cylinder. Pressure from the closed-end of the cylinder will supply pilot pressure to open the valve.

Reverse this procedure if there is a pilot-operated check valve mounted on the opposite cylinder port.

In this example, a pilot-operated check valve controlling the rod-side of a cylinder will be tested.

Step 4.   Install pressure gauge (1) in parallel with the connector at the pilot pressure port of the pilot-operated check valve.

Step 5.   Start the prime mover. Inspect diagnostic equipment connectors for leaks.

Step 6.   Allow the system to warm up to approximately 130°F. (54°C.). Observe pyrometer (2).

1. Pressure gauge

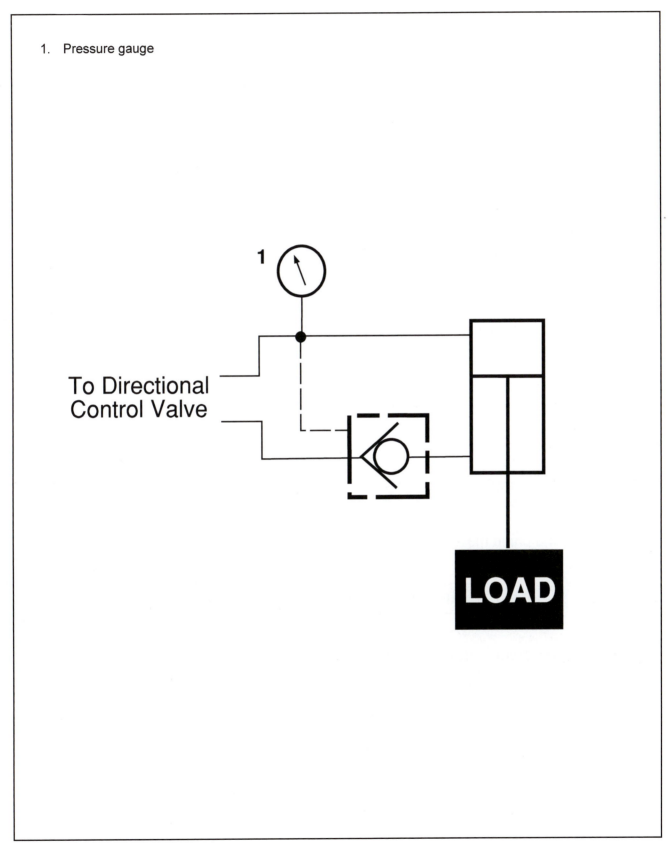

Figure 12-4   Placement of diagnostic equipment

Step 7.   Place the directional control valve in the neutral position. Record on the test worksheet the pressure indicated on pressure gauge (1).

NOTE: *If there is a gradual pressure build-up in the pilot pressure line refer to "Analyzing the Test Results" at the end of this test procedure. If there is a nominal pressure in the pilot pressure line continue with the test procedure.*

Step 8.   Activate the directional control valve and direct pump flow to the closed-end of the cylinder.

Step 9.   Record on the test worksheet the pilot pressure indicated on pressure gauge (1).

NOTE: *If the pilot pressure, registered on pressure gauge (1), increases to the value of the pressure relief valve setting, and the cylinder rod fails to move or moves erratically, refer to "Analyzing the Test Results" at the end of this test procedure.*

Step 10.   Allow the cylinder rod to travel through its entire stroke. Observe the pilot pressure while the rod is traveling. Note any deviation in pilot pressure which affects cylinder rod performance.

NOTE: *If there is a pilot-operated check valve in the opposite cylinder port, connect pressure gauge (1) in the opposite pilot pressure line and repeat this test procedure.*

Step 11.   Shut the prime mover off and analyze the test results.

Step 12.   At the conclusion of this test procedure, remove the diagnostic equipment. Reconnect the transmission lines and tighten the connectors securely.

Step 13.   Start the prime mover and inspect the connectors for leaks.

ANALYZING THE TEST RESULTS

---

1.   **Diagnostic Observation:** There is gradual pressure build-up when the directional control valve is in the neutral position which causes the cylinder rod to drift.

Diagnosis: In a pressure manifold system it is common for pressure to build across the spool of a closed-center directional control valve. If the rod-to-bore ratio of the cylinder is great enough, the cylinder rod could drift out (extend) due to pressure equalization on both sides of the piston.

A pilot-operated check valve will not necessarily correct cylinder drift. If the pressure attempts to equalize on both sides of the cylinder there may be enough pressure in the pilot-line to open the check valve.

To correct this problem, a closed-center directional control valve spool can be substituted with a float-center spool. A float-center spool will prevent residual pressure from opening a pilot-operated check valve. This is because it allows the pilot-line of the check valve to connect with the tank return-line when the valve is in the neutral (float) position.

2. **Diagnostic Observation:** Pressure gauge (1) indicates system pressure in the pilot line but the cylinder rod fails to move.

   Diagnosis: Any of the following problems could prevent the cylinder rod from moving:

   a. Pilot ratio of pilot-operated check valve incorrect. Refer to the manufacturer's specifications for pilot ratio recommendations.

   b. Cylinder rod overload. Remove load and repeat the test procedure.

   c. Pilot piston leaking internally. Refer to "Pilot Piston Seal Direct-Access Test Procedure" on page 12-9 of this chapter.

   d. Cylinder piston and/or rod seals too tight. Refer to the seal manufacturer's specifications for proper seal installation procedures.

   e. Load too heavy for pilot ratio. Install a pressure gauge at the cylinder port (between the pilot-operated check valve and the cylinder port). Record the "load-generated" pressure at the cylinder port.

WARNING! Follow all recommended safety procedures when disconnecting transmission lines from pilot-operated (load-holding) check valves.

---

3. **Diagnostic Observation:** Erratic cylinder rod operation.

   Diagnosis: If the pilot pressure line is too long, it will cause pressure transients in the pilot-line which will cause the pilot piston to fluctuate. This will cause erratic cylinder rod travel.

   Sudden load changes could cause the pilot pressure to fluctuate.

   In dual pilot-operated check valve applications, a sudden load change could begin a chain reaction. The pilot-operated check valves will open and close alternately causing violent load pulsations. The load pulsations become more violent with increases in actuator speed.

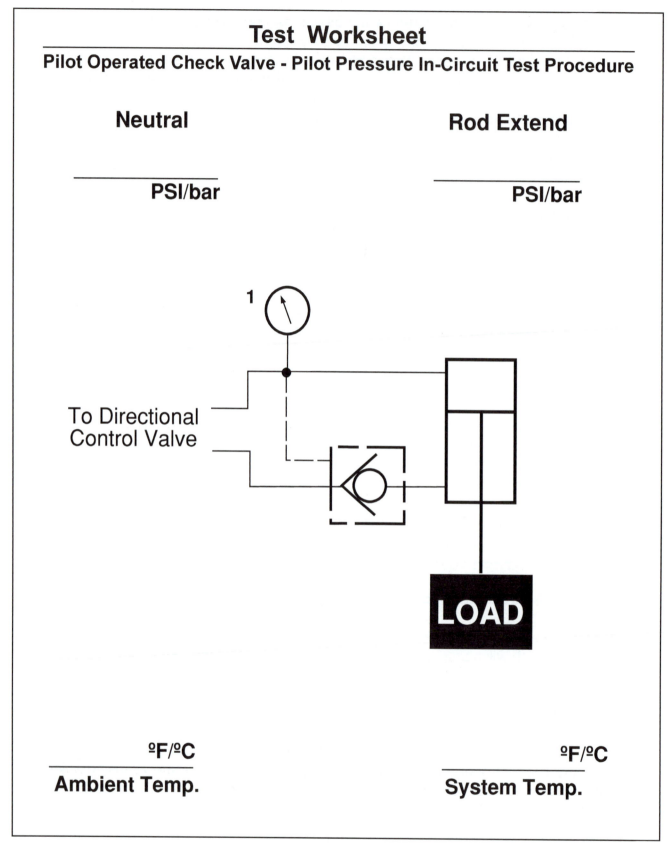

Test Worksheet

Pilot Operated Check Valve - Pilot Pressure In-Circuit Test Procedure

Neutral

_____
PSI/bar

Rod Extend

_____
PSI/bar

To Directional
Control Valve

LOAD

_____
ºF/ºC
Ambient Temp.

_____
ºF/ºC
System Temp.

Figure 12-5   Test Worksheet - Pilot operated check valve - pilot pressure in-circuit test
procedure

chapter thirteen

Safety with Accumulators

Accumulator Precharging Procedure

What will this test procedure accomplish?

Accumulators are an integral part of many hydraulic systems. System application determines the specific function of an accumulator in a hydraulic system.

These functions include:
    a. Energy storage (potential).
    b. Develop flow.
    c. Maintain pressure.
    d. Cushion shock.
    e. Absorb pulsations.

Basically, an accumulator is a device which stores potential energy in the form of compressed gas (precharge) which is used to exert a force on a relatively incompressible fluid.

This procedure will:
    a. Determine the condition of the bladder or piston seals inside the accumulator.
    b. Determine existing precharge pressure.
    c. Provide steps on the proper gas precharge procedures.
    d. Provide safety guidelines for working with gas charged accumulators.

 **-WARNING-** Do not work on or around hydraulic systems without wearing safety glasses which conform to ANSI Z87.1-1989 standard.

YOU WILL NEED THE FOLLOWING EQUIPMENT
TO CONDUCT THIS PRECHARGE PROCEDURE

CAUTION! The pressure and flow ratings of the diagnostic equipment which will be used to conduct this test procedure must be equal to, or greater than, the pressure and flow ratings of the system being tested. Refer to the system schematic for recommended pressure and flow data.

1. Accumulator Precharge Assembly
2. Accumulator Precharge Hose Assembly
3. Nitrogen Cylinder Pressure Regulator
4. Nitrogen Cylinder- Nitrogen is a commercially available bottled gas.

A fully charged nitrogen cylinder is normally supplied with an internal pressure of approximately 2000 PSI (138 bar), which is usually sufficient for precharging most accumulators.

-WARNING-

1. There is an inherent danger when servicing gas charged devices. Read <u>ALL</u> manufacturer's information regarding safety before attempting any service procedures.

2. Have a person trained in accumulator service procedures provide training on how to service accumulators.

3. Observe the system pressure gauge. Release any residual pressure trapped in the system by an accumulator, counterbalance valve, pilot-operated check valve, suspended load on an actuator, intensifier, or pressurized reservoir.

4. A check valve is usually located between the accumulator and the pump. This prevents the energy stored in the accumulator from discharging into the pump outlet when the prime mover is shut down. Reversing oil through the pump outlet port could damage the pump, high pressure filter elements, and/or suction screens.
If the system pressure gauge is located between the pump and a check valve, it will not indicate if there is pressure in the system downstream of the check valve. If the system is not equipped with an automatic or manual accumulator bleed-off system, release the pressure trapped in the system by operating an actuator associated with the accumulator. The power unit must be shut off to bleed-off the system pressure.

5. Accumulators should be charged with "oil pumped" or "dry" nitrogen only. Never use compressed air or oxygen for precharging accumulators, unless recommended by the manufacturer. A mixture of oxygen and oil has the tendency to ignite due to the heat caused by the rapid (adiabatic) compression of gas.

6. Cylinders designed for transporting high pressure gases have a threaded metal cover on top of the cylinder to protect the shutoff valve from accidental damage while handling. Do not remove the cover unless the cylinder is firmly supported. Replace the cover immediately after accumulator precharging, and before attempting to move the cylinder.

7. Keep hands, face, and entire body away from the accumulator precharge valve. Depressing the valve core by hand will release a potentially high velocity gas through the precharge valve to atmosphere. Gas or oil at high velocity can easily pierce the skin or come into contact with the eyes, causing serious injury, limb amputation, or even death. If a piston seal, bladder, or diaphragm have failed, hot oil at high velocity, could exhaust from the valve.

PREPARATION

To conduct this procedure safely and accurately, refer to Figure 13-1 for correct placement of pre-charging equipment.

PROCEDURE

Step 1. Shut the prime mover off.

Step 2. Lock the electrical system out or tag the keylock switch.

Step 3. Move the nitrogen cylinder into position near the accumulator and support it firmly.

Step 4. The accumulator precharging valve is protected by a bracket which is attached to the accumulator, or a cover which screws onto the charging valve. Remove the bracket or cover to gain access to the charging valve.

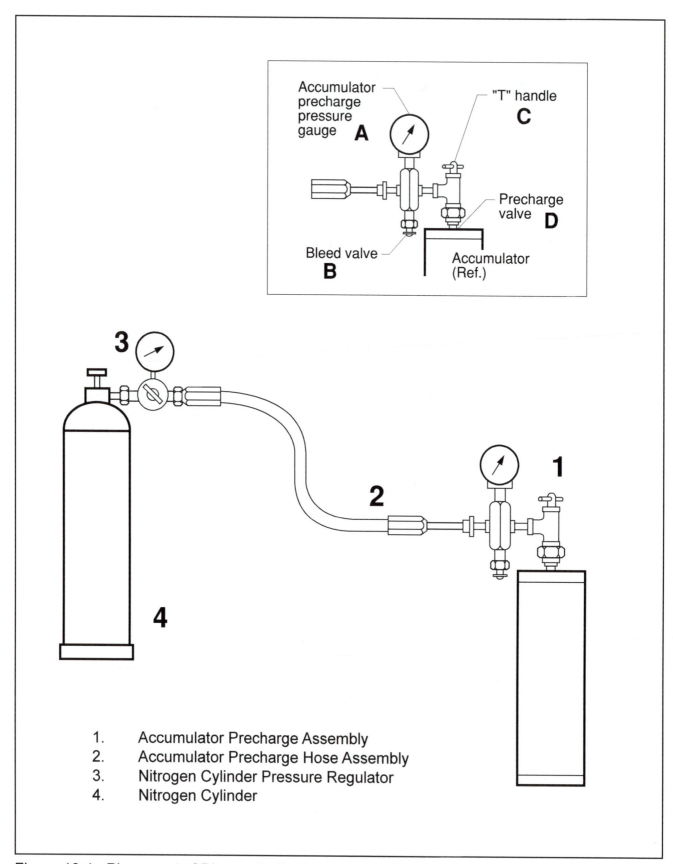

Figure 13-1   Placement of Diagnostic Equipment

Step 5.   Before installing precharge assembly (1) on precharge valve stem (D), first determine if the nitrogen chamber in the accumulator is contaminated with oil. This will occur if the piston seal, bladder, or diaphragm is damaged. Depress the valve core with a suitable tool keeping hands, face, and body out of the line of velocity. If oil discharges from the precharge valve, remove the accumulator for repair or replacement. If there is no indication of oil exhausting from the precharge valve, continue with the charging procedure.

Step 6.   Install accumulator precharge assembly (1) on the accumulator precharge valve. Tighten it securely.

Step 7.   Connect one end of accumulator precharge hose assembly (2) to nitrogen cylinder pressure regulator (3). Tighten it securely.

NOTE: *The precharge hose assembly must be equipped with the proper gland nut to adapt it to the pressure regulator on the nitrogen cylinder. The standard thread connection for a U.S. nitrogen cylinder is 29/32-14 with left-hand thread.*

Step 8.   Connect the opposite end of accumulator precharge hose assembly (2) to the valve stem on accumulator precharge assembly (1). Tighten it securely.

Step 9.   Open bleed valve (B) on accumulator precharge valve assembly (1). Open nitrogen cylinder pressure regulator valve (3) very slightly and bleed the air out of charging hose assembly (2). Once the charging hose is bled, close valve (3) and bleed valve (B).

Step 10.   A special "T" handle device (C) is fitted to accumulator precharge valve assembly (1) (See inset of Figure 13-1). Screwing the "T" handle in (clockwise), will push the valve core open allowing the precharge pressure in the accumulator to enter the precharge assembly. The pressure, indicated on the precharge assembly pressure gauge (A), will be existing nitrogen precharge pressure. Screw the "T" handle in and record the pressure.

Step 11.   If the pressure, indicated on the accumulator precharge pressure gauge, is consistent with the manufacturer's specifications, remove accumulator precharge assembly (1) as follows:

NOTE: *If the precharge pressure is low, proceed with Step 12.*

    a.   Screw the "T" handle out (counter-clockwise).
    b.   Release the residual pressure trapped in the accumulator precharge assembly and precharge hose, by opening the bleed valve located on the precharge assembly.
    c.   Remove the precharge hose from the precharge assembly.
    d.   Remove the precharge assembly from the precharge valve.
    e.   Install the valve cap on the precharge valve.
    f.   Install the protective cover over the precharge valve.

    g. Remove the precharge hose from the nitrogen cylinder.

    h. Install the protective cover over the shutoff valve on the nitrogen cylinder.

    i. Store the nitrogen cylinder safely.

At what precharge pressure should the accumulator be charged to?

Generally, an accumulator must be precharged to the pressure recommended by the manufacturer. If there is no information regarding precharge pressure recommendations, apply the following rule-of-thumb: precharge the accumulator to 50% of pressure relief valve or compensator setting.

If the accumulator is new or the bladder has been replaced, pour some oil into the oil port at the bottom of the accumulator. The fluid will act as a cushion and will lubricate and protect the bladder as it unwinds and contours to the shell. The initial 50 PSI (3.45 bar) precharge should be introduced slowly. Failure to follow these recommendations could result in premature start-up bladder failure.

High pressure nitrogen will freeze if it expands rapidly. This will cause the deflated bladder to become brittle and possibly rupture. In accumulators using anti-extrusion poppet valves, the bladder could be forced under the poppet causing it to tear or cut in the shape of the poppet.

If the nitrogen precharge pressure is too high, it will drive the bladder into the anti-extrusion poppet valve which could cause fatigue failure of the poppet return spring, or the bladder could be pinched and cut if it is trapped underneath the poppet valve as it is forced shut. Excessive precharge pressure is the most common cause of bladder failure.

Step 12.   Adjust nitrogen cylinder pressure regulator valve (3) on nitrogen cylinder (4), to approximately 100 PSI (6.9 bar) higher than recommended precharge pressure.

Step 13.   Open and close nitrogen cylinder pressure regulator valve (3) on nitrogen cylinder (4) and slowly charge the accumulator. Stop when the nitrogen pressure, indicated on the accumulator precharge assembly pressure gauge, is approximately 100 PSI (6.9 bar) higher than the recommended nitrogen precharge pressure.

NOTE: *The expansion of the nitrogen as it flows from the bottle to the accumulator will cause the gas to cool.*

Step 14.   When the required precharge pressure has been reached, allow time for the temperature of the gas to equalize. Check the pressure indicated on the accumulator precharge assembly pressure gauge.

Step 15.   Close nitrogen cylinder pressure regulator valve (3) on nitrogen cylinder (4) securely.

Step 16.   Open bleed valve "B" very slightly, and gradually bleed down to the recommended pre-charge pressure.

Step 17.   Turn the "T" handle on accumulator precharge assembly (1) out (counter-clockwise) until it stops. This will close the valve core in the precharge valve and will prevent the nitrogen from escaping when accumulator precharge assembly (1) is removed.

Step 18.   Open the bleed valve on accumulator precharge assembly (1), and release residual pressure trapped in the charging hose and precharge assembly.

Step 19.   Disconnect accumulator precharge hose (2) from accumulator precharge assembly (1).

Step 20.   Install the valve cap. Tighten it securely.

Step 21.   Disconnect accumulator precharge hose assembly (2) from nitrogen cylinder pressure regulator (3).

Step 22.   Remove accumulator precharge assembly (1) from the accumulator precharge valve.

Step 23.   Make up a small solution of soapy water. Pour a small amount onto the accumulator precharge valve and check for leaks. A leak will cause the solution to bubble.

Step 24.   Install the valve cap on the precharge valve stem. Tighten it securely.

Step 25.   Install the protective cover over the precharge valve. Tighten it securely.

Step 26.   Install the protective cover over the shutoff valve on the nitrogen cylinder.

Common causes of accumulator failure:

PISTON TYPE-
Excessive Precharge Pressure
Excessive nitrogen precharge pressure will cause the piston to contact the bottom of the cylinder when the system pressure is low.

Low Precharge Pressure
Low nitrogen precharge will allow the piston to be driven up into the gas end-cap which could cause damage. It will usually remain in this position with no further damage after initial contact.

Incorrect Mounting - Contamination
The optimum mounting position for an accumulator is vertical with the oil port down. If an accumulator is mounted horizontally, it acts like a reservoir. Solid contaminants entering with the fluid settle out and collect at the bottom of the cylinder. The shape of the accumulator and position of the oil port makes it difficult for the contaminants to leave with the fluid. The accumulation of contamination will cause premature piston seal failure, and/or cylinder wall damage as the piston rubs against the contaminants trapped in the cylinder. The extent of damage and piston seal life, depend on fluid cleanliness and cycle rate.

Incorrect Mounting - Clamping

A piston-type accumulator is usually mounted with clamps which clamp around the outside diameter of the cylinder. If the clamps are too tight, the cylinder will deform and cause the piston to jam. If the piston does not jam, minor cylinder deformity could cause uneven or accelerated seal and/or bore wear.

BLADDER TYPE-
Excessive Precharge Pressure

Excessive nitrogen precharge pressure can force the bladder into the poppet valve assembly which could cause fatigue failure of the spring and poppet valve assembly. It could also cause damage to the bladder should it get caught under the poppet valve as it is forced shut.

Low Precharge Pressure

Low precharge pressure will cause the bladder to be crushed into the top of the steel shell which could extrude it into the precharge valve. A single low precharge cycle is sufficient to cause the bladder to rupture or puncture.

Incorrect Mounting

Mounting an accumulator in the horizontal position can damage the bladder because it is forced to ride against the top of the shell while floating on the fluid.

The optimum mounting position for an accumulator is vertical with the oil port down. If an accumulator is mounted horizontally, it acts like a reservoir. Contaminants entering with the fluid, settle out and collect at the bottom of the shell. The shape of the accumulator and position of the oil port make it difficult for the contaminants to leave with the fluid. The accumulation of contamination could cause premature bladder failure because the bladder rubs against the contaminants trapped in the shell. The extent of damage, and bladder life depends on fluid cleanliness, cycle rate, and maximum/minimum system pressures.